OPTICAL PHYSICS

BY

MAX GARBUNY

CONSULTANT, WESTINGHOUSE RESEARCH LABORATORIES
PITTSBURGH, PENNSYLVANIA
LECTURER, SUMMER COURSES
UNIVERSITY OF CALIFORNIA EXTENSION,
LOS ANGELES, CALIFORNIA

1965

ACADEMIC PRESS   New York and London

ACADEMIC PRESS, INC.
111 Fifth Avenue, New York, New York 10003

United Kingdom Edition published by
ACADEMIC PRESS, INC. (LONDON) LTD.
Berkeley Square House, London W1X 6BA

LIBRARY OF CONGRESS CATALOG CARD NUMBER: 65-19999

Third Printing, 1970

PRINTED IN THE UNITED STATES OF AMERICA

To the memory of my mother

PREFACE

Recent years have seen the ascendency of a field which can be described as the interactions of radiation and matter in the optical spectrum, here quite generally defined as the region bounded by microwaves and x-rays, containing the infrared, visible, and ultraviolet wavelengths. To a major extent, the growth in importance of this field is due to the maturity which certain of its experimental techniques have reached, especially those of generating and detecting monochromatic or continuum radiation over this entire range of frequencies. These practical advances in turn have had two major consequences. First, they stimulated —and made possible—more basic inquiries into such diverse topics as higher order coherence, the generation of stimulated emission from inverted populations, solid state spectra, and phenomena of photoconductivity. Second, there has emerged a wealth of applications, from detectors of far infrared radiation to sources of ultraviolet laser beams. The purpose of this book is to describe by means of a unified treatment the various interaction phenomena, their experimental and theoretical exploration, and their applications.

The title "Optical Physics" is chosen so as to convey concisely the two symmetric aspects of the field: The study of dynamic interactions between light and matter and the conclusions on structure and state that can be derived from optical phenomena. Of course, the name "Optical Physics" has been used for more limited and quite diverse subjects, such as the radiative phenomena of gases, the optical properties of solids, or the statistical behavior of light beams. However, these various fields are embraced under the definition adopted here, so that the area to be treated in this book is extensive and diverse. Furthermore, the intent is to present aspects of theory, phenomenological description, and application in about equal measure.

The text material is divided into three major parts, the processes of emission and absorption, the phenomena of propagation, and finally the secondary effects caused by radiation. The first chapter introduces concepts fundamental to the rest of the book, particularly the relationship between structure and radiation. The next three chapters deal with emission and absorption following a sequence of increasingly detailed models: Thermal radiation continua from solids and plasmas, first from the viewpoint of thermodynamics, then of quantum statistics; monochromatic radiation, first from the classical, then from the quantized oscillator; the line shapes for various conditions and degrees of resonance broadening; the spectra from structures in which the complexity is increased stepwise, viz., atoms with single and multiple valence electrons, spins, multipoles, molecules, and finally solids. The next two chapters are concerned with propagation. Chapter 5 treats propagation phenomena in general with regard to the optics of various materials. In Chapter 6 aspects of propagation which are characteristic of coherent light are discussed. This includes criteria and measurement of coherence with such practical applications as intensity interferometry, the theory and practice of optical masers, and nonlinear optics. The last chapter deals with various photoeffects, their applications in detectors, and the theory of fluctuations as the ultimate limit to detection. As this last chapter was being written, it was possible to report on the solution of a problem which had been under attack for many decades: The opening of the entire optical spectrum to photon detectors.

This work is intended as a textbook, even though many of its subjects are still in the forefront of evolving research. For this reason, certain basic theories are briefly stated, their availability in diverse texts notwithstanding. This is particularly true of such subjects as quantum statistics, atomic spectra, and the electromagnetic theory of light. If in such discussions I have followed rather closely the classic treatments—for example, Born's "Atomic Physics," Heitler's "Quantum Theory of Radiation," and Becker-Sauter's "Theorie der Elektrizität," this was done because I could not improve on perfection; in these cases I have cited the source. In all other instances, whenever possible, I have relied on the original research paper, as quoted at the end of each chapter. Nevertheless, the bibliography is in no way complete. It represents only a portion of the available literature; but it is that portion which I have used, or at least read, for this book. Finally, certain ideas and experiments which had been carried out in my Optical Physics Section at the Westinghouse Research Laboratories have been mentioned in this text, although they have been briefly published elsewhere. One area of special interest, the theory and applications of infrared radiation, was initially intended to form a somewhat larger portion of the book. However, the appearance of several good texts on this subject, such as the "Elements of Infrared Technology" by Kruse, McGlauchlin, and McQuistan, made a more extensive discussion of this field unnecessary.

I wish to acknowledge my indebtedness, first of all, to the Westinghouse Research Laboratories for encouragement and cooperation extended to me in writing this book. Several of my colleagues have assisted in the arduous task of proofreading and with helpful comments, especially Dr. M. Gottlieb and T. P. Vogl, who read this entire manuscript, and Dr. R. D. Haun, Dr. H. F. Ivey, A. H. Boerio, and Mrs. S. Banigan. The treatment of fluctuations and noise in the seventh chapter has been strongly influenced by suggestions from Dr. J. W. Coltman. I have had the benefit of illuminating discussions with Prof. E. Wolf regarding the chapter on coherence phenomena. A number of authors deserve acknowledgment who sent me communications on their work, photographs of spectra, and other material which would be difficult to obtain otherwise, notably Prof. A. Hadni, Dr. R. Tousey, Prof. G. H. Dieke, and Dr. T. P. Hughes. Last but not least, thanks are due to my wife for helping with corrections and for charitably forgoing many a long weekend on behalf of this book.

MAX GARBUNY

Pittsburgh, Pennsylvania
February, 1965

CONTENTS

xi

CHAPTER 4   Spectra of Atoms, Molecules, and Solids

CHAPTER 5   Phenomena of Propagation

CHAPTER 6 Interactions of Coherent Radiation with Matter

CHAPTER 7 Secondary Effects of Light and Processes of Detection

1

FUNDAMENTALS OF RADIATION
AND ITS INTERACTION WITH MATTER

1.1 Introduction and Orientation

1.1.1 CLASSICAL OPTICS

It is possible to divide the study of optical radiation, historically as well as didactically, into two different parts which may be referred to as classical and modern optics. The first of these consists of optics in its original and more restricted sense, namely, insofar as it is concerned with the understanding of the nature and propagation properties of visible light. This discipline—as old as scientific thought itself—represents one of the original divisions of physics according to the human senses, viz., that dealing with the perceptions of vision. Characteristically, throughout most of its history, it remained a science isolated by itself, having little interaction with the other fields of physics. Here belong such major subdivisions as geometric optics, which had its beginnings among the treatments of early Greek mathematicians and reached its highest sophistication with the mathematical theorems of Fermat (1601–1665) and Malus (1755–1812); and the various manifestations of physical optics, such as interference, first observed by Boyle (1626–1691) and Hooke (1635–1703), diffraction (Grimaldi, 16th century; Fresnel, 1816), and polarization.* All these phenomena, including those connected with the velocity of light in vacuo and other media, could be explained by a system of differential equations, stated by Maxwell (1873) in terms of interrelated and time-variant electric and magnetic fields.

1.1.2 ELECTROMAGNETIC SPECTRUM

Now this electromagnetic theory of light marks not only the culmination, but, in a sense, also the conclusion of the "pure" or classical optics which it is able to explain. Thus began the age of modern optics where

---

* See also historical introduction and beginning of Chapter 8 in Born and Wolf.[1]

1

the emphasis is placed on the interaction between radiation and matter, particularly in atomic processes. The reason for this lies in the consequences of Maxwell's equations themselves. These emphatically state that the nature and propagation properties of light are explainable in terms of another discipline of physics, viz., that of electromagnetism. Therefore, the interaction between light and matter, especially the processes governing the emission and absorption of radiation, cannot be understood unless the structure of matter is known to the extent to which it determines the electromagnetic behavior. This, it will be seen, involves an entirely new area of study, concerned with electrons, ions, atoms and their structural bonding. Again, the electromagnetic theory of radiation introduces the fact that light, to the extent to which it is perceptible by the human eye, is merely part of an infinite continuum, the electromagnetic spectrum.

These disclosures of the theory did not find the scientific world altogether unprepared. Already in 1800, W. Herschel, endeavoring to measure the caloric content of the solar spectrum, had found that heat radiation was received in the invisible portion beyond the red. This *infrared* emission was found in succeeding experiments to exhibit the salient properties of wave propagation and other characteristics of visible light, although both generation and detection were accomplished at that time exclusively by thermal processes. The existence of a spectrum beyond the visible violet was discovered almost simultaneously with infrared. In 1801 J. W. Ritter found that radiant energy in that region produced blackening in silver chloride, and subsequent interference experiments by Young (1804) allowed the conjecture that this *ultraviolet* radiation was of the same nature as visible light.

Altogether, it is now known that the electromagnetic spectrum (see Fig. 1.1) encompasses radio- and microwaves, infrared, visible, and ultraviolet light, *x* (or Roentgen-) rays, gamma, and cosmic radiation. All of these have in common that they propagate through space as transverse electromagnetic waves; and that in vacuo their speed of propagation which is the product of their frequency and their wavelength, is the same for all, the speed of light. We write

$$\lambda\nu = c \tag{1.1}$$

where λ is the wavelength, ν the frequency, and c the propagation velocity of light in vacuo which is almost exactly 3×10^{10} cm/sec. The various parts of the electromagnetic spectrum differ in wavelength and frequency, and this, in turn, leads to profound differences in their generation and their interaction with matter. Thus, the spectral regions are distinguished

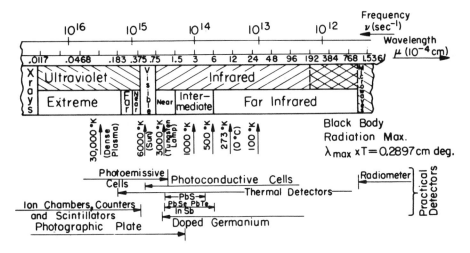

The optical spectrum.

FIG. 1.1. The electromagnetic spectrum at and around the optical region. Position of black body radiation maxima is indicated on wavelength scale for various absolute temperatures. Spectral range of operation is indicated for various practical detectors.

by the methods of generating or observing their wavelengths, and their limits are fixed by convention rather than by sharp discontinuities of the pertinent physical phenomena. Radiation at lowest frequency is produced and received by electron beam tubes in combination with extended resonant structures of capacitors and inductances between which the energy oscillates (see Fig. 1.2). In the microwave region, at successively higher frequencies, the structures shrink, and the interplay between electric and magnetic fields is supported by single components, viz., cavities and waveguides. At yet shorter wavelengths, we have to turn to the oscillators provided by nature: molecules rotating and vibrating in the infrared; electrons performing transitions across atomic fields in the visible and ultraviolet regions. The mechanism by which radio frequency oscillations are produced is still recognizable, at least in principle, on the molecular scale; but the radiation is now generated, not in a single oscillator, but typically by an enormous number of individual vibrators, and this engenders totally different phenomena and techniques of experimentation. Finally, at still higher frequencies, radiation is encountered which is produced either by charged particle bombardment, as in the case of x rays and certain cosmic rays, or by nuclear transitions, as in the case of gamma radiation and certain other cosmic rays. Taken as a group, these radiations are characterized by the high energy involved in their generation and by the fact, to be described later, that they behave like corpuscles rather than waves in their interaction with matter.

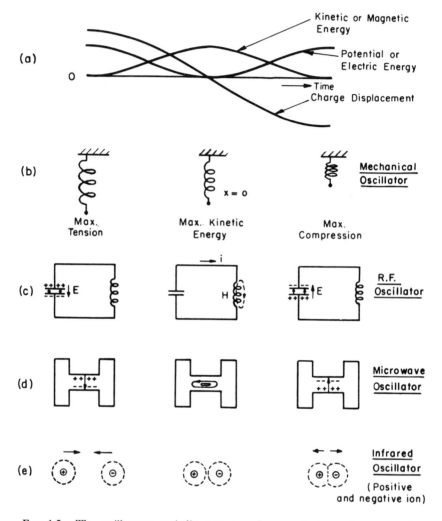

FIG. 1.2. The oscillator as periodic converter of energy from one form to another. Mechanical oscillators (b) change from potential to kinetic energy and back. Their electric analogues shown for various wavelength regions (c–e) oscillate between electrostatic and magnetic field energy. Phase relationship of the various parameters is shown in (a).

1.1.3 The Optical Spectrum

It has become apparent in the preceding paragraph that the electromagnetic spectrum consists of three major portions: (1) the very energetic (corpuscular type) emission from nuclear and bombardment processes at the high-frequency end; (2) the radio and radar waves generated in extended circuit structures at the long wavelength end; and (3) in the intermediate region, the radiations of light which exhibit, as will as be seen, both wave and particle character, and which have their origin in atoms and molecules. This third group, which includes (see Fig. 1.1) the infrared, visible, and ultraviolet wavelength regions, constitutes the optical spectrum in the wider sense. The justification for this collective name is based, not merely on the close relationship of the emission processes, but on common experimental techniques, such as the use of lenses and mirrors for focusing purposes and of prisms and gratings for spectroscopy, although the various spectral regions within this group must be accommodated by proper choice of materials and design of components.

TABLE 1.1

Conversion Factors and Units

Lengths

$$1 \text{ micron } (\mu) = 10^{-3} \text{ mm} = 10^{-4} \text{ cm}$$
$$1 \text{ angstrom } (\text{Å}) = 10^{-8} \text{ cm} = 10^{-4} \, \mu$$
$$1 \text{ millimicron } (\text{m}\mu) = 10^{-7} \text{ cm} = 10 \text{ Å}$$
$$1 \text{ centimeter } (\text{cm}) = 0.3937 \text{ in.} = 1 \times 10^4 \, \mu = 0.010936 \text{ yard}$$
$$1 \text{ mile (mi) U.S. Statute} = 1.609 \text{ km} = 1.609 \times 10^5 \text{ cm}$$
$$= 1.69 \times 10^{-13} \text{ light year} = 5280 \text{ ft}$$
$$= 0.868 \text{ mile (nautical)}$$
$$1 \text{ foot (ft)} = 30.480 \text{ cm} = 1.894 \times 10^{-4} \text{ mi (statute)}$$

Temperatures

Kelvin (absolute, °K) and centigrade (°C) scales:

$$T(°\text{K}) = t(°\text{C}) + 273.18$$

Centigrade and Fahrenheit (°F) scales:

$$t(°\text{C}) = \tfrac{5}{9}[t'(°\text{F}) - 32]$$

Energy Units

$$1 \text{ gram-calorie (gm-cal)} = 4.186 \text{ joules} = 3.968 \times 10^{-3} \text{ Btu}$$
$$1 \text{ erg} = 10^{-7} \text{ joule} = 2.3889 \times 10^{-8} \text{ gm-cal}$$
$$= 1 \text{ dyn-cm}$$
$$1 \text{ joule} = 10^7 \text{ ergs} = 0.23889 \text{ gm-cal}$$
$$= 2.778 \times 10^{-7} \text{ kw-hr} = 1 \text{ watt-sec}$$
$$1 \text{ ev} = 1.602 \times 10^{-12} \text{ erg}$$

Figure 1.1 shows that the optical spectrum ranges in wavelength from about 0.1 cm, as the conventionally defined limit of the far infrared, to about 10^{-6} cm, at the short wavelength end of the ultraviolet. Since the experimentally significant extent of the electromagnetic spectrum covers more than 20 orders of magnitude in wavelength, one uses for convenience diversified units of length (see Table 1.1). The units important for the optical spectrum are the micron ($= 10^{-4}$ cm), the millimicron ($= 10^{-7}$ cm), and the angstrom ($= 10^{-8}$ cm). It is sometimes convenient to represent the electromagnetic spectrum in steps of octaves (viz., intervals in which the frequency and wavelength vary by a factor of two). The visible spectrum, then, spans about one octave, viz., 0.38–0.75 micron, the ultraviolet five, and the infrared about ten octaves. Furthermore, it has been the practice to subdivide the ultraviolet and infrared regions each into three parts, such as "near," "intermediate," and "far" infrared, as indicated in Fig. 1.1. These distinctions were originally made because of differences in techniques and phenomena; however, as more is learned about the various processes, their applicability is broadened beyond the initial limits, and the subdivisions become less meaningful. In fact, for the same reasons, the domain of the far infrared is already partially overlapped by that accessible to the generation of microwaves. A similar situation exists near 100 Å where the same region is claimed by x rays and the extreme ultraviolet. In such cases, the radiation is usually named for the method of its generation.

1.1.4 MODERN OPTICS

To the extent of the validity of Maxwell's equations, the phenomena of classical optics remain unchanged throughout the spectrum. In other words, if the linear dimensions of the apparatus (e.g., slits, gratings, mirrors) are scaled in constant proportion to the wavelength, such effects as interference and diffraction remain invariant with the spectral region. Conversely, if the size of the structure is fixed, such as in atoms and molecules, the effect on incident electromagnetic waves will depend qualitatively and quantitatively on wavelength. For this reason, interaction phenomena between radiation and matter are, in general, spectrally selective. The sixteen octaves of the optical spectrum affect matter in different ways, and the possible combinations with various materials give rise to a multitude of effects. Clearly, these phenomena reveal as much or more about the state of matter as they do about the nature of radiation and radiative interactions. This, then, is the subject of modern optics which had its beginning in the latter part of the nineteenth century.

Using the tools of radiometry and spectroscopy, the new approach moved quickly into a pivotal position for the development of physics: the discovery of the quantized nature of energy (Planck, 1900, and Einstein, 1905); the theory of atom structure and spectral lines (Bohr, 1914); the theory of the periodic system (Bohr, 1921); the exclusion principle for the spinning electron (Pauli, 1926); Raman effect (1928) and the understanding of the molecular bond; all these were results due to the exploration of matter by its emitted or absorbed light. More recently, the finer features of the radiation phenomena in excited gases and plasmas were brought into focus. Measurements of emission, absorption, and dispersion with instruments, often of high spectral resolution, were used to determine the dynamics between light and atoms, viz., the energy distribution, lifetimes, and transition probabilities. This area of modern optics is properly called *optical physics*, the name implying the study of interaction phenomena between optical radiation and matter. The field now includes in equal measure the phenomena occuring in solids, either intrinsically or owing to the presence of impurities. There are topics which lie at the outer fringes of this field, in which interest is centered, not on the radiation phenomena as such, but on the energy structure of a given material such as a superconductor. However, quite often such studies lead to the discovery of new means of detecting or exploring the radiation itself, and so both fields grow in cooperative fashion.

1.1.5 APPLICATIONS

It is not our purpose to describe in detail the enormous number of applications which follow from the basic science of classical and modern optics and which, in fact, continue to be in a state of development. The field of geometrical optics, for example, aside from continuous improvements of instruments invented long ago, is forever surprising us with new ideas, methods, and systems, such as the new art of fiber optics. Again, multiple-beam interferometry is used to image and control submicroscopic surface irregularities by comparing them with the wavelength of light or rather with fractions thereof; in turn, the capabilities of physical optics are steadily advanced by the improvements of surface quality which it made possible. Although we shall briefly discuss the sources of radiation in the various wavelength regions, we shall not dwell on extraneous applications: the therapeutical and diagnostic value of ultraviolet and infrared; the use of infrared for power purposes such as drying; or even the application of spectroscopy to chemical analysis. What does concern us, however, are specific applications of modern optics, namely, those which are the exclusive domain of this field and which have been apt in the past to stimulate its fundamental study. Here belong such practical developments as line-sources or plasma radiators, masers and other sources of coherent radiation, and the vast field of photoelectric, photoconductive, and other means for the detection of radiation. In short, our concern is the generation, propagation, and detection of optical radiation and the physical processes on which the art is based.

1.2 Classical Model of Structure and Radiation

It has been brought out in the introduction that the subject of our treatise is largely the interaction between optical radiation and matter. The procedure that we shall follow is to first describe basic concepts of what constitutes matter, insofar as it affects radiation, and then develop in detail the connection with optical processes. Thus we shall deal with the physical models accounting for the origin of optical radiation; the phenomena associated with its propagation through a medium, viz., reflection, refraction, scattering, polarization, etc.; and the processes responsible for absorption, viewing these either as the inverse of the generation processes, or as mechanisms of attenuation, or as the phenomena used in detection. The media with which these interactions occur will be mainly solid or gaseous. The optics of the liquid state has few aspects which are different from either that of solids or that of gases, and their practical significance is relatively small. Important and common to all these interactions are the following considerations.

Almost by definition, at least within the scope of our subject, electromagnetic radiation interacts with matter only through the electric charges which make up matter, or more accurately, through the motion of such charges. This may be only indirectly accounted for, or implicitly described, in the physical models and quantitative relations of which the theory of electromagnetism consists. As an example, the theory of metal optics relates such parameters as reflectance and absorptance to conductivity, dielectric constant, and permeability, through which the role of atomic charges is only indirectly represented. It is, then, in the last analysis, electrons and ions which individually or cooperatively interact with electromagnetic fields in the phenomena of radiation.

Now the various types of structural bonds which link charges together, thereby determining the properties of matter, comprise also the most important factor in determining what kind of interaction can occur with radiation and at what frequencies the phenomenon takes place. In other words, the structural state of the charges, whether they are free or bound, independent of each other or coupled, will decide to what extent there will be reflection, refraction, or absorption for an electromagnetic wave at the interface between two media, and if radiation is emitted and absorbed monochromatically or in a continuum. This effect of structure on radiation is quite important for the differentiation of optical phenomena, and is implicit in such classifications as thermal radiation and line spectra, metal and semiconductor optics, or selective and nonselective detection. We will, therefore, in our treatment of optical physics, benefit from an early distinction between the various structures capable of interacting with electromagnetic radiation. We will, in fact, begin with a discussion of pertinent models in terms of mechanical analogues. This is useful because the motion of electric charges is, to a first approximation, quite well described by such simple pictures. With this background the processes by which energy is exchanged with the electromagnetic field will then be discussed as the various phenomena are taken up in detail.

1.2.1 The Free Electron Model

A simple ideal case is that of free charges, which are not bound by forces to specific centers. Such is the situation, at least approximately, for the conduction electrons of a metal, if the effect of collisions is negligible. Furthermore, it is possible to consider electrons, although they may be bound to fixed atom sites, as free in certain processes such as scattering, provided the coupling can indeed be ignored in the interaction. Free charges will vibrate in response to an external periodic field, written in complex presentation as $E_0 e^{i\omega t}$, where $\omega = 2\pi\nu$ is the angular frequency and t the time. Although the charges are not bound by local centers, their own motion will produce reaction forces in the field. These are, however, small enough to be ignored for most purposes. Under these conditions, the equation of motion is simply

$$M\ddot{x} = eE_0 e^{i\omega t} \tag{1.2}$$

where $\ddot{x}$ is the second derivative of position with respect to time t and M and e are mass and charge of the particle, respectively (see Table 1.2 for values of atomic and other constants). If initial velocity and position are assumed to be zero, the solution is

$$x = -\frac{eE_0}{M\omega^2} e^{i\omega t} \tag{1.3}$$

TABLE 1.2

VALUES OF FUNDAMENTAL CONSTANTS

F	Faraday	96,487.3 $\pm$ 0.4 coul gm-equiv^{-1}
N	Avogadro's number	(6.0228 $\pm$ 0.0011) $\times$ 10^{23} mole^{-1}
h	Planck's constant	(6.624 $\pm$ 0.002) $\times$ 10^{-27} erg-sec
m	Electron mass	(9.10710 $\pm$ 0.00022) $\times$ 10^{-28} gm
e	Electronic charge	(4.80296 $\pm$ 0.00006) $\times$ 10^{-10} esu
		(1.602095 $\pm$ 0.00002) $\times$ 10^{-19} coul
e/m	Specific electronic charge	(1.758896 $\pm$ 0.000032) $\times$ 10^7 emu-gm^{-1}
a_0	First Bohr radius	(0.5291483 $\pm$ 0.0000024) $\times$ 10^{-8} cm
σ	Stefan-Boltzmann constant	(5.6697 $\pm$ 0.0009) $\times$ 10^{-5} erg-cm^{-2}deg^{-4}sec^{-1}
k	Boltzmann's constant	(1.38053 $\pm$ 0.00006) $\times$ 10^{-16} erg-deg^{-1}
c_1	First radiation constant	(4.99208 $\pm$ 0.00011) $\times$ 10^{-15} erg-cm
c_2	Second radiation constant	(1.43879 $\pm$ 0.00006) cm-deg
$\lambda_{max} T$	Wien displacement law constant	(0.289779 $\pm$ 0.000012) cm-deg
λ_0	Wavelength associated with 1 ev	(1.239805 $\pm$ 0.000012) $\times$ 10^{-4} cm ev
ν_0	Frequency associated with 1 ev	(2.418061 $\pm$ 0.000022) $\times$ 10^{14} sec^{-1} (ev)$^{-1}$
E_0	Energy associated with 1 ev	(1.602095 $\pm$ 0.000022) $\times$ 10^{-12} erg (ev)$^{-1}$
v_0	Speed of 1-ev electron	(5.981888 $\pm$ 0.00030) $\times$ 10^7 cm-sec^{-1}
$(R_0/F) \times 10^{-8}$	Energy associated with 1°K	(8.61632 $\pm$ 0.00042) $\times$ 10^{-5} ev
T_0	"Temperature" associated with 1 ev	(11605.6 $\pm$ 0.6) deg K
c	Velocity of light	(299,792.5 $\pm$ 0.2) km-sec^{-1}
R_∞	Rydberg constant for infinite mass	109737.303 $\pm$ 0.024 cm^{-1}
R_0	Gas constant per mole	(8.31436 $\pm$ 0.00038) $\times$ 10^7 erg-mole^{-1} deg^{-1}
V_0	Standard volume of perfect gas	(22414.6 $\pm$ 0.6) cm^3 mole^{-1}

It is seen that free particles vibrate with the frequency of the external field. Thus there is no resonance, although the amplitude decreases with the square of increasing frequency. The optical behavior of metal electrons bears this out to a certain extent. In the general case of metal optics, however, Eq. (1.2) has to be modified to account for electron collisions with the ions. This is accomplished by assuming a damping force, $-R\dot{x}$, which has the effect (see Chapter 4) that resistivity and absorptance at optical frequencies are no longer zero as is implicit in (1.2) and (1.3).

1.2.2 THE CLASSICAL OSCILLATOR

A different situation exists when charges are bound by internal forces to fixed locations, such as the positive and negative ions in the lattice configuration of an insulator or the ions in a polar molecule. A useful concept, in fact, historically the first approach to an understanding of the interaction phenomena involving bound charges, is that of the oscillator. Classical physics understood the structure of atoms, molecules, and crystals in terms of quasi-elastic forces. In this picture then, electrons and ions are held together by spring-type bonds such that they are capable of vibrating around positions of equilibrium. The various systems and the analytical methods for treating them are, of course, available from elementary mechanics. We shall begin with the "free" oscillator, leaving the discussion of its interaction with a periodic driving force to Chapter 3. The most important model on the basis of which it is possible to understand the interaction of light and matter classically, is that of the harmonic oscillator.

A charged particle of mass M is bound to a position of equilibrium, at $x = 0$. When displaced from it, the particle experiences a linear restoring force $-Gx$, where G, the "stiffness," is the force per unit length displacement. Assuming that there are no other forces exerted on the particle, we obtain for Newton's equation of motion

$$M\ddot{x} + Gx = 0$$

The solution of this homogeneous, second-order differential equation is the simple, undamped harmonic oscillation

$$x = x_0 e^{2\pi i \nu_0 t} \tag{1.4}$$

where

$$\nu_0 = \frac{1}{2\pi}\sqrt{\frac{G}{M}}$$

represents the resonant frequency. We thus arrive at a simple model of a light source; a charge vibrating according to Eq. (1.4) will induce an electromagnetic field oscillating at frequency ν_0. We have to refine this model, however, since Eq. (1.4) describes an infinite, unattenuated process in which no energy is spent on radiation. The Newtonian equation of motion must, therefore, in addition contain a "frictional" or "damping" term $R\dot{x}$ (which with constant R is proportional, but opposite, to velocity) to account for energy transfer since no light emission would be possible otherwise. In the absence of a driving force, the "free" equation of motion (1.2) then contains 3 terms,

$$M\ddot{x} + R\dot{x} + Gx = 0$$

The integration of this yields a harmonic oscillation, with the amplitude decaying in time as $e^{-(R/2M)t}$. Thus, with an attenuation constant $\gamma = R/M$

$$x = x_0 e^{-(\gamma/2)t} e^{i2\pi\nu_0 t} \qquad (1.5)$$

The oscillation with exponentially decreasing amplitude is shown in Fig. 1.3a. The mechanical analog to (1.5) is, of course, a spring in which

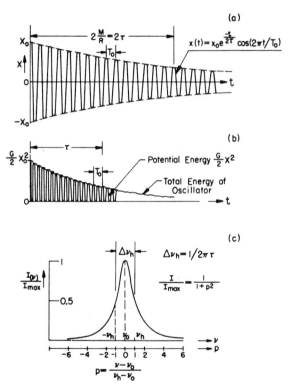

FIG. 1.3. Characteristics of classical damped oscillator. Top curve (a) shows harmonic oscillations with period T_0 and decay time $2\tau_0$. Curves (b) present energies associated with (a). Potential energy is $\frac{1}{2}Gx^2$, hence vibrates at twice the frequency of (a), while total energy is dissipated in small periodic pulses following the envelope $e^{-t/\tau}$ with a Q of $2\pi\tau/T_0 \approx 47$. Broadening of spectral lines due to damping follows quantitatively frequency response curve (c) of damped oscillator, shown in terms of $I(\nu)/I(\nu_0)$ vs. ν or (lower scale) $p = (\nu - \nu_0)/(\nu_h - \nu_0)$.

damping is taken into consideration and which is released from the extended position x_0 at time $t = 0$. The resonant frequency is now

$$\nu_0 = \frac{1}{2\pi} \sqrt{\frac{G}{M} - \left(\frac{R}{2M}\right)^2} \cong \frac{1}{2\pi} \sqrt{\frac{G}{M}} \tag{1.6}$$

which can be simplified to the form of Eq. (1.4) whenever the damping term $R/2M$ is small compared with the resonant frequency.

The significance of this model for our purposes lies first in the fact that it assigns approximately correct wavelength regions to the resonance spectra of transparent insulators and gaseous molecules. If G is representative of the forces that hold atoms and ions together in molecules, absorbed layers, and solids, and if M is their (effective) mass, then the values are such that the resonant frequency falls into the infrared region —somewhere between 10^{12} and 10^{14} sec^{-1}. The existence of these resonances in the infrared reveals itself in absorption and emission or in reflection and refraction. Equation (1.6) indicates that the resonant frequency is proportional to $M^{-1/2}$, a trend which is rather well confirmed by observation (Chapter 5). Indeed, assuming that under given conditions the same binding forces obtain for electrons as for the 10^4–10^5 times heavier ions, this simple classical model places electronic resonances correctly into the visible or ultraviolet regions around 10^{15} sec^{-1}.

A second aspect of this model is associated with the damping term in Eq. (1.5). The energy of an oscillation is proportional to the square of the amplitude (see Fig. 1.3b). Thus the time τ during which the oscillator energy is spent to $1/e$ of the initial value is given by

$$\tau = 1/\gamma = M/R \tag{1.7}$$

Oscillatory systems are also characterized by means of quality factor Q which is defined as

$$Q = 2\pi \frac{\text{stored oscillator energy}}{\text{energy spent per period, } T_0} \tag{1.8}$$

Hence

$$Q = \frac{2\pi}{1 - e^{-T_0/\tau}} \cong 2\pi \frac{\tau}{T_0} = 2\pi\nu_0\tau \tag{1.9}$$

The Q-value represents, therefore, 2π times the number of periods T_0 during which the oscillation process lasts before its total energy has dropped to $1/e$ of its initial value. In other words, the Q-value is the oscillation time expressed in radians. Conversely, a periodic, resonant driving force would have to be applied for Q radians to bring the oscillator within $1/e$ of its equilibrium value.

Now the damping force, besides limiting the lifetime of the oscillation, has another—although mathematically equivalent—effect. If the term γ in Eq. (1.5) is zero, the oscillation is unlimited in time and resonates with zero bandwidth, i.e. at exactly one point in the frequency scale. If, however, there is damping and therefore a finite lifetime, the oscillation process must be considered as a superposition of Fourier components such that a broadening of the frequency response appears. Basically, the oscillation time τ corresponds by itself to a frequency $1/\tau$, and this is the amount which one may expect for the detuning or broadening of the pure resonance frequency. Thus the oscillator response, i.e. the power emitted or absorbed per unit frequency interval, becomes a continuous function $I(\nu)$. We present here merely the result of the Fourier analysis which yields for the oscillator response the expression

$$I(\nu) = A \, \frac{1}{4\pi^2(\nu - \nu_0)^2 + \gamma^2/4} \tag{1.10}$$

in which it has been assumed that $\gamma \ll \nu_0$. Equation (1.10) shown normalized in Fig. 1.3c, contains the factor A which quite generally accounts for the relationship between the radiation intensity and the amplitude of the vibrating charge, but which does not affect the spectral profile of the damped oscillation (in the emission process, $A = 16\pi^4\nu_0^4 e^2 x_0^2/3c^3$, as will be shown in Section 3.1). The maximum of the intensity distribution, at $\nu = \nu_0$, equals $4A/\gamma^2$. If we now solve for that frequency deviation $\nu_h - \nu_0$, at which the intensity has fallen off to half the maximum value, we obtain

$$\nu_h - \nu_0 = \pm \, \gamma/4\pi \tag{1.11}$$

and for the "halfwidth" of the spectral distribution

$$2(\nu_h - \nu_0) = \Delta\nu_h = \frac{\gamma}{2\pi} = \frac{1}{2\pi\tau} \tag{1.12}$$

$$\Delta\omega_h = \gamma = 1/\tau \tag{1.13}$$

The last equation expresses the halfwidth due to damping in units of angular frequency: $\Delta\omega_h = 2\pi \, \Delta\nu_h$. On this scale, the halfwidth is just the reciprocal of the oscillation time constant τ.

Equation (1.10) for the spectral profile can be simplified with the use of Eq. (1.11):

$$\frac{I}{I_{\max}} = \frac{1}{1 + p^2}$$

Here the intensity, normalized to the maximum value at ν_0, is presented as a function of $p = (\nu - \nu_0)/(\nu_h - \nu_0)$. The lower scale for the abscissa of Fig. 1.3c is given in terms of this parameter p.

Correlation to the halfwidth provides an additional definition of the Q-value, valid if the latter is large enough. One obtains from (1.9) and (1.12)

$$Q = 2\pi\nu_0\tau = \frac{\nu_0}{\Delta\nu_h} \tag{1.14}$$

The damping terms, R or γ, may be representative of any power-draining processes which limit the life of the oscillation. Damping by the radiation process itself, which causes in atomic oscillators the so-called "natural halfwidth," presents merely a lower limit under the special conditions that no other broadening effects are present. We mention, however, in anticipation of a detailed discussion later, that the natural lifetime can be "artificially" prolonged by driving a multiplicity of oscillators in the same phase (*stimulated emission*). Left to itself each radiating atom, producing typically a line width of 10^{-4} Å at $\lambda = 10^4$ Å, represents a sharply monochromatic oscillator with a Q-value of 10^8. The spectra of gases at the usual glow discharge conditions or of certain ions at low concentrations in host lattices show considerably broader line widths, although one may still speak of isolated oscillators. It is under conditions of strong coupling, as in high pressure discharges or in radiating solids, that the individual oscillators lose their identity and merge into a continuum.

1.2.3 THE OSCILLATOR CONTINUUM

The manner in which the selective response of isolated oscillators broadens, with increasing degree of coupling, into a nonselective continuum is well understood on the basis of classical mechanics. At sufficiently large distances from each other, radiating atoms may be considered as undisturbed oscillators, all vibrating at the same frequency with a small but finite spectral width determined by the factors mentioned in the previous section. Assume that two of these atoms diminish their distance so that they begin to exert a force on each other, for instance, of electrostatic nature. Generally we say that the two oscillators are coupled if the motion of one exerts a force on the other. This has the result that the stiffness G, in Eq. (1.4), assumes different values depending on the relative phases, and that the system of the two atoms as a whole responds to two frequencies which diverge increasingly as the two particles interact in tighter coupling. Figure 1.4 illustrates this with the

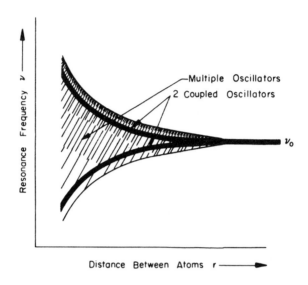

FIG. 1.4.   Effect of oscillator coupling on the broadening of frequency. The example used here is that of a gas at varying density, each molecule or atom representing an individual oscillator. At low gas densities (large distances r) coupling is negligible and all atoms radiate (or absorb) within the same narrow frequency interval. As two atoms are brought closer together, they form a coupled system resonant at two diverging frequencies (black regions). When more atoms are added to the coupled system, additional frequencies appear, favoring the upper region (indicated by degree of shading). Frequency distribution at smallest r represents density of states for solid or correspondingly dense gas.

two heavy black lines which represent the response of two coupled oscillators at varying distance. We have now to complete the picture of the closely packed oscillator structure by adding additional splittings for additional bonds so that ultimately a continuous frequency region of response results as indicated by the lighter shading in Fig. 1.4. If we were actually performing this process, we would notice that certain frequency regions are more often covered than others, i.e. that the density of frequency states is, in general, not uniform but itself a function of position in the spectrum.

It is obvious that this "density function" has to play an important role in the interaction phenomena which we will encounter. To compute it, one could, in fact, apply the method just indicated, viz., that of adding the effect of N individual oscillators (Landé). However, a much simpler approach exists. To discuss this, it is best to start with the one-dimensional model of the vibrating string (see Fig. 1.5). Such a structure is capable of maintaining standing waves which we assume, for simplicity, to be restricted to a plane. Now if the string consists of N atoms, it should be able to entertain N different types of vibration, owing to the fact that each atom can oscillate with respect to its environment with one degree of freedom in position. On the other hand, it is known from the theory of vibrations that only certain, so-called "proper" modes can exist, namely those in which the ends of the string present nodes, i.e. points of

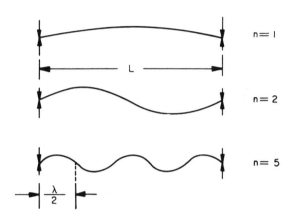

FIG. 1.5. Standing waves in vibrating string. Only those oscillations are stable which subdivide length L by an integer number of half waves, shown here for $n = 1$, 2, and 5. All harmonics are allowed until half wavelengths are equal to interatomic distance.

zero movement, forming in this manner standing waves. This means that the string of length L can maintain only an integral number of proper half wavelengths:

$$n \frac{\lambda}{2} = L \tag{1.15}$$

An upper limit $n = N$ is, of course, reached when the half wavelengths have become as small as the distance between atoms. Thus there exist exactly N proper vibrations which fully account for the N degrees of freedom in position.

We now proceed to compute the density function in the three-dimensional case by counting the number of possible vibrations in a solid. A cube of side L will replace the string used for the one-dimensional case. Here, however, we must also admit those plane standing waves for which directions of propagation may form angles α, β, and γ with the x, y, and z coordinates, respectively. Again, the standing waves must intersect the sides of the cube in integral numbers, and this requires (see Fig. 1.6) that

$$n_1 \frac{\lambda}{2} = L \mid \cos \alpha \mid$$

$$n_2 \frac{\lambda}{2} = L \mid \cos \beta \mid \qquad (1.16)$$

$$n_3 \frac{\lambda}{2} = L \mid \cos \gamma \mid$$

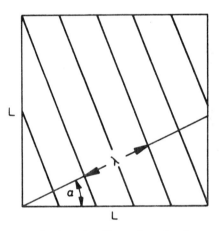

Fig. 1.6.   Vibrational modes in cube. Shown here for the two-dimensional case are proper oscillations which intersect sides with 2 and 5 half waves, respectively.

Furthermore, since the squares of the cosines must add up to 1, it follows from (1.16) that

$$n_1{}^2 + n_2{}^2 + n_3{}^2 = \frac{4L^2}{\lambda^2} \qquad (1.17)$$

Thus, each triplet of integer numbers n_1, n_2, n_3 determines the wavelength and direction of one proper vibration or "mode." However, it is possible to generate vibrations of the same wavelength by a multiplicity of such triplets. More precisely, the question regarding the density of oscillations is answered by determining the number of possible modes, i.e. the number of possible triplets, in the wavelength interval between λ and $d\lambda$ such that

$$\frac{4L^2}{(\lambda + d\lambda)^2} \leqslant n_1{}^2 + n_2{}^2 + n_3{}^2 \leqslant \frac{4L^2}{\lambda^2} \qquad (1.18)$$

Now the form of Eqs. (1.17) and (1.18) suggests the representation of the triplets as points in n-space such that each set of 3 integers stands for unit volume at coordinates n_1, n_2, n_3. Equation (1.18) determines then the points in a spherical shell of radius $2L/\lambda$ and thickness $d\lambda$, and since the integers are all positive numbers, the sphere is limited to one octant of space. Furthermore, because each triplet occupies unit volume, the number dZ of proper vibrations within $d\lambda$ is simply given by the volume of the octant shell, in an approximation improving with the magnitude of n:

$$dZ = \tfrac{1}{8}4\pi\left(\frac{2L}{\lambda}\right)^2 d\left(\frac{2L}{\lambda}\right) = 4\pi V \frac{d\lambda}{\lambda^4} \qquad (1.19)$$

where the cube volume $V = L^3$ has been introduced. The density function can also be expressed in terms of the number of proper vibrations in the frequency interval between ν and $\nu + d\nu$, with $\lambda = c/\nu$ and $|d\lambda| = c|d\nu|/\nu^2$:

$$dZ = \frac{4\pi V}{c^3} \nu^2 d\nu \qquad (1.20)$$

where c is the velocity of wave propagation in the oscillator medium. In this equation dZ represents the number of different modes or, in other words, the "density of states" within a frequency interval $d\nu$. Primarily because the oscillator system is three-dimensional, this number increases with the square of the frequency.

Integration of Eq. (1.20) yields the total number of possible vibrations between $\nu = 0$ and ν, representing the volume of the spherical octant with radius $2L/\lambda$. If there are in the volume N oscillators, each with three degrees of freedom, the total number of oscillations cannot exceed $3N$. This, in turn, defines a maximum frequency ν_M with which the system can vibrate:

$$Z_{\max} = 3N = \frac{4\pi}{3} \frac{V}{c^3} \nu_M{}^3 \qquad (1.21)$$

This theory of vibrational states is applicable to a wide range of problems. First, the results contained in Eqs. (1.18)–(1.21) are not restricted to cubes, but are valid for any volume V (Weyl[3]). Applied to quasi-elastic vibrations, the theory concerns the interaction of infrared and visible radiation with solids. Since dZ is, in effect, the number of degrees of freedom the results apply also to the theory of specific heat (Debye[4]). Furthermore, the electromagnetic field enclosed by a cavity, just as the elastic medium in solids, can give rise to standing waves. It was, in fact, for the purpose of computing the possible vibrational modes of light in a radiating enclosure that Jeans[5] devised and applied the procedure presented here. Significantly there exists in this case, however, no definite number of oscillators (or minimum distance between nodes) so that frequencies are not subject to an upper limit as given by Eq. (1.21). We will see that certain difficulties arise as a consequence of this fact for which the classical model has no remedy.

1.3 Quantized Structure and Radiation

1.3.1 THE QUANTIZED OSCILLATOR

The main difficulty encountered by the purely classical model of the radiating oscillator was this:

According to the just-derived relation (1.20) for the density of states, there can exist a number $dZ = 4\pi\nu^2\, d\nu/c^3$ of possible standing waves within $d\nu$ per unit volume of the radiating cavity. Now according to the classical law of equipartition (of thermal energy at absolute temperature T), each degree of freedom will assume a mean energy $\frac{1}{2}kT$ (where k is Boltzmann's constant, see Table 1.2). Since each oscillator—or oscillating mode—has, according to classical statistics, one degree of freedom in position and another in momentum, each has a mean energy of kT. Furthermore, because we are dealing with electromagnetic waves, another degree of freedom has to be allotted to the direction of the electric vector; in other words, a factor of 2 results from the fact that each mode can exist in two states of polarization. Thus, we can assign to the various oscillator frequencies of the radiating cavity an energy density distribution:

$$u(\nu)\, d\nu = 2kT\, dZ = \frac{8\pi\nu^2}{c^3}\, kT\, d\nu \qquad (1.22)$$

This, in fact, is the Rayleigh-Jeans law of thermal radiation which successfully describes the spectral distribution for long waves. In particular, it correctly presents the vanishing of energy for decreasing ν. For high values of ν, however, the law is in fundamental conflict with experience. This is obvious since no upper limit can be assigned to ν (see remarks at end of previous section), and thus the energy contained in a volume filled with radiation [Eq. (1.22)] would tend to infinity. The experimental facts are, however, that the intensity distribution has a maximum at a frequency which depends on temperature.

For a violent contradiction, there had to be radical interpretation. The scientific event that assumed this role was the quantum postulate of Planck[6,7]: Oscillators cannot assume a continuous set of energy values, but exist at discrete, so-called "quantized" levels of energy*

$$E_n = n\epsilon_0 = nh\nu, \qquad n = 0, 1, 2, \cdots \qquad (1.23)$$

where h is Planck's constant (see Table 1.2). In other words, the oscillator energy can only stepwise change from one state to another, each step being proportional to the frequency ν.

This assumption immediately modifies the statistics of the equipartition law. Since at large values of ν the oscillators have to assume large quanta of energy, the probability that such high frequency states are at all occupied rapidly becomes zero. The mean energy per mode of high frequency, instead of being kT, is in fact much smaller. Chapter 2 will show quantitatively how these considerations lead to the correct description of the spectral energy distribution by Planck's radiation law. Here, we shall continue with a number of basic rules related to the original quantum concept.

1.3.2 THE CORPUSCULAR NATURE OF LIGHT

The radical break with the classical views on the interaction between oscillators and the electromagnetic field was completed by Einstein.[8,9] In his paper on the corpuscular nature of light, Einstein proposed that light itself is emitted in quanta of energy as "photons":

$$\epsilon = E'' - E' = h\nu \qquad (1.24)$$

The classical model of the radiating oscillator (see Fig. 1.2) is therefore modified such that the oscillators, which themselves can exist only in discrete levels E', E'', etc., give off energy in sudden steps as light quanta. The decay time τ of the classical oscillator [defined by Eq. (1.5) and (1.7)] is now interpreted statistically as the mean life of the upper energy state; in other words, after a time τ, all but $1/e$ of the original states have emitted their light quantum (see Section 1.3.5).

Photons, furthermore, are not merely quantized in energy. They are corpuscular in the sense that the radiating atoms emit them as "wave packets" of statistically determined direction and with a momentum p the amount of which is quantized:

$$p = \frac{h}{\lambda} \qquad (1.25)$$

where, again, h is Planck's constant and λ is the wavelength.

---

* The existence of a zero point energy for $n = 0$ is discussed in Section 4.5.3.

The experimental verification of Einstein's quantization conditions for energy and momentum of photons came in the case of the former [Eq. (1.24)] through the photoelectric effect then already discovered (Chapter 7); while in the case of the latter [Eq. (1.25)], commensurate with its difficulty, proof had to wait a relatively long time—until the discovery of the Compton[10,11] effect. The impulse nature of light quanta manifests itself, according to (1.25), to an increasing extent as the wavelength becomes shorter. Thus, it was through the momentum transfer to electrons by x rays, which after scattering showed a frequency shift, that the Compton effect confirmed Eq. (1.25).

1.3.3 Matter Waves

An important implication of (1.25) results, furthermore, when this equation is written in the form

$$\lambda = \frac{h}{p} \tag{1.26}$$

The implication is simply this: since light waves of wavelength λ behave like corpuscles of momentum p, particles of momentum p may behave like waves of wavelength λ (de Broglie[12,13]). Experimental evidence for the existence of such "matter waves" was, in fact, obtained not long after de Broglie's theoretical work. Diffraction patterns could be produced by electron beams after reflection from metal surfaces (Davisson and Germer[14,15]). Not only was the appearance of these patterns equal to that produced by x rays under similar conditions, but the wavelength (which can be computed from such patterns and the lattice parameters of the reflecting crystal) confirmed Eq. (1.26). Since then, electron diffraction methods have replaced x-ray techniques in the exploration of surface structure. Finally, matter waves could be demonstrated by reflection from crystal surfaces with much heavier particles than electrons, viz., with molecules and helium atoms (Stern,[16] Estermann and Stern[17]).

1.3.4 Dualism of Wave and Particle Model

We have cited this evidence for matter waves primarily to show that there exists a mutually complementary dualism between wave and corpuscular concept: both aspects are valid, for phenomena in which the wave nature is prominent, as well as for phenomena in which the particle character is manifest. More precisely, in principle, the same radiation process has features to be described, albeit with varying difficulty, either by the mathematical formalism of the time-and-space expanded wave or by the particle which appears at statistically determined time and position. If the associated pictures appear to be irreconcilable, the conflict lies, not with their physical basis, but with the implicit expectation that submicroscopic events can be demonstrated by macroscopic analogues. In this sense, then, we need, and have to accept, both models. Which one is to be emphasized will depend on the occasion, although sometimes both may serve equally well. The phenomena at long wavelengths of the electromagnetic spectrum are best described by the classical oscillator, while those at high frequency are best represented by the corpuscular picture; in the infrared and visible region, both models have a common domain of usefulness and both have to be consulted.

1.3.5 THE UNCERTAINTY PRINCIPLE

It was seen that the two descriptions, in terms of electromagnetic waves or of discrete particles, cannot be expected to yield more than a formal agreement since they approach their subject, as it were, from two opposing points of view. Nevertheless, there exists, due to Born,[18,19] a successful attempt to unify the two models by statistical description. According to this interpretation, quantum theory, in predicting the behavior of large numbers of particles (or the probable course of an individual), arrives at statistical distributions for which the contours are provided precisely by the wave envelopes of electromagnetic theory. This view proved to be quite fruitful when combined with another quantum rule, which is closely related to those discussed previously—Heisenberg's[20] uncertainty principle. We may introduce this subject by comparing the classical and quantized oscillator models.

The classical viewpoint is that the radiating oscillator loses its energy gradually according to an $e^{-t/\tau}$ law so that all but $1/e$ of the initial energy is spent after a time τ. Furthermore, as a result, the resonant frequency ν broadens to a halfwidth $\Delta\nu_h$ [cf. Eqs. (1.7), (1.10), and (1.12)].

How can quantum theory present this behavior statistically? First of all, photons are observed, the energies of which scatter around a mean $h\nu$ with a halfwidth $h\,\Delta\nu_h$. The radiation is due to quantized upper states of radiating centers which can spontaneously drop to lower states with a transition probability γ per unit time, each thereby emitting a quantum of light. Just when this event occurs, is, of course, uncertain for the individual center, but on the whole the population should behave classically and decay as $e^{-\gamma t}$. The uncertainty Δt for the time of photon emission is thus in the average

$$\overline{\Delta t} = \frac{\int_0^\infty te^{-\gamma t}\, dt}{\int_0^\infty e^{-\gamma t}\, dt} = -\frac{d}{d\gamma} \ln \left(\int_0^\infty e^{-\gamma t}\, dt \right) = \frac{1}{\gamma} \qquad (1.27)$$

In order that the two models agree, the halfwidths have to agree and this also holds for the times of decay, i.e. $\tau = 1/\gamma$. Equation (1.12) for the classical oscillator now assumes the role of a relation between uncertainties of energy and time for the quantized case:

$$\overline{\Delta \epsilon}\, \overline{\Delta t} = \frac{h}{2\pi} \qquad (1.28)$$

Equation (1.28) represents one form of Heisenberg's uncertainty principle. In the context in which (1.28) has been derived, it merely states that the product of the width and the lifetime of the upper energy level is equal to Planck's constant, at least by order of magnitude. Actually, however, the validity of Heisenberg's principle is quite general. Whenever a system exists, or is observed, during a time Δt, its energy is, in principle, only determinable within $\Delta \epsilon$. Furthermore, there exist more fundamental derivations of Heisenberg's uncertainty relation,* and these state an inequality:

$$\Delta \epsilon\, \Delta \tau \geqslant \frac{h}{4\pi} \qquad (1.29)$$

Another, quite related, presentation of Heisenberg's uncertainty principle is stated in terms of position (x) and momentum (p_x) variables:

$$\Delta x\, \Delta p_x \geqslant \frac{h}{4\pi} \qquad (1.30)$$

The position and momentum of a particle is again undetermined to the extent prescribed by (1.30). It is sometimes advantageous to describe events in six-dimensional phase space in which a point has, in addition

---

* $\Delta \epsilon$ and Δt represent rms-deviations (cf. Chapter 7) such as $[\langle (\Delta \epsilon)^2 \rangle]^{1/2}$. In general, $\Delta \epsilon\, \Delta t \sim h$, but the exact coefficient of h in the uncertainty relation depends on the averaging process performed, as well as the probability distribution of the particle.

to the three position coordinates x, y, z, three momentum coordinates p_x, p_y, p_z. Equation (1.30) then asserts that the position of a particle in phase space can be determined only within $\sim h^3$ as the smallest volume element in phase space which can have a physical meaning.

We may, for example, ask for the number of photons in an enclosure of volume V that differ from each other in the amount and direction of their momentum $\mathbf{p}$. Here we use the relation

$$| \mathbf{p} | = (p_x{}^2 + p_y{}^2 + p_z{}^2)^{\frac{1}{2}} = h\nu/c \qquad (1.31)$$

We have then, in the interval $dp = (h/c) \, d\nu$ a phase space volume of $4\pi V h^3 \nu^2 \, d\nu/c^3$. This volume contains a finite number of smallest distinguishable elements h^3, and hence, of photon modes:

$$dZ = \frac{4\pi V \nu^2 \, d\nu}{c^3} \qquad (1.32)$$

This, by order of magnitude, agrees with the number of standing waves which followed before from purely classical considerations [see Eq. (1.20)].

1.4 Processes of Interaction

Starting with the premise that all optical phenomena are ultimately based on the interaction between the electromagnetic field and the electric charges of which matter consists, we have in this chapter evolved the ground rules which govern the bonding between the charges. It is seen that there are several major aspects under which this material has to be treated. First, an important differentiation exists between those interactions which can be described with classical, quasi-mechanical models and, conversely, those processes which are best understood with quantized structures. Another aspect concerns selectivity of response to frequency such as represented by the two extreme cases of the pure harmonic oscillator and the coupled oscillator continuum. Another example, which we will encounter in Chapter 3 is the distinction according to phase relationship of the oscillations. In the context of the theory, these classifications serve as guiding threads. Derived, as they were, for quasi-mechanical models, they nevertheless are valid for the interaction with the electromagnetic field.

The major subject of the following chapters is the detailed treatment of the interactions between matter and radiation. To a certain extent, this treatment follows the subdivision advanced here. Thus, the phenomena of emission and absorption are first discussed for continua, viz., as thermal radiation of solids (Chapter 2). The interaction with isolated oscillators is next treated as emission and absorption of spectral lines (Chapter 3). This sequence has the advantage of using the thermodynamics of heat radiation also as a foundation for the monochromatic case, namely for the interrelation between processes of line emission and absorption. The concept of monochromatic radiation is, of course, an idealization; the emission of light is always intimately associated with processes which tend to broaden, or at least alter, the pure harmonic oscillator. In fact, the physical mechanisms responsible for relative line widths varying from 10^{-14} to 10^{-1}, i.e., from an almost pure spectral line to almost a continuum, warrant sufficient interest to be treated in a special subchapter of their own (Section 3.4). In Chapter 4, radiating structures of increasing complexity are discussed: electron states determined by four quantum numbers, hyperfine structure, the spectroscopy of molecules and, finally, the spectra of the solids. While such interactions as the generation of light are treated mainly as quantized processes, the propagation phenomena of radiation are best understood on the basis of the wave picture. This includes reflection, refraction, absorption, and polarization in gases, insulators, and metals (Chapter 5). The next chapter deals with the more recent field of coherent radiation and lasers in which aspects of emission and propagation or of the quantum and the wave picture have to be treated together. Finally, in Chapter 7, the secondary effects of light in solids are taken up, together with their application in detectors and the fluctuation processes which present the ultimate limit of performance.

REFERENCES

1. M. Born and E. Wolf, "Principles of Optics." Pergamon Press, New York, 1959.
2. A. Landé, *Ann. Physik* [4] **50**, 89 (1916).
3. H. Weyl, *Crelles J.* **141**, 163 (1912).
4. P. P. Debye, *Ann. Physik* [4] **39**, 789 (1912).
5. J. H. Jeans, *Phil. Mag.* [6] **10**, 91 (1905).
6. M. Planck, *Verhandl. Deut. Phys. Ges.* **2**, 237 (1900).
7. M. Planck, *Ann. Physik* [4] **4**, 553 (1901).
8. A. Einstein, *Ann. Physik* [4] **17**, 132 (1905).
9. A. Einstein, *Ann. Physik* [4] **20**, 199 (1906).
10. H. A. Compton, *Bull. Natl. Res. Council* **20**, 10 (1922).
11. H. A. Compton, *Phys. Rev.* **21**, 483 (1923).
12. L. de Broglie, *Nature* **112**, 540 (1923).
13. L. de Broglie, *Ann. Phys. (Paris)* [10] **3**, 22 (1925).

14. C. DAVISSON and L. H. GERMER, *Phys. Rev.* **30**, 705 (1927).
15. C. DAVISSON and L. H. GERMER, *Proc. Natl. Acad. Sci. U.S.* **14**, 318 (1928).
16. O. STERN, *Naturwissenschaften* **17**, 391 (1929).
17. I. ESTERMANN and O. STERN, *Z. Physik* **61**, 95 (1930).
18. M. BORN, *Z. Physik* **37**, 863 (1926).
19. M. BORN, *Z. Physik* **38**, 803 (1926).
20. W. HEISENBERG, *Z. Physik* **43**, 172 (1927).

GENERAL BIBLIOGRAPHY

W. HEITLER, "The Quantum Theory of Radiation," 3rd ed. Oxford Univ. Press, London and New York, 1954.

A. LANDÉ, Optik und Thermodynamik, *in* "Handbuch der Physik" (H. Geiger and K. Scheel, eds.), 1st ed., Vol. 20, p. 453. Springer, Berlin, 1928.

L. R. KOLLER, "Ultraviolet Radiation." Wiley, New York, 1952.

R. A. SMITH, F. E. JONES, and R. P. CHASMAR, "The Detection and Measurement of Infrared Radiation." Oxford Univ. Press, London and New York, 1957.

2

THE EMISSION AND ABSORPTION
OF HEAT RADIATION

An important interaction between optical radiation and matter, both as to relevancy and physical significance, is that of generation and absorption of radiation. The next two chapters will deal with this subject, following an obvious dividing line: by first treating spectral continua in heat radiation and discussing subsequently the laws governing monochromatic emission and absorption. It will be seen that thermal radiation is not limited to any particular wavelength region or to a particular type of spectrum since all electromagnetic radiation represents flux of energy and thus can, at least in principle, generate heat, or in turn be generated by it. If, nevertheless, the concept of heat radiation is used, it is because it defines energy emitted or absorbed solely owing to the thermal equilibrium of the system. It follows that this subject matter belongs as much to the theory of heat as of optics. This is reflected by the fact that most of the radiation laws can be derived either by means of thermodynamics or, conversely, by means of the statistics of quanta. The fact that emission and absorption are collectively treated is, in itself, an indication that these two opposite processes are thermodynamically related. The second law will indeed show an intimate connection between the probability of generating a photon with that of absorbing it in the same mode. Much of optical radiation theory revolves around this fact so that it is actually not possible to treat one process without the other.

Heat radiation is, of course, not limited to the continuous spectra of solids, liquids, and dense gases. Isolated atoms and molecules can be thermally excited as well and will in that case emit spectral lines and bands, respectively. However, gases present special more complicated cases of the thermodynamic relationships involved, and their practical significance as thermal emitters is small. We will, therefore, study the laws of heat radiation mainly in connection with solid matter. The importance of gaseous interactions with radiation will emerge in the next chapter in terms of monochromatic absorption and emission in which thermal excitation plays only a subordinate role.

2.1 Geometrical Relationships in Radiation Theory

There exist certain relationships which, since they follow directly from solid geometry, apply to radiation theory in general. They are sometimes referred to as physical laws, although this is correct only in the sense that compliance with geometry is in itself a fact of physical significance. Thus, the starting point of these considerations is an axiomatic statement: if a radiation field can be considered sufficiently uniform in space and time, the energy passing through a surface element will be proportional to the area, the time, and the solid angle of the radiation cone with which the element intercepts such flux. We will now use this statement to relate the powers radiated from the surface of one body to that received by another. The results of such computations are, of course, not limited to radiation which originates at a surface, but apply to reflected, scattered, or transmitted energy as well. It is, at any rate, an idealization to consider radiation as being emitted or absorbed by surfaces since these interactions require depths of at least several atom layers.

2.1.1 Brightness and Lambert's Cosine Relation

We define as brightness or radiance[1] B_0 of a surface the amount of energy radiated normally from the surface per second per square centimeter per unit solid angle. Furthermore, we define as radiation power, or radiant flux density, W, the energy emitted totally over all angles, i.e. the solid angle 2π, again per second and per square centimeter. Usually, we shall limit ourselves to a stated wavelength interval between λ and $\lambda + d\lambda$. The units of such spectral brightness are, then, watts/(cm² · steradian · μ). To express radiation power W in terms of brightness B_0, one has to know the dependence of radiation on angle.

Assume that a point P receives radiation emitted normally from a small surface area S (i.e. emitted toward the left of Fig. 2.1). This radiation

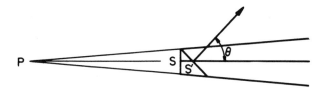

Fig. 2.1. Geometrical relationships in Lambert's law.

fills a cone or pencil of rays. Now the solid angle of this cone remains unchanged, if radiation were to originate from a larger surface area S' which intersects the cone with an angle θ between its normal and the axis. Thus, if there exists no preferential direction in emission (or reflection), that is to say, if radiation is perfectly diffuse, S' transmits the same power $B(\theta)$ toward P as its projection S along the cone axis. This is the basis of Lambert's law

$$B(\theta) = B_0 \cos \theta \qquad (2.1)$$

which is valid for all "black" surfaces (see Section 2.2.2), but also holds approximately for other radiating, or diffusely reflecting, surfaces. Similarly, if radiation is emitted from a surface element at P (toward the right of Fig. 2.1), the power transmitted through any surface intersecting the same solid angle remains the same and obeys Eq. (2.1.).

Lambert's cosine relation explains why incandescent spherical and cylindrical bodies appear to the eye as flat disks or ribbons. There are, however, exceptions. The sun, for instance, because of the radial density distribution of its luminous mass, does not emit diffusely, and accurate photometric evaluation of its image shows the existence of a darker edge.

2.1.2 Radiant Flux

Assuming Lambert's law, the radiation power W (or radiant flux density) is obtained from B_0 by integration over a half-sphere with unit radius (see Fig. 2.2). We divide the unit half-sphere into zones of width $d\theta$ and circumference $2\pi \sin \theta$. The total radiation density into solid angle 2π is then

$$W = 2\pi B_0 \int_0^{\pi/2} \sin \theta \cos \theta \, d\theta = \pi B_0 \qquad (2.2)$$

Hence, the average flux density per unit solid angle equals half the brightness:

$$B_{av} = \frac{W}{2\pi} = \frac{B_0}{2} . \qquad (2.3)$$

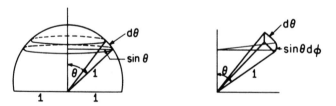

Fig. 2.2.   Geometrical relationships for the derivation of radiant flux density.

The power radiated within a solid angle $d\Omega$ ($= \sin\theta\, d\theta\, d\varphi$) will, at a distance R, be distributed over an area dA normal to the direction of the beam:

$$dA = R^2\, d\Omega = R^2 \sin\theta\, d\theta\, d\phi \qquad (2.4)$$

Aside from the reduction of the flux density by $1/R^2$, there may be, in addition, attenuation due to absorption and scattering, although such effects will be ignored in this section. If the radiation source consists of an element dA_1 with brightness B_0, the power radiated from element dA_1 to element dA_2 is given by

$$dP = B_0 \cos\theta\, dA_1\, d\Omega$$

or

$$dP = \frac{B_0 \cos\theta_1 \cos\theta_2\, dA_1\, dA_2}{R^2} = \frac{W \cos\theta_1 \cos\theta_2\, dA_1\, dA_2}{\pi R^2} \qquad (2.5)$$

where θ_1 and θ_2 are the angles which the normals of the surface elements dA_1 and dA_2, respectively, form with the direction of propagation.

If the projections of the dA's are summed to form A_1 and A_2, viz., $A_1 = \int \cos\theta\, dA_1$ and, if W is uniform over A_1, one obtains a simplified form for the radiant flux, i.e., the power transmitted from A_1 to A_2:

$$P = \frac{B_0 A_1 A_2}{R^2} = \frac{W A_1 A_2}{\pi R^2} \qquad (2.6)$$

where it has been assumed that the linear dimensions of A_1 and A_2 are small compared with the distance R (see also Fig. 2.3).

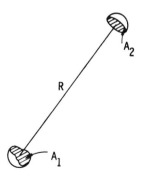

FIG. 2.3. Power radiated from A_1 to A_2 for irregularly shaped surfaces.

2.1.3 Energy Density

Because radiation is propagated at finite velocity, there exists in any radiation field a finite amount of radiation energy per unit volume. This is called energy density or simply radiant density u [joules/cm³]. Assume first the simple case that uniform radiation B_x [watts/cm²] is propagated in the x-direction with velocity c. The energy density is then

$$u = \frac{B_x}{c} \qquad (2.7)$$

Of greater interest, however, is the case of the field enclosed by radiating walls of uniform brightness B_0 since it is for these conditions that the laws of temperature radiation are derived. The relationship between the densities of energy and flux follows from geometrical considerations[2] on the basis of Fig. 2.4. Here, a small volume V_2 is shown near the center of a surrounding spherical wall, presumed to be the radiating enclosure.* All linear dimensions of V_2 shall be small compared with the radius R of the wall, although still large compared to the wavelength of the radiation. Each surface element dA_1 of the enclosure then contributes to V_2, according to Eqs. (2.5) and (2.7), an energy content

$$dU = dA_1 \frac{B_0}{R^2 c} \int_{(V_2)} l \cos \theta_2 \, dA_2 = dA_1 \frac{B_0}{R^2 c} V_2 \qquad (2.8)$$

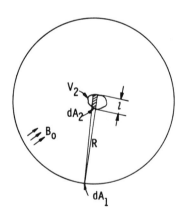

Fig. 2.4.   Brightness and energy density in black body.

---

* The restriction of V_2 to the center serves only mathematical convenience. Because of Lambert's law, whatever holds for the center in terms of energy density is valid everywhere in the enclosure.

where l is the length of the section, and θ the angle of incidence, with which the rays from dA_1 intersect V_2 at dA_2. Integrating Eq. (2.8) over the whole spherical surface $A_1 = 4\pi R^2$ and dividing by V_2 then yields the energy density

$$u = \frac{4\pi B_0}{c} \tag{2.9}$$

In turn, if we consider the energy density u in an enclosure as given, we can now derive the radiant flux density W over a solid angle 2π. Assuming a small opening in the enclosure, we find with Eqs. (2.2) and (2.9)

$$W = \pi B_0 = \frac{c}{4} u \tag{2.10}$$

This relationship between density of flux and energy is, of course, plausible from simpler, if less precise, arguments.

2.2 Radiation Laws of Thermodynamic Origin

We shall now turn from the geometrical considerations, which are typical for electromagnetic rays as such, to laws specifically valid for heat radiation. Thus, we are here concerned with the dependence of emitted and absorbed power on the nature and temperature of the interacting media as well as the wavelength. Since these relationships are part of the theory of heat, they must, to a large extent, be accessible to its methods. The conventional starting point is, in fact, the second law of thermodynamics or, perhaps more precisely, one of its consequences. This is the empirically irrefutable principle that bodies in an isolated system will tend to reduce their temperature differences, thereby approaching an irreversible equilibrium in which all temperatures are equal. Thus, the radiation laws introduced thermodynamically do not employ, for their derivation, any particular model of a light source such as discussed in Chapter 1. Yet this approach will suffice, first to relate power emitted to that absorbed (Section 2.2.1); then to determine such power, radiated over the entire spectrum, as a function of temperature; and finally to derive a number of laws, albeit with limitations of validity, which govern the distribution of intensity over the electromagnetic spectrum. In all this, we will use only the simple picture of a radiation field enclosed by walls at uniform temperature. Such a field has, according to Section (2.1.3), a finite energy density, and it is this quantity that the various theories of thermal radiation yield first, so that subsequently the flux densities can be computed with Eq. (2.10).

2.2.1 Kirchhoff's Law

Assume that two bodies (1) and (2) are at the same temperature in radiation equilibrium with each other. The first body (1) transmits a power $P_1 = W_1 A_1 A_2 / \pi R^2$ [Eq. (2.6)] to the other which shall have an absorption coefficient Σ_2, i.e. which will absorb the fraction $\Sigma_2 P_1$ of the incident flux. Similarly, the first body with an absorption coefficient Σ_1 will receive from (2) a power $P_2 = W_2 A_1 A_2 / \pi R^2$ of which it will absorb the fraction $\Sigma_1 P_2$. Now, the thermodynamic statement of the previous paragraph demands that these two amounts of absorbed power be equal. Otherwise, although starting from the same temperature, one of the bodies would heat up spontaneously at the expense of the other which would get colder. Thus, we have

$$\Sigma_2 W_1 = \Sigma_1 W_2 \tag{2.11}$$

So far, we have neglected the fractional powers $r_2 \Sigma_2 (W_1 A_1 A_2 / \pi R^2)^2$, etc., returned to the emitters by reflection (with coefficients r) as well as other effects of the surrounding system. It is, of course, conceivable that under specialized conditions a thermal equilibrium may be obtained, although Eq. (2.11) is not fulfilled. The point is, however, that the temperatures must remain equal regardless of these external conditions. Thus, Eq. (2.11) must obtain at least sometime; however, since the quantities in (2.11) do not depend on the rest of the system, it must obtain at all times. This leads to Kirchhoff's law:

$$\frac{W_1}{\Sigma_1} = \frac{W_2}{\Sigma_2} = W_0 \tag{2.12}$$

which states that the ratio of emitted power and absorption coefficient is the same for all objects and equal to a value W_0 which, while still dependent on wavelength and temperature, is not dependent on the shape or material of the radiator. Kirchhoff's law in the form of Eq. (2.12) makes a statement only with respect to the overall balance of heat radiation between two (or more) bodies. It will be easily seen that a more encompassing statement is obtained, if reference is made to separate intervals of frequency and space.

For this purpose, we imagine a small test body within a radiating enclosure, the whole system having reached temperature equilibrium.

The power density in the cavity is uniform and isotropic. We can then state that an equal amount of radiation is incident on the body as emerges from it by the processes of emission, reflection, and transmission. Furthermore, this must be true for any separate interval dv of the spectrum. For if this were not the case, the radiation density at such a frequency would either decrease or increase at the expense of the rest of the spectrum. Any other object in the enclosure which selectively absorbs within an interval v and $v + dv$ would then get spontaneously warmer or colder in contradiction to the second law of thermodynamics. Similar reasoning shows that the equality between incoming and outgoing radiation must hold also for any subdivision of the solid angle Ω and the radiating surface A. Thus, if K_v is the power density incident on the body, B_v its emissive power or brightness and Σ_v its "absorbing power," i.e. the fraction absorbed of the incident power, then the fraction $(1 - \Sigma)$ is reflected or transmitted and the power balance at the surface requires that

$$2K_v \cos \theta d\Omega \, dA \, dv = (1 - \Sigma_v)2K_v \cos \theta \, d\Omega \, dA \, dv + 2B_v \cos \theta \, d\Omega \, dA \, dv \quad (2.13)$$

where the factor of 2 accounts for two directions of polarization at any interval dv. From Eq. (2.13) another form of Kirchhoff's law follows:

$$\frac{B_v}{\Sigma_v} = K_v \quad (2.14)$$

The ratio of emissive power to "absorbing power" at any frequency is for all bodies equal to the surface brightness K_v of the surrounding walls with which they are in thermal equilibrium. The difference between this derivation and that of Eq. (2.12) is not merely that validity is extended to separate intervals of frequency and space, but lies, in addition, in the reference of (2.14) to the volume, rather than the surface, of the body which may transmit, reflect, absorb, and emit radiation.

2.2.2 Spectral Emissivities and Black Body Radiation

As a consequence of Kirchhoff's law, there must exist a function $K(v, T)$, valid for all forms and materials of radiating bodies, such that their spectral radiance is

$$B(v, T) = \Sigma(v, T)K(v, T) \quad (2.15)$$

In other words, the spectral brightness equals the product of a general function $K(\nu, T)$ and a specific coefficient $\Sigma(\nu, T)$ called spectral emissivity or radiant emissivity, which according to Kirchhoff's law is equal to the absorptivity, i.e. the fraction of incident power absorbed at the same frequency and temperature. In the inside of an enclosure, radiation is completely absorbed, if the walls are impervious to it. This condition of "blackness" implies that $\Sigma(\nu, T) = 1$ for all frequencies and temperatures and

$$B(\nu, T) = K(\nu, T) \qquad (2.16)$$

This ideal state of emission, in which the emissivity is everywhere unity, is called "black body radiation." A black body is experimentally realized by a cavity held at constant temperature T. The radiation is observed through an opening, the linear dimensions of which are small compared with that of the cavity (although still large in comparison with the wavelength to avoid diffraction effects).

The Σ-values of surfaces can reach unity only within a few percent, even if a type of black body action is obtained by virtue of a porous or rough structure. One speaks of "gray" surfaces, if their emissivities are independent of wavelength so that they radiate, at lower intensity levels, with the spectral distribution of a black body. Metals in bulk often radiate gray, and since they are, under such conditions, impervious it follows that their reflectivity is also independent of wavelength. Most materials, of course, have emissivities which vary strongly with wavelength (i.e. they have "color"). Gases, in particular, may be considered as having zero emissivity everywhere in the spectrum except in certain discrete regions. The dependence of emissivity on temperature is usually small and can be ignored for most solids under practical conditions.

2.2.3 The Stefan-Boltzmann Law

It will be seen from the aforesaid that black body radiation summed over all wavelengths can only depend on temperature. A formula for $W(T)$ can be derived from principles of thermodynamics.* This

---

* This derivation, which is outlined briefly in the following, is in itself an example of a very general procedure in thermodynamics, namely, of drawing conclusions merely from the fact that its variables must form total differentials. Assume a medium of total energy content U at pressure P and consider its temperature T and volume V

relation was first found experimentally by Stefan (1879) and then derived thermodynamically from the radiation pressure by Boltzmann (1884). The Stefan-Boltzmann law of radiation has the form

$$W(T) = \sigma T^4 \tag{2.21}$$

Here $W(T)$ is the total power radiated by a black body per cm² into a solid angle 2π. Thus, $W(T)$ is proportional to the fourth power of the absolute temperature. The constant of proportionality (Stefan-Boltzmann constant) has the value $\sigma = 5.67 \times 10^{-5}$ erg cm^{-2} deg^{-4} sec^{-1}. The radiation emitted from a body of surface area A and emissivity Σ is then given by

$$W_A(T) = A\Sigma\sigma T^4 \tag{2.22}$$

Equations (2.5) and (2.6) can now be used to compute the radiation power transmitted from one surface to another. As an example, if radiation from an element of A_1 at angle θ fills a solid angle $d\Omega$ and arrives perpendicular on dA_2 at distance R, the transmitted radiation power is

$$dP = \frac{1}{\pi} \Sigma_1 \, \sigma T_1^4 \, dA_1 \cos \theta \, d\Omega \tag{2.23}$$

---

as independent variables. The first and second law of thermodynamics can then be expressed by the total differential of the entropy

$$dS = \frac{dU + PdV}{T} = \frac{1}{T}\frac{\partial U}{\partial T} \, dT + \frac{1}{T}\left(\frac{\partial U}{\partial V} + P\right) dV \tag{2.17}$$

where now the condition of integrability demands that

$$\frac{\partial}{\partial V}\left(\frac{1}{T}\frac{\partial U}{\partial T}\right) = \frac{\partial}{\partial T}\left[\frac{1}{T}\left(\frac{\partial U}{\partial V} + P\right)\right] \quad \text{or} \quad \frac{\partial U}{\partial V} = T\frac{\partial P}{\partial T} - P \tag{2.18}$$

The meaning of this very general relation, Eq. (2.18), for an electromagnetic radiation field of volume V within enclosing reflecting walls at temperature T is that U represents the total contained radiation energy, i.e., $U = V \cdot u(T)$, and is, furthermore, based on the interpretation of P as a radiation pressure. Such a pressure results, in fact, as a consequence of Maxwell's equations or from statistical mechanics, namely, if the radiation field is treated as a gas of photons having a density $n = u/h\nu$ and a momentum $mc = h\nu/c$ so that

$$P = nm\frac{c^2}{3} = nh\frac{\nu}{3} = \frac{u}{3} \tag{2.19}$$

Introducing the radiation pressure into Eq. (2.18) yields

$$u = \frac{T}{3}\frac{du}{dT} - \frac{u}{3} \quad \text{or} \quad \frac{du}{u} = 4\frac{dT}{T}$$

hence,

$$u = aT^4 \tag{2.20}$$

which with Eq. (2.10) leads to $W = \sigma T^4$, where $\sigma = c \cdot a/4$.

Suppose dA_2 has an emissivity Σ_2. The power absorbed by dA_2 is then

$$dP_2 = \frac{1}{\pi} \Sigma_1 \Sigma_2 \sigma T_1^4 \, dA_1 \cos \theta \, d\Omega = \frac{1}{\pi R^2} \Sigma_1 \Sigma_2 \sigma T_1^4 \cos \theta \, dA_1 \, dA_2 \qquad (2.24)$$

If the radiating source is a sphere of radius r_1, the projections of all surface elements amount to πr_1^2. At a distance R, a disk of radius r_2 will intercept an area $dA_2 = R^2 d\Omega = \pi r_2^2$, where R has been assumed to be large compared with radii r_1 and r_2. Hence, in this example, the power absorbed by the disk is

$$P_2 = \frac{\pi r_1^2 r_2^2}{R^2} \sigma \Sigma_1 \Sigma_2 T_1^4$$

Equations (2.21)–(2.24) are used also to determine radiation equilibrium conditions, such as the temperature which a body assumes when the radiation it emits balances the power which it receives in some manner. Unless the temperature of the body is much higher than that of its environment, account must be taken of the radiation it receives from its environment.

2.2.4 WIEN'S DISPLACEMENT LAW

We now return to the radiation function $K(\nu, T)$, which has been introduced by Eq. (2.15). From a historical point of view, it may well be said that the difficulties, and the ultimate success, in the determination of this one function marked one of the most important turning points in the evolution of scientific thought. The Stefan-Boltzmann law had, by purely thermodynamic reasoning, uncovered one aspect of the radiation function, namely, that when integrated over the entire spectrum, its value is proportional to the fourth power of the temperature:

$$\int_0^\infty K(\nu, T) \, d\nu = \frac{1}{\pi} W(T) = \frac{\sigma}{\pi} T^4 \qquad (2.25)$$

Whatever $K(\nu, T)$ turns out to be, it has to fulfill Eq. (2.25). Progressing beyond the derivation indicated in the footnote to Section 2.2.3, Wien (1893) proved thermodynamically another relationship for $K(\nu, T)$:

$$K_\nu = \frac{\nu^3}{c^3} F\left(\frac{T}{\nu}\right) \qquad (2.26)$$

where $F(T/\nu)$ is a still-undetermined function of the single variable T/ν. Thus neither Eq. (2.25) nor (2.26) give a description of the functional relationship between the radiation flux density and the two variables of temperature and frequency, yet they reveal certain properties that such a function must have, and these properties are of interest in their own right. The most important feature of Wien's relation, Eq. (2.26), is that it implies a statement concerning the spectral location of a radiation maximum. That such a maximum must exist follows, not only from experimental evidence, but from the fact that the flux density must be zero for infinitely large, and infinitely small, frequencies. The manner in which the maximum shifts with temperature is more simply represented by the wavelength rather than the frequency scale and is the reason why Wien's statement is referred to as *displacement law*.

The conversion from units of frequency to those of wavelengths must, of course, leave the spectral interval and the corresponding value for the intensity unchanged. This condition determines the relationship between K_ν and the wavelength function for the radiation density $B_\lambda = B(\lambda, T)$:

$$B_\lambda \, d\lambda = K_\nu \, d\nu \tag{2.27}$$

where with $\nu = c/\lambda$ and $d\nu = -c \, d\lambda/\lambda^2$, Eq. (2.26) now has the form

$$B_\lambda = \frac{c}{\lambda^5} F\left(\frac{\lambda T}{c}\right) \tag{2.28}$$

The location of the radiation maximum is determined by the condition that the derivative of Eq. (2.28) with respect to λ is zero.
Thus

$$\frac{\partial B_\lambda}{\partial \lambda} = 0 = -\frac{5c}{\lambda^6} F\left(\frac{\lambda T}{c}\right) + \frac{T}{\lambda^5} F'\left(\frac{\lambda T}{c}\right)$$

$$F'\left(\frac{\lambda T}{c}\right) - 5 \frac{c}{\lambda T} F\left(\frac{\lambda T}{c}\right) = 0 \tag{2.29}$$

It is seen that the variables λ and T do not appear in Eq. (2.29) except as a product. It follows that, whatever the form of the function F, its solution must have the form

$$(\lambda T)_{\text{max}} = C_0 \tag{2.30}$$

Thus, the spectral radiation profile peaks at a wavelength which is inversely proportional to the absolute temperature of the black body radiator. In other works, Eq. (2.30) confirms the basic observation that the dominant color of a glowing object turns toward shorter wavelengths for higher temperatures.

Another conclusion[2] to be drawn from Eq. (2.26) results from integration of the radiation flux over the entire spectrum:

$$\int_0^\infty K_\nu \, d\nu = \frac{1}{c^3} \int_0^\infty \nu^3 F\left(\frac{T}{\nu}\right) d\nu$$

Introducing $x = \nu/T$ as a new variable of integration and a function $G(x) = F(1/x)$, one obtains

$$\int_0^\infty K_\nu \, d\nu = \left\{ \frac{1}{c^3} \int_0^\infty x^3 G(x) \, dx \right\} \cdot T^4 \tag{2.31}$$

where the value in brackets on the right-hand side of (2.31) amounts to a simple constant. Since the left-hand side of the equation is a measure of the total radiation power $W(T)$, as defined in Section 2.2.3, one obtains again the Stefan-Boltzmann law:

$$W(T) = \sigma T^4$$

verifying the statement made in the beginning of this section that the function K_ν has to satisfy Eq. (2.25).

2.3 Quantum Models and Statistics

The methods and results presented in the last section are more or less indicative of the efforts made by theoreticians, notably during the end of the nineteenth century, to conclude as much as possible of the heat radiation principles by the use of thermodynamics alone. An attack, thus limited to the tools of the theory of heat, cannot be expected to penetrate to the heart of the radiation mechanism, either for localized emitting centers or for the continua of radiating solids. It is, in particular, not possible to derive by these means the complete form of the radiation function $K(\nu, T)$, Eq. (2.15). Basically the reason for this limitation is that while thermodynamics can draw general conclusions on heat exchange by radiation, it cannot reveal details of the spectral radiation distribution without specific assumptions about the manner in which the interaction between radiation and matter depends on frequency. In other words, for further progress detailed models are needed such as described in Chapter 1. Furthermore, since heat radiation involves a large number of interacting centers, statistical methods are indicated. Thus we are concerned in the following with the application of various statistical treatments to specific oscillator models. The conclusions arrived at by thermodynamics must, of course, remain valid and are, in fact, used to complement the statistical approach.

2.3.1 Planck's Radiation Law and Its Approximations

There exist two—apparently unrelated—model concepts on the basis of which one can undertake to compute the black body radiation functions $u(v, T)$, or from this, $K(v, T)$.

The first of these models[2] considers radiating matter, such as the walls of a black body cavity, as made up of dipole oscillators. Positive and negative charges thus describe linear harmonic oscillations at various proper frequencies and, as a result, electromagnetic radiation is emitted or, conversely, such motion is induced by the absorption of an incident electromagnetic wave. The radiation density in thermal equilibrium is then just that for which the power absorbed by the oscillators equals that emitted by them.

The second approach[3] is based on the procedure of Rayleigh-Jeans which has been described in Section 1.2.3. In this model, the electromagnetic field within a radiating enclosure is compared to an elastic medium capable of maintaining a number dZ of standing waves or proper modes within a frequency interval between v and $v + dv$. Allowing for two polarizations for each proper vibration, this number of modes per unit volume follows from Eq. (1.20):

$$dZ = \frac{8\pi v^2}{c^3} \, dv \tag{2.32}$$

The energy density per unit frequency interval is then obtained by assigning a mean energy $\bar{\epsilon}_v$ to each mode:

$$u_v = \frac{8\pi v^2}{c^3} \, \bar{\epsilon}_v . \tag{2.33}$$

The same expression is also obtained by the first method, although we shall forego here the derivation by electromagnetic theory. The value $\bar{\epsilon}_v$ is, of course, conceptually—although not quantitatively—different for the two cases: it represents the mean energy of a dipole, e.g., an oscillating electron, in the first instance; but that of an electromagnetic field mode in the second case. At any rate, $\bar{\epsilon}_v$ must be available from statistical considerations. Indeed, classical Boltzmann statistics has an immediate answer in the equipartition law which sets $\bar{\epsilon}_v$ equal to kT at any frequency. The resulting dilemma, because of its importance for physics in general, has already been discussed in the first chapter, along with its subsequent solution. First, the introduction of kT for $\bar{\epsilon}_\nu$ into Eq. (2.33) yields the radiation law of Rayleigh-Jeans

$$u_\nu = \frac{8\pi\nu^2}{c^3} kT$$

which, while in experimental agreement for small ν, fails in the high frequency end of the spectrum since u_ν grows as a monotonic function of ν. Furthermore, similar objections against the unqualified application of the equipartition law apply to other, albeit related, situations such as the specific heat of solids, especially at low temperatures.

As a remedy for the failure of previous radiation laws, Planck postulated that the dipole oscillators of radiating matter could exist only in discrete energy states, $n\epsilon_0$ ($n = 0, 1, 2 \ldots$), rather than in a continuum of energy values. This leads to an apparently innocuous modification for computing the mean oscillator energy (see, e.g., Born[3], p. 238). Namely, unlike the classical method which proceeds by integrating over a continuous range of energy values [according to Eq. (2.34)], the condition of quantized states requires a summing over discrete values [in Eq. (2.35)]. In either case, the averaging process demands that the individual energy values ϵ be weighted by the Boltzmann probability factor $e^{-\epsilon/kT}$

Thus we have in classical statistics, with $\beta = 1/kT$:

$$\bar{\epsilon} = \frac{\displaystyle\int_0^\infty \epsilon e^{-\beta\epsilon}\, d\epsilon}{\displaystyle\int_0^\infty e^{-\beta\epsilon}\, d\epsilon} = \frac{\left[\dfrac{e^{-\beta\epsilon}}{\beta^2}(\beta\epsilon + 1)\right]_0^\infty}{\left[\dfrac{e^{-\beta\epsilon}}{\beta}\right]_0^\infty} = \frac{1}{\beta} = kT \tag{2.34}$$

while the Planck postulate leads to

$$\bar{\epsilon} = \frac{\sum_{n=0}^\infty n\epsilon_0 e^{-\beta n\epsilon_0}}{\sum_{n=0}^\infty e^{-\beta n\epsilon_0}} = \frac{-\dfrac{d}{d\beta}\sum_{n=0}^\infty e^{-\beta n\epsilon_0}}{\sum_{n=0}^\infty e^{-\beta n\epsilon_0}}$$

$$= \frac{\epsilon_0 e^{-\beta\epsilon_0}(1 - e^{-\beta\epsilon_0})^{-2}}{(1 - e^{-\beta\epsilon_0})^{-1}} = \frac{\epsilon_0 e^{-\beta\epsilon_0}}{1 - e^{-\beta\epsilon_0}} = \frac{\epsilon_0}{e^{\epsilon_0/kT} - 1} \tag{2.35}$$

It is this latter value of $\bar{\epsilon}$ which has now to be introduced into Eq. (2.33). The meaning of the "energy quantum" ϵ_0 becomes immediately clear when we consult the general statements derived from thermodynamics such as Wien's displacement law. Thus Eq. (2.33) with the value of $\bar{\epsilon}$ from (2.35) must be compatible with Eq. (2.26) since the latter is valid regardless of the actual radiation mechanism. A comparison shows that ϵ_0 must be proportional to frequency:

$$\epsilon_0 = h\nu \qquad (2.36)$$

where h, the constant of proportionality, is Planck's "quantum of action" with the numerical value

$$h = 6.624 \times 10^{-27} \text{ erg. sec}$$

Combining Eqs. (2.33), (2.35), and (2.36), we obtain the energy density of the black body

$$u_\nu = \frac{8\pi h\nu^3}{c^3} \frac{1}{e^{h\nu/kT} - 1} \qquad (2.37)$$

This is Planck's radiation law. It is seen, first of all, that this function tends to vanish, as it should, for very small and very large values of ν, the physical reason being that for very small ν the number of modes goes to zero, while for very large ν the probability of exciting these modes disappears. Direct observation yields, of course, not the energy density Eq. (2.37), but either the brightness $K(\nu, T)$ or the radiant flux density $W(\nu, T)$ (over solid angle 2π) of the black body radiator. These can be obtained from Eq. (2.37) with Eq. (2.10). Thus,

$$W_\nu(\nu, T) = \pi K_\nu(\nu, T) = \frac{c}{4} u_\nu = \frac{2\pi h\nu^3}{c^2} \frac{1}{e^{h\nu/kT} - 1} \qquad (2.38)$$

Equation (2.38) refers to unit frequency interval. To convert this into terms of wavelengths, we apply Eq. (2.27) with the relations $\nu = c/\lambda$ and $| d\nu | = c | d\lambda |/\lambda^2$ so that

$$W_\lambda(\lambda, T)\, d\lambda = \frac{2\pi c^2 h}{\lambda^5} \frac{1}{e^{ch/\lambda kT} - 1}\, d\lambda = \frac{c_1 \lambda^{-5}}{e^{c_2/\lambda T} - 1}\, d\lambda \qquad (2.39)$$

where

$$c_1 = 3.74 \times 10^{-12} \quad \text{watt-cm}^2 \qquad c_2 = 1.438 \quad \text{cm-deg}$$

The agreement of Eq. (2.39) with measurements is such that there can be no question about its validity or the fact that it completely describes all observable phenomena of black body radiation. The equation is, of course, dependent on h and would degenerate into the Rayleigh-Jeans law for $h \to 0$. However, Planck's h can be independently and more accurately determined by other connections since it is, together with the electron mass m and charge e, one of the basic atomic constants.[4] Figure (2.5) shows a family of curves $W_\lambda(\lambda, T)$. The spectral radiation power $W_\lambda(\lambda)$ is plotted for various values of the temperature T. It is seen, and in fact follows from an inspection of the radiation equations, that the power emitted at any one wavelength increases with temperature; in other words, the curves W_λ for various temperatures do not cross. Furthermore, it is seen that each curve has a maximum which shifts towards shorter wavelengths for higher temperatures. The geometric locus of all these maxima is given by Wien's displacement law:

$$\lambda_{\max} T = 0.2897 \quad \text{cm. deg}$$

These features had, of course, already been predicted by thermo-dynamics. The radiation law, Eq. (2.38), was adjusted so as to conform to Wien's general relation, Eq. (2.26), except for constant factors. Therefore, statements derived from this latter equation, such as the Stefan-Boltzmann law, Eqs. (2.21) and (2.31), and that concerning the location of the radiation maxima, must also be valid for Planck's radiation law. However, thermodynamics derived the laws with undetermined

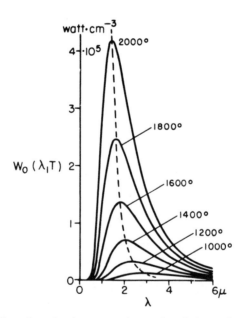

Fig. 2.5. Radiant flux density per unit wavelength interval emitted by the black body at various temperatures over the halfsphere.

functions; hence, there resulted constants, σ and C_0, about which nothing else was known other than that they were constants. The completely determined radiation law supplies the missing link to the universal constants. Thus one obtains from Eq. (2.38):

$$W(T) = \frac{2\pi h}{c^2} \int_0^\infty \frac{\nu^3 e^{-h\nu/kT}}{(1 - e^{-h\nu/kT})} \, d\nu = \frac{2\pi h}{c^2} \int_0^\infty (e^{-h\nu/kT} + e^{-2h\nu/kT} + \cdots)\nu^3 \, d\nu$$

$$= \frac{2\pi^5 h}{15 c^2} \left(\frac{kT}{h} \right)^4 = \sigma T^4$$

with

$$\sigma = \frac{2\pi^5 k^4}{15 c^2 h^3} = 5.672 \times 10^{-5} \quad \text{erg cm}^{-2} \text{ deg}^{-4} \text{ sec}^{-1}$$

and from Eq. (2.39) and $(\partial W/\partial \lambda)_T = 0$

$$C_0 = \frac{ch}{4.96k}$$

Furthermore, two other radiation laws are contained in Eq. (2.39) in the limit of very large, and very small wavelengths, respectively. In the first instance, for $ch/\lambda kT \ll 1$, we can expand

$$e^{ch/\lambda kT} = 1 + \frac{ch}{\lambda kT} + \cdots$$

so that we obtain the Rayleigh-Jeans law in terms of λ and $d\lambda$:

$$W_\lambda(\lambda, T) \, d\lambda = \frac{2\pi c}{\lambda^4} kT \cdot d\lambda, \qquad \lambda \gg \frac{ch}{kT} \tag{2.40}$$

In the second instance, for $ch/\lambda kT \gg 1$, we can ignore unity in the denominator of Eq. (2.39) and obtain a radiation law which (except for constant factors) was stated by Wien in 1896:

$$W_\lambda(\lambda, T) \, d\lambda = \frac{2\pi c^2 h}{\lambda^5} e^{-ch/\lambda kT} \, d\lambda, \qquad \lambda \ll \frac{ch}{kT} \tag{2.41}$$

In the foregoing, we have dealt with energy density and flux of radiation in vacuum either within, or emergent from, a black body cavity. If instead of vacuum, we consider media in which the velocity of light is smaller by a factor of n, we have for energy and radiant flux density, instead of Eqs. (2.37) and (2.38), respectively,

$$u_\nu = \frac{8\pi h n^3 \nu^3}{c^3} \frac{1}{e^{h\nu/kT} - 1} \tag{2.42}$$

$$W_\nu(\nu, T) = \frac{2\pi h n^2 \nu^3}{c^2} \frac{1}{e^{h\nu/kT} - 1} \tag{2.43}$$

Similarly, Eqs. (2.39), etc., are changed by substituting c/n for c.

2.3.2 Bose-Einstein Distribution of Photon Energies

Planck attempted, in his theory of heat radiation, to introduce the postulate of quantized energy levels while otherwise leaving the classical picture of the radiation mechanism as undisturbed as possible. As described in the preceding section, this method led to complete success. Nevertheless, the viewpoint adopted in either of the two models (viz., that of the dipole oscillator in equilibrium or that using standing waves of the enclosed electromagnetic field), represents a hybrid of classical and quantum concepts which is difficult to reconcile with the picture of corpuscular photons; yet this latter picture has been equally successful in its own right.

Therefore, it is tempting to obtain Planck's radiation law as the result of a statistical mechanics of photons, just as Boltzmann's energy distribution law is obtained from a statistical mechanics of molecules. In the following, we shall outline the procedure and results of such "quantum statistics" and refer, for a more complete treatment, to more specific texts.[3,5,6]

We imagine again a black body cavity of volume V at temperature T. However, we now assume that the radiation field consists of discrete photons which are emitted in random directions, and that its total energy U is thus given by the sum of all the quanta which are, at any instant, in transit through the cavity. Hence,

$$U = \sum_j N_j h\nu_j = \sum_j N_j \epsilon_j \tag{2.44}$$

where we have summed over j discrete values of energy $\epsilon_j = h\nu_j$, each populated by N_j photons. Now it is the aim of this type of statistics to compute the energy distribution, i.e. the values N_j for the various ϵ_j in thermal equilibrium. This model of particles with a thermal distribution of energies, albeit equal velocities, suggests treating it as a "photon gas" following as far as possible the methods of classical statistics. There are, however, important differences. First, the thermal equilibrium of energies, unlike in ordinary gases, does not establish itself by collisions of the photons among each other. This would require a dominant process of photon-photon scattering with change of wavelengths, in contradiction to experience. Rather is the thermalizing mechanism based on the emission and absorption of the photons by matter which, in this respect, acts as a catalyst. Now such a process has the important consequence that the total number of photons in the enclosure is not fixed, and the usual boundary condition of particle statistics, $\Sigma_j N_j = N$, does not hold.

Another important difference with respect to ordinary particles basically affects the rules of statistical counting. One deals, in fact, with a new statistics, named after their discoverers, the Bose-Einstein statistics of light quanta. The equilibrium distribution of particles over the various energy states is, as usual in statistics, computed by the general method of counting the number of different, possible arrangements; that distribution for which such number is largest, subject to certain boundary conditions, is then the most probable state, i.e., that of equilibrium. Now an important feature of the Bose-Einstein statistics is that there are restrictions on counting arrangements as different or distinguishable.

This means that in order to compute the probability of a distribution, one first subdivides the available volume and momentum range into intervals; in other words, one subdivides phase space (see Section 1.3.5) into cells. Classical statistics considered the size of cells as arbitrary. On the other hand, quantum statistics takes account of the fact that the minimum size of distinguishable cells is h^3 (Section 1.3.5). There exists, therefore, in a volume V in a frequency interval between ν and $\nu + \varDelta\nu$, a number G_j of distinguishable cells which according to Eq. (1.32) is

$$G_j = \frac{8\pi p_j^2 \, \varDelta p_j V}{h^3} = \frac{8\pi \nu_j^2 V}{c^3} \, \varDelta\nu_j \tag{2.45}$$

This number results from a subdivision of a spherical shell in phase space of finite thickness $\varDelta p$ into cells of size h^3, allowing for a factor of 2 for polarization. The number G_j corresponds, of course, to the number of standing waves in the classical picture; each cell represents a different mode (or location and direction of the photon); but all G_j cells are characterized by the same amount of momentum $|p_j|$ and energy $\epsilon_j = h\nu_j$, within the limits of the respective intervals, $\varDelta p_j$ and $\varDelta\nu_j$.

Furthermore, quantum statistics makes another distinction which is perhaps even more important. In classical statistics, particles are considered distinguishable; their interchange between two different points in phase space represents a new distribution. Photons, however, derive their entire identity from their position and momentum; they are otherwise indistinguishable, and their interchange between two different cells does not create a different arrangement.

Under these rules, we now proceed to count the number of distinguishable arrangements with which N_j photons can be distributed over G_j cells with energy ϵ_j. For this purpose, we designate the photons by the symbols a_1, a_2, ..., a_{N_j}. We arrange these symbols in a row, but separate them into G_j cells by $G_j - 1$ partitions which we call $b_1, b_2, ...,$ b_{G_j-1}. We thus obtain

$$\underset{a_1 a_2 a_3 b_1 a_4 a_5 b_2 b_3 a_6 \;\cdots\; b_{G_j-1} a_{N_j}}{\overset{\circ \;\;\circ\;\;\circ\;\;|\;\;\circ\;\;\circ\;\;|\;\;|\;\;\circ\;\;\cdots\;\;|\;\;\;\;\circ}{}}$$

indicating that there are 3 particles in the first cell, 2 in the second none in the third, etc., and 1 in the last. Now if we interchange the position of these symbols in all possible ways, we have covered all possible permutations of the particles over the cells. From the theory of combinations it is known that $G_j + N_j - 1$ symbols can be arranged in $(G_j + N_j - 1)!$ different sequences. This latter product contains, however, $(G_j - 1)! N_j!$ permutations of the partitions and photons, each among themselves, and these do not present distinguishable arrangements according to the rules of Bose-Einstein statistics. Thus, the number Γ_j of possible and distinguishable arrangements of N_j photons over G_j cells is given by the ratio

$$\Gamma_j = \frac{(G_j + N_j - 1)!}{(G_j - 1)! N_j!} \tag{2.46}$$

For a fixed number G_j of cells, Γ_j depends only on the number of photons N_j in the j-shell. What number N_j is appropriate in the equilibrium state, follows now from the total number Γ of arrangements which results as the product Π of the Γ_j's in shell 1, 2, 3, ..., j:

$$\Gamma = \prod_j \Gamma_j = \prod_j \frac{(G_j + N_j - 1)!}{(G_j - 1)! N_j!} \tag{2.47}$$

This number Γ of possible arrangements is a measure of the likelihood with which a given distribution $N_j(\epsilon_j)$ will occur and it is, therefore, called the probability of such distribution. Thus, to find the state of equilibrium, we have to solve for the maximum of Eq. (2.47), subject to the condition of restraint [Eq. (2.44)] that the total energy is fixed:

$$U = \sum_j N_j \epsilon_j = \text{constant}$$

The calculation is simplified by solving for a maximum of $\log \Gamma$, rather than Γ, because of the existence of Sterling's theorem:

$$\log N! = N \log N - N \tag{2.48}$$

We may point out here, in passing, that in this manner we are directly computing the maximum value of entropy, since the latter is defined statistically by the Boltzmann relation

$$S = k \log \Gamma \tag{2.49}$$

We thus have

$$\log \Gamma = S/k = \sum_j [(G_j + N_j) \log (G_j + N_j) - G_j \log G_j - N_j \log N_j] \tag{2.50}$$

to which we apply the method of Lagrangian multipliers:

$$\frac{\partial}{\partial N_j} (\log \Gamma + \beta U) = 0. \tag{2.51}$$

This yields with Eqs. (2.44) and (2.51)

$$\log \frac{G_j + N_j}{N_j} + \beta \epsilon_j = 0 \tag{2.52}$$

We thus have for the most probable distribution of the N_j:

$$N_j = \frac{G_j}{e^{-\beta \epsilon_j} - 1} \tag{2.53}$$

Returning now to continuous functions, we have from Eq. (2.44)

$$N_j h \nu_j = U(\nu) \, \Delta \nu = u(\nu) V \, \Delta \nu \tag{2.54}$$

so that with Eq. (2.45) and reduction to unit volume, Eq. (2.53) assumes the form

$$u(\nu) \, d\nu = \frac{8\pi \nu^3 h \, d\nu}{c^3} \frac{1}{e^{h\nu/kT} - 1} \tag{2.55}$$

where we have set the Lagragian parameter β equal to $-1/kT$. This follows, as before, on the basis of a comparison with thermodynamics—either because of the general validity of Wien's displacement law or by comparing the statistical definition of entropy, Eq. (2.49), with its thermodynamic origin, $dS = dQ/T$, where dQ stands here for the differential of heat content in the photon gas at constant volume.

Equation (2.55) represents Planck's radiation law, derived on the basis of the photon particle concept by statistical means.

2.4 Sources of Heat Radiation Continua

Before turning to the physics of monochromatic emission, it seems appropriate to interrupt the theoretical treatment by discussing the practical sources of heat radiation continua. Not all radiation continua, of course, are of thermal origin: energy exchanged by a charged particle in an unbound starting or ending state (in so-called *free-bound, bound-free*, or *free-free* transitions) is always continuous, but may not be thermal. Thus the *bremsstrahlung* of the x ray region and the recombination of electrons and positive ions, at low gas densities, are examples of such nonthermal mechanisms. Nor does heat radiation manifest itself exclusively in the form of continua; line spectra can be generated thermally, although we may regard them as still dominated by the Planck curve as an envelope such that the emissivity is zero everywhere, except in selected wavelength regions. Nevertheless, even if we restrict ourselves to optical emission which is continuous and—at least predominantly—thermal, there exists a very large number of different sources, and we can only describe briefly the most representative examples. We shall subdivide these into natural and laboratory sources. Another important criterion of classification is that of the operating temperature. In particular, the usefulness of thermal emitters depends in most cases on the maximum temperature which they can tolerate. An upper limit is dictated by the processes of decomposition, melting, or evaporation. Solids may operate at temperatures of 3000–$4000°K$, liquids at 5000–$6000°K$, while gases are not at all limited in that respect. This is the reason why gases play an increasingly competitive role as thermal emitters. At high pressures and densities, produced by magnetic confinement of ions and electrons, for instance, the oscillator continuum case discussed in Section 1.2.3 prevails, and these characteristics can be combined with those of very high temperatures from about $10^4°K$ in arcs to $10^9°K$ in nuclear reactions.

2.4.1 THE PRACTICAL REALIZATION OF BLACK BODY CAVITIES

Black body radiators have been defined in Section 2.2.2 by their property of having a radiant emissivity $\Sigma = 1$, independently of wavelength and temperature. Therefore, if the temperature of the radiator is known, brightness and radiant flux density follow exactly Planck's law and its derivations, and thus the number of quanta emitted in each wavelength interval is also known. Therein lies the significance of black bodies for all those measurements in which a response per photon or per erg of light power is to be determined, as in detector calibration. The practical realization of a black body demands a cavity so constructed that (1) the temperature within is uniform and known and (2) the opening is large compared to λ^2, yet small in comparison to the surface of the walls. This latter condition may be relaxed to the requirement that there re-emerge no reflections by the cavity walls of any radiation that may enter the cavity from the outside.

Figure 2.6 shows black body constructions for various typical temperatures of operation: (a) indirectly heated emitters in air for infrared calibrations; (b–d) cavities in vacuum, heated by the passage of current

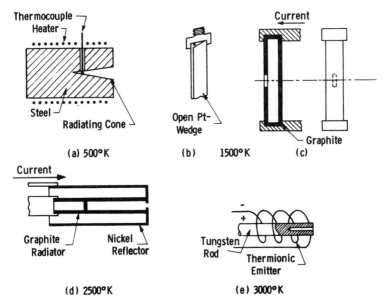

FIG. 2.6. Embodiment of black body cavities for various temperature regions. (a) and (b) Mendelhall type wedges, indirectly and directly heated, respectively; (c) and (d) graphite cylinders heated by current passage, (c) for observation in radial, (d) in axial direction; (e) electron bombardment heated tungsten rod (Dorgelo).

through their walls, for the near infrared and visible region; and (e) electron bombardment heated radiators for the visible and possibly the near ultraviolet.

Since a much larger surface is heated than observed, black radiating cavities are not ordinarily used as light sources.

2.4.2 Solid Radiators of Known Emissivity

For purposes of infrared spectroscopy or for other requirements of continuum radiation, certain calibrated sources are in common use.

2.4.2.1 *The Globar*

The Globar is a silicon carbide rod of several centimeters length and several millimeters radius, conventionnally heated by the passage of a current up to about $1500°K$. The Globar emits nearly "gray," with an emissivity of about 80% throughout much of the infrared spectrum.

2.4.2.2 *The Nernst Glower*

This is essentially a hollow cylinder, several centimeters long and 2–3 millimeters in diameter, composed of a mixture containing the oxides of zirconium, yttrium, and other rare earths. Like the Globar, it operates by the dissipation of ohmic heat. However, to be in a state of sufficient conductivity, the Nernst glower has first to be preheated to about $1100°K$ by means of a small gas flame or, preferably, by glowing wires of platinum. The source has the advantage of operating at temperatures near $2000°K$, and thus it yields a large amount of radiation power. Disadvantages result from the fact that the glower is quite fragile, that it needs preheating, and that the emissivity for wavelengths larger than 3μ decreases towards zero.

2.4.2.3 *Metal Ribbon Sources*

Heated wires, such as tungsten filaments, radiate at relatively high temperatures and sometimes, for certain wavelength regions, as gray emitters. Tungsten filaments reach the highest operating temperatures; and while in ordinary incandescent lamps the spectrum is limited to the transmission region of glass, namely, between about 0.3 and 3.0μ, accommodation with an inert atmosphere and a quartz window extends the spectrum. Table 2.1[7] lists for well-aged tungsten some values of spectral emissivities at specific wavelengths and of the total emissivities for the entire spectrum at various pertinent temperatures. Emissivity values are cited usually for smooth and polished materials since black radiation always can be approached with sufficient surface roughness.

TABLE 2.1[a]

Temperature °K	Spectral emissivity $\Sigma_{0,655\mu}$	Spectral emissivity $\Sigma_{0,467\mu}$	Total Σ
300	0.470	0.505	0.024
500	0.466	0.498	0.042
1000	0.456	0.486	0.105
1500	0.445	0.476	0.192
2000	0.435	0.469	0.259
2500	0.425	0.462	0.301
3000	0.415	0.455	0.334
3300	0.409	0.451	0.344

[a] A. G. Worthing and D. Halliday, "Heat." Wiley, New York, 1948.

2.4.3 THERMAL RADIATION FROM PLASMAS

In this section we shall discuss the emission from gases at high temperatures—from $\sim 6000°K$ and upward—and at high pressures, in the order of an atmosphere or more. We are concerned here with temperatures large enough to break up the bonds between electrons and ions and thus to create a state of partial or complete ionization, although as a whole the number of positive and negative particles is equal and the space charge is zero. Such a state is called a *plasma*. In principle any kind of energy is capable of generating a plasma—the mechanical energy of shock waves, for instance, or the nuclear energy in fusion or fission. Our discussion of light sources, however, will deal mainly with electrical excitation as the most important practical case in which the pertinent radiation phenomena, at any rate, are represented.

Chemical energy, although easily available in flames and explosions, is quite limited in the temperatures it can produce, and this for a reason which is, to some extent, interesting for the understanding of the heat exchange mechanism in all plasmas. We shall briefly explain this in terms of the energy processes affecting 1 gram of gas by means of the following equation:

$$Q_0 = \int_{T_0}^{T} C_v \, dT + \int_{T_0}^{T} q_{\text{dis}} \, dT + \int_{T_0}^{T} R(T) \, dT \qquad (2.56)$$
$$\quad (1) \qquad\qquad (2) \qquad\qquad (3)$$

The implication of Eq. (2.56) is that the energy Q_0 generated per gram of gas distributes itself over three processes such that a temperatue T results: (1) the heating itself, at a rate per degree which is given by the specific heat $C_v(T)$; (2) ionization or dissociation to a fraction $q_{dis}(T)$ which assumes here the role of a latent heat; and (3) radiation, convection, expansion, and other losses to the outside. The energy expended on $q_{dis}(T)$ is negligible until fairly high temperatures are reached; but after this, it rises sharply. For example, it requires five times as much heat to dissociate hydrogen as that needed to bring its temperature from $300°K$ to the onset of dissociation, near $3000°K$. Since chemical reactions always involve dissociated molecules, the energy developed suffices only to raise the flame or explosion temperature to the onset of dissociation, typically $2000–5000°K$. However, if the chemical energy Q_0 in a flame is "augmented" by an electrical discharge, higher temperatures can be reached (Karlovitz[8]).

Because the energy of ionization is in the order of five to ten times larger than that of dissociation, the latent heat absorption process does not set in until correspondingly higher temperatures are reached. Even at moderate energy densities, therefore, temperatures in the column of electric arcs vary from about $6000°K$ for atmospheric air to about $18,000°K$ in helium which has the highest potential for single ionization (see Chapter 4). Furthermore, the input energy Q_0 in electrically produced plasmas is unlimited, in principle, if proper energy transfer schemes can be found, and in this lies, of course, a second difference with respect to chemically generated plasmas. In fact, there unfolds a multitude of possible alternatives which will assume increasing importance along with the plasma processes produced by other means.

2.4.3.1 *Thermal Equilibrium and Thermal Radiation*

Certain concepts introduced in the beginning of this chapter deserve further interpretation in the case of plasmas. When we speak of the temperature of a very hot gas, we imply that (1) such temperature is measurable and (2) a thermal equilibrium prevails. Now as we have seen (Section 2.2.1), a perfect equilibrium between atomic states, electrons, and photons exists in the black body cavity. Such conditions are also realized in the core of very dense gas masses—inside stars, exploding wires, nuclear explosions, and certain electrically generated plasmas, for instance. It follows that the radiation within such systems must be given by Planck's law for the radiant flux density, Eq. (2.38).

A different situation exists in the more commonly observed plasmas which are transparent for much of the spectrum and which, if they emit a continuum at all, also exhibit a superimposed line spectrum (for a general review of this field, see Finkelnburg and Peters[9]). Nevertheless, one speaks even in this case of a thermal equilibrium between atoms, ions, and electrons, provided the interactions between them are predominantly thermal. By this is meant that the particle energies attributable to the equipartition law [or to quantum statistics, as in Eq. (2.35)] must be overwhelmingly large compared with the average contributions made, during the time between collisions, by such outside sources as electric fields and radiation. More precisely stated, we shall recognize an equilibrium as thermal if the various energy states follow the Boltzmann distribution (or its quantum-statistical counterpart); that is, if

$$n_i = g_i A e^{-E_i/kT} \qquad (2.57)$$

where n_i is the population density of the state i, g_i its *statistical weight*, which is analogous to the term G_j in Eq. (2.45) but is here either 1 or a small integer; E_i its energy above the ground state; and A a constant. The plasma temperature T is thus defined, and made accessible to measurement, by Eq. (2.57). Obviously, a thermal plasma of known components is fully described with respect to all its energy exchange processes by the temperature T, which in equilibrium must be the same for all components and their energy states. It follows that if particles of widely differing mass are to be brought into thermal equilibrium, as is the case for the electrons and ions of the thermal plasma, the number of collisions must be particularly large so that the ineffective energy transfer is compensated by the frequency of the encounters. Thus thermal plasmas can develop only for small electron mean free paths and the gas densities large enough to achieve them.

2.4.3.2 *Optical Path Length and Plasma Brightness*

We have surmised already that even if all the components of a gas are in thermal equilibrium, its radiation in general will be quite different from that of a black body. Nevertheless, under equilibrium conditions Kirchhoff's law, Eq. (2.14), must be valid, and from this we can draw certain conclusions regarding the brightness B_ν of the plasma (Finkelnburg[10]).

For this purpose we consider a gas layer of thickness dl which reduces the incident brightness B by absorption, but also increases it by emission. We have for the contribution of absorption

$$(dB_\nu)_{\text{abs}} = -\Sigma_\nu B_\nu \, dl \qquad (2.58)$$

where Σ_ν is the absorption coefficient per unit length (Σ_ν may include a negative part due to stimulated emission, as will be explained in Section 3.3.1). The contribution of emission is, according to Eq. (2.15)

$$(dB_\nu)_{em} = \epsilon_\nu \, dl = \Sigma_\nu K_\nu \, dl \tag{2.59}$$

where K_ν is the spectral brightness of black radiation, as defined in Eq. (2.38), and where we have also introduced the emission per unit plasma length, $\epsilon_\nu = \Sigma_\nu K_\nu$. The sum of the two contributions (2.58)and (2.59) yields the differential equation

$$\frac{dB_\nu}{dl} = \Sigma_\nu (K_\nu - B_\nu) \tag{2.60}$$

The total brightness from a plasma mass of thickness l is then obtained by integration with the boundary condition that for vanishing thickness, $l = 0$, $B_\nu = 0$:

$$B_\nu(l) = K_\nu(1 - e^{-\Sigma_\nu l}) \tag{2.61}$$

It is seen that as the *optical path length* $\Sigma_\nu l$ in the plasma increases, the radiance changes from

$$B_\nu = \Sigma_\nu l K_\nu \qquad \text{for} \quad \Sigma_\nu l \ll 1 \tag{2.62}$$

to

$$B_\nu = K_\nu \qquad \text{for} \quad \Sigma_\nu l \gg 1 \tag{2.63}$$

Thus even for very small Σ_ν, the plasma will emit black radiation throughout the spectrum provided its dimension l is large enough. Of course, in the general case of gases Σ_ν changes violently with frequency, and thus the thicknesses needed for blackness will vary by corresponding orders of magnitude. However, when the density of the atoms in the plasma is increased, two effects will result: (1) the average absorption coefficient per unit frequency interval will increase (more or less linearly); and (2) as lines broaden and merge into a continuum, Σ_ν will vary much less with frequency and finally become a constant. Thus plasmas even of small extent may behave as black bodies.

2.4.3.3 *Models for Emission and Absorption Mechanisms in Plasmas*

In order to compute the absorption coefficient per unit length, Σ_ν, or the brightness ϵ_ν emitted from 1 cm³ of plasma within 1 steradian, it is necessary to know density and distribution of the electric charges, bound or free, and the laws by which they interact with radiation. The detailed models for the energy exchange between charges and radiation will be discussed in Chapter 3. In particular, the energy emitted during a time dt by a charge e which undergoes an acceleration (d^2x/dt^2) is given by [cf. Eq. (3.15)]:

$$dE = \frac{2e^2}{3c^3} \left(\frac{d^2x}{dt^2}\right)^2 dt \tag{2.64}$$

This equation can then be used to compute the radiation, for instance, from a free electron which describes a hyperbola in the Coulomb field of a positive ion and thereby changes its velocity in amount and in direction. A classical approach of this nature which in addition bridged the gap to the quantized states by the so-called *correspondence principle*, was used by Kramers[11] to compute the absorption coefficient for x rays owing to free and bound electrons. The Kramers theory is also applicable to the optical spectrum of various plasmas, provided that there exist reasonable assumptions regarding the energy distributions of the free and bound electrons. Models for the mechanism by which a plasma emits a conti-nuum were developed qualitatively by Finkelnburg[12] and quantitatively by Unsöld[13,14]. Figure 2.7 shows the various energy states, assumed to be hydrogenlike which an electron can occupy under plasma conditions (cf. Maecker and Peters[15]). As explained in Section 3.2.1 in greater detail, there exists a series of quantized energy states from a normal, unexcited ground state E_1 onward. In an isolated and unperturbed atom, this *term series* converges towards a limit (conventionally set at zero), and at this point the electron is freed and the atom ionized. The term scheme in Fig. 2.7, however, depicts in addition the effect of high plasma density in that all the levels are "detuned" and broadened so that upward from a certain energy state, E_g, they are close enough to overlap to and form a "pseudocontinuum." Of course, it is conceivable that this pseudocontinuum extends down to the ground state so that the electron assumes no discrete energy levels throughout the bound range.

The various features of the emitted spectrum result from the panorama of the possible transitions between the levels. Jumps of the bound-bound variety from higher to lower discrete states give rise to the line spectrum. On the other hand, electrons of positive energy, which approach ions with a certain kinetic energy, $mv^2/2$, produce continuous radiation. Such free-bound transitions occur, of course, even in low density non-thermal plasmas but the continuum intensity is very much enhanced in hot dense plasmas because of the larger number of ions. As the pressure becomes high, the pseudocontinuum, which is really the analog of the band structure in solids, plays a dominant role: all transitions starting or ending in it contribute to the continuous radiation. Finally, the free-

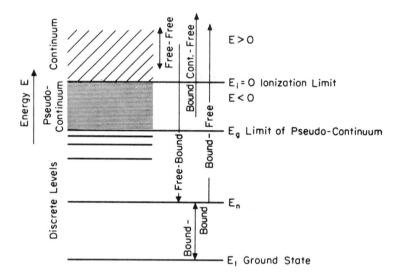

FIG. 2.7. Hydrogenlike energy levels and transitions in Finkelnburg-Unsöld theory of continua. At high densities upper terms near ionization limit broaden and overlap, thus forming a "pseudocontinuum." Transitions involving energies above the ionization limit start or end, or both, with free electrons and give rise to a spectral continuum even at lowest densities. Such processes are called, respectively, *free-bound*, *bound-free*, and *free-free* transitions.

free electron exchanges with radiation simply amount to a conversion of (continuous) kinetic energy into light and vice versa—in hyperbolic paths around the ion, the electron must obviously pass through a central field, but the net energy of the ion is unaffected.

In his calculation, Unsöld uses Kramers' formula for the absorption coefficient per (hydrogenlike) atom in the nth orbit

$$a_n = \frac{1}{n^2} \frac{64\pi^4}{3\sqrt{3}} \frac{me^{10}}{ch^6 n^3} \frac{1}{\nu^3} \qquad (2.65)$$

This has to be summed over all atoms, taking account of their distribution over the various states E_n. Applying the Boltzmann relation Eq. (2.57) with the proviso that the statistical weight of the hydrogen states equals n^2, one has for the occupancy numbers N_n relative to the density N_1 of the ground state

$$\frac{N_n}{N_1} = n^2 \exp(-(E_1 - E_n)/kT) \qquad (2.66)$$

Obviously, the values E_n above E_g are no longer discrete, and the sum is replaced by an integral. In this manner, Unsöld obtained an absorption coefficient in which the contribution of free-free transitions stood in a ratio to that of the free-bound and that of the overlapping lines of the pseudocontinuum, as follows:

$$
\begin{array}{ccccc}
\text{free-free} & & \text{free-bound} & & \text{pseudocontinuum} \\
1 & : & e^{h\nu/kT} - 1 & : & e^{h\nu/kT}(e^{|E_g|/kT} - 1)
\end{array} \qquad (2.67)
$$

The effect of increased pressure is a growing value of E_g so that the contribution of the smeared-out lines becomes correspondingly larger. From the absorption coefficient follows, by means of Kirchhoff's and Planck's law, the brightness ϵ_ν attributable to 1 cm length of plasma:

$$
\epsilon_\nu = \Sigma_\nu K_\nu = \gamma \frac{32\pi^2}{3\sqrt{3}} \left(\frac{e^2}{h_c}\right)^3 Z_{\text{eff}}^2 e^{-(E_1 - E_g)/kT} . P \qquad (2.68)
$$

where P is the gas pressure, Z_{eff} the effective ion charge, and γ the statistical weight of the ground state which is unity for hydrogen, but may be a small integer for the more complex atoms. An expression equivalent to Eq. (2.68) introduces the ion and electron densities per cm^3, n_i and n_e, so that one obtains, after substituting numerical values, the simplified expression

$$
\epsilon_\nu = 5.41 \times 10^{-46} Z^2 n_i n_e / T^{1/2} \qquad (2.69)
$$

Equations (2.68) and (2.69) state that the value $\Sigma_\nu K_\nu$ does not contain a term for the frequency; in other words, the emission is constant in that part of the spectrum for which the derivation of (2.68) is valid, up to a limit $h\nu_g = E_g$. In this region, then, the absorption coefficient $\Sigma_\nu = \epsilon_\nu / K_\nu$ is proportional to the inverse Planck function. For $\nu > E_g/h$ the radiation power ϵ_ν of the continuum decreases as $\exp(- h\nu/kT)$, while the absorption coefficient Σ_ν changes accordingly as $1/\nu^3$.

What a measurement of plasma brightness actually yields is the value $B_\nu(l)$ which contains Σ_ν according to Eq. (2.61). Agreement between the Kramers-Unsöld theory and observation on plasmas of the types described in the following section can be considered as satisfactory[15] (i.e. within a factor of 2) in view of the approximations in the theory and the uncertainties in the measurement of the experimental parameters. Figure 2.8 shows the emission from a water-stabilized plasma [see Section 2.4.3.4(c)] at various high pressures, as calculated by Peters.[16] It is seen that a plasma layer of 2 mm at 1000 atm emits a spectral brightness $I_\nu = B_\nu(l)$ which approaches the black body radiation curve, K_ν, over most of the spectrum. The limit described by Eq. (2.63) is therefore

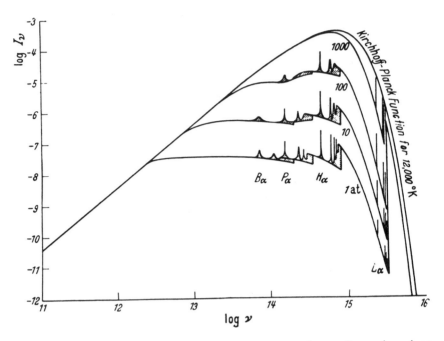

FIG. 2.8. Spectral continuum emitted by water-vapor plasma. Curves have been computed by T. Peters [Z. *Physik* **135**, 573 (1953)] for conditions in a water-stabilized arc with an emitting thickness of 2 mm at 12,000°K at a variety of pressures. Hydrogen emission lines are superimposed on continuum. At highest pressures, black-body radiation is approached (reprinted from *Zeitschrift für Physik*).

nearly reached. Experiments on the arc confirmed this conclusion. In the case of atoms for which the assumption of hydrogenlike states is not justified, e.g., for the inert gases, the Kramers-Unsöld theory agrees only in limited aspects with experiment.[17-19]

2.4.3.4 *Practical Sources of Thermal Plasmas*

The development of plasma radiators has been motivated by the need for light sources of high brightness. Various forms of electrical energy conversion proved to be the most suitable means, and the overwhelming majority of development efforts have been concentrating on such sources as arcs, sparks, and exploding wires (in this order of importance). Some of these have been in commercial use for a long time, and we can mention here only typical examples of schemes which are of interest because of the physical phenomena involved or because of their importance for physical measurement. There are, in addition, conversion systems other than electrical, notably shock wave tubes. An example of the latter is the ballistic piston compressor (Lalos[20]) which can produce pressures of 10^4 atm and temperatures up to at least $2 \times 10^{4}°K$ for a duration of milliseconds. The highest plasma temperatures are generated by nuclear fission and fusion. We shall not dwell on these sources, but refer to the literature on nuclear engineering.

(a) *Atmospheric Carbon Arc.* The plasma sources in longest use and of greatest simplicity are electric arcs operated in the atmosphere, especially between carbon electrodes. The crater which forms in the anode has a temperature of about $4000°K$, and because its characteristics are relatively well known, it serves as an intensity standard for the comparison with other plasma emitters. The arc plasma proper has temperatures in the order of $6000°K$ nearly independent of current. The main drawback of the atmospheric carbon arc is its instability which has to be minimized by such measures as automatic adjustment of electrode distance, application of short gap with pointed cathode, special circuitry to maintain constant currents, and stabilization methods such as those described in the next paragraphs.

(b) *Stabilized Arcs.* An important principle of stabilization, introduced by Schönherr,[21] consists of operating the arc in a rotating gas or vapor mass. A jet of air or other gas is blown tangentially along the inner wall of a cylinder so that a whirling motion of the contained mass results. The arc is maintained in the axial region over lengths of several meters.

Similarly, stabilization is possible by a rotating water jet (Gerdien[22] arc) in which the arc creates its own vapor atmosphere. Such systems[23] have been used with 1500 amp current at temperatures of $50,000°K$.

(c) *High Pressure Arcs.* High density plasmas, of the type discussed in Section 2.4.3.2, are realized in a variety of devices.

The simplest of these is the high pressure arc source, first described by Schulz,[24] which contains an inert gas at 20 to 40 atm and an electrode configuration such as shown in Fig. 2.9. The most effective gas appears to be xenon, although mercury is sometimes added in commercial versions of the tube. Quartz envelopes of about an inch inner diameter are able to sustain pressures of this magnitude. Typical xenon arc tubes were found[25] to exceed the intensity of other arc sources in the ultraviolet by a factor of 10 to 100. Between 2200 and 3200 Å, where $\nu > E_g/h$, the radiance of the xenon continuum is about 2 watts/100 Å $\times$ steradian $\times$ cm², and the spectral behavior is that of a gray emitter with $\Sigma = 0.058$

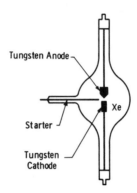

FIG. 2.9.   High pressure xenon arc source.

at $6600°K$, although the actual gas temperature is much higher. At 2500 Å the radiance is 23 times that of the carbon arc crater.

Higher pressures and closer approach to black body radiation continua are possible with the water-stabilized arc source developed by Peters[16] (see Fig. 2.10). The arc chamber consists of two steel cylinders butted against an isolating and transparent plastic ring. This assembly is spun at about 100 revolutions per second around its axis so that a suitable quantity of water contained in the cylinders will be pressed against the walls. A hollow channel is thus formed in the axial region, and it is this sharply defined water cylinder which contains the arc plasma and which allows a variation of the pressure by a change of the dissipated arc power.

(d) *Plasmas Excited by High-Frequency Discharges.* Nearly thermal plasmas can be produced also in radio-frequency and microwave fields. A characteristic of this method is the absence of electrodes with the resulting advantages of improved power transfer and greater efficiency.

It is necessary to distinguish between high impedance and low impedance devices. The former (see, e.g., Cobine and Wilbur[26]) operate by means of high field breakdown effects which, other things being equal, are less likely to produce thermal equilibrium conditions than ordinary arcs. Low impedance radio-frequency plasmas are maintained by means of induced currents (from 100 to 1000 amp upward) in sufficiently ionized gases (Babat[27]).

An important advance in this field was made with the so-called *induction coupled plasma torch* (Reed[28,29]). The device uses a stream of argon (with or without additional gases) which passes axially through the induced field of a radio-frequency coil. A plasma of apparently thermal distribution is obtained with temperatures in the order of

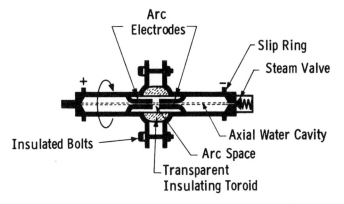

FIG. 2.10.   Scheme of water stabilized high pressure arc of T. Peters [*Z. Physik* 135, 573 (1953)].

10,000 to 20,000°K, and it is a particular advantage of the method that these temperatures are maintained over certain distances of convection. This feature makes the plasma torch valuable for heat transfer purposes, but it may conceivably also serve as an effective source of radiation continua.

(e) *Spark Sources.* In the preceding paragraphs we have treated intense spectral continua sources which could be maintained stationary over long times. It is obvious that considerably larger intensities can be produced, if the condition of continuous operation is traded for power developed in pulsed duty cycles. A characteristic of pulsed methods is that energy gathered during relatively long times (seconds to minutes) is released in periods which are 10^{6}–10^{7} times shorter (10^{-6}–10^{-7} sec).

The most important conversion mechanism of pulsed electrical power into light is the spark discharge. The powers available to the radiation mechanism with a capacitor discharge are by orders of magnitude larger than in the stationary arc case (typically, for a capacity $C = 20$ μf, $U = 10$ kv, $P_s = 10^9$ watts). The current near the peak power is determined by

$$i_{\max} = U\left(\frac{C}{L}\right)^{1/2} \quad \text{amp} \qquad (2.70)$$

the ohmic resistance of the spark being negligible. For highest efficiency, the inductance L should be as small as possible. Perhaps the most radical method of reducing the inductance is realized in the coaxial spark-gap capacitor developed by Fischer[30] (see Fig. 2.11). Circuit time constants

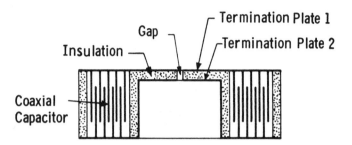

FIG. 2.11.   Coaxial spark source of H. Fischer [*J. Opt. Soc. Am.* **47**, 981 (1957); Air force Cambridge Rearch Center Report TN-59-185 (1959)] for discharges of high speed and large light power.

as low as 36×10^{-9} sec at $L = 1.1 \times 10^{-9}$ henry could be achieved with variations of this arrangement.[31]

A review of the entire field of pulsed light sources is given by Vanyukov and Mak.[32]

2.4.4 NATURAL SOURCES OF HEAT RADIATION

No object is exempt from being a source of thermal radiation. Therein lies an important application because, even in the "dark" of night, a terrain or any scene in which differences of temperature exists can be made visible by a conversion of infrared into visible radiation. The earth's atmosphere has, of course, a modifying effect on the spectral distribution of any light passing through it. Aside from scattering, various atoms and molecules of the air absorb, radiate, and reradiate much of the optical spectrum between 0.01 and 1000μ. In the ultraviolet, below wavelengths of 1800 Å, oxygen is a most effective absorber. Therefore spectrographic work in this region has to be performed in vacuum. Ozone, at a 25-km height in the ionosphere, absorbs ultraviolet for $\lambda < 2900$ Å and this, therefore, represents the lower wavelength limit of the optical spectrum which the atmosphere admits from sun and star light. In the infrared above 2.5μ, water vapor and carbon dioxide produce intermittant spectral regions of strong absorption, until in the far infrared, from about 15μ upward, the atmosphere is completely opaque save for occasional intervals of limited range transmission, particularly near 1 mm. We have therefore the result that the atmosphere of the earth, except for the region between 0.3 and 15μ, behaves not unlike a black body at 200–300°K rather than as a transparent dielectric medium.

The sun, too, emits to a certain extent as a black body. Nevertheless, the spectrum is complicated by the fact that various gaseous layers at different densities and temperatures contribute to the radiation. The detailed description of how the sun's brightness is distributed over the spectrum, or how it varies geometrically over the disk is therefore quite involved. The radiant flux density over the total optical spectrum amounts to 6.13×10^3 watts/cm^2 at the surface, corresponding to an effective black body temperature of 5713°K. The values refer to the so-called *photosphere*, defined as those layers of the sun from which the visible and infrared rays emerge directly. However, the photosphere is an atmosphere in which the temperature increases with depth. For this reason, the rim of the sun is viewed through gas masses which are colder than the center. Because of the difference in absorption the rim is darker, an effect which increases for shorter wavelengths. Monochromatic absorption by atoms of the outer layers is responsible for the solar part of the so-called *Fraunhofer* lines. The spectrum is further modified by layers surrounding the photosphere: the much hotter *chromosphere*, followed by the extremely rarified *corona* which extends to a height

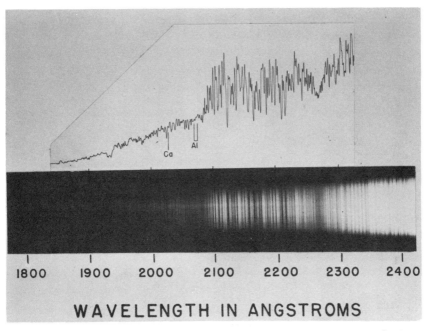

FIG. 2.12. Sun spectrum and densitometer trace between 1800 and 2400 Å photographed from rocket at 115-km altitude [F. S. Johnson, H. H. Malitson, J. D. Purcell, and R. Tousey, *Astrophys. J.* **127**, 80 (1958)]. (Official United States Navy Photograph.)

of about 10^6 km and has a temperature of about 10^6 °K. The earth receives from the sun 0.133 watts/cm² of optical radiation power, a part of which is absorbed, scattered and reradiated by the atmosphere. At ground level the amount received as scattered light from the atmosphere over a solid angle of 2π is about equal to that of direct sunlight in normal incidence.

Solar spectra photographed from rockets or high altitude balloons are, of course, largely free of absorption effects of the earth's atmosphere. Figure 2.12 shows the sun's spectrum between 1800 and 2400 Å taken from a rocket at 115-km altitude (Johnson et al.,[33] U.S. Naval Research Laboratory). The photograph and the densitometer trace above it show a transition of the continuum with its solar Fraunhofer lines to an absorption region beginning rather abruptly near 2085 Å. Presumably the absorption is caused by the chromosphere which also gives rise to certain emission lines due to particle bombardment (e.g., the Lyman series). The width of the emission lines offers a clue regarding the location and mechanism of their generation as explained in Section 3.4.

REFERENCES

1. Report of the Working Group on Infrared Backgrounds, Part II 2389-3SR. Univ. of Michigan Willow Run Laboratories, Ann Arbor, Michigan (1959).
2. M. PLANCK, "The Theory of Heat Radiation" (transl. by M. Masius). Dover, New York, 1955.
3. M. BORN, "Atomic Physics," 5th ed. Hafner, New York, 1951 (7th ed., 1962).
4. J. W. M. DuMOND and E. R. COHEN, Rev. Mod. 20, 82 (1948).
5. L. D. LANDAU and E. M. LIFSHITZ, "Statistical Physics." Pergamon Press, New York, 1958.
6. R. H. FOWLER, "Statistical Mechanics," 2nd ed. Cambridge Univ. Press, London and New York, 1955.
7. A. G. WORTHING and D. HALLIDAY, "Heat." Wiley, New York, 1948.
8. B. KARLOVITZ, Int. Sci. Technol., 6, 36 (1962).
9. W. FINKELNBURG and T. PETERS, in "Handbuch der Physik" (S. Flügge, ed.), Vol. 28, p. 79. Springer, Berlin, 1957.
10. W. FINKELNBURG, J. Opt. Soc. Am. 39, 185 (1949).
11. H. A. KRAMERS, Phil. Mag. [6] 46, 836 (1923).
12. W. FINKELNBURG, Z. Physik 88, 297, 763, and 768 (1934).
13. A. UNSÖLD, Ann. Physik [5] 33, 607 (1938).
14. A. UNSÖLD, Z. Astrophys. 8, 32 (1934).
15. M. MAECKER and T. PETERS, Z. Physik 139, 448 (1954).
16. T. PETERS, Z. Physik 135, 573 (1953).
17. H. N. OLSEN, Phys. Rev. 124, 1703 (1961).
18. K. LARCHÉ, Z. Physik 136, 74 (1953).
19. H. SCHIRMER, Z. Physik 136, 87 (1953).
20. G. T. LALOS, Rev. Sci. Instr. 33, 214 (1962).
21. D. SCHÖNHERR, Elektrotech. Z. 30, 365 (1909).

22. H. Gerdien and A. Lotz, *Wiss. Veröff. Siemenskonz.* **2**, 489 (1922).
23. F. Burhorn, H. Maecker, and T. Peters, *Z. Physik* **131**, 28 (1951).
24. P. Schulz, *Ann. Physik* [6] **1**, 95 (1947).
25. W. A. Baum and L. Dunkelman, *J. Opt. Soc. Am.* **40**, 782 (1950).
26. J. D. Cobine and D. A. Wilbur, *J. Appl. Phys.* **22**, 835 (1951).
27. G. I. Babat, *J. Inst. Elec. Engrs.* (*London*) **94**, Part III (1947).
28. T. B. Reed, *J. Appl. Phys.* **32**, 821 (1961).
29. T. B. Reed, *Int. Sci. Technol.* **6**, 36 (1962).
30. H. Fischer, *J. Opt. Soc. Am.* **47**, 981 (1957).
31. H. Fischer, Air Force Cambridge Research Center Report TN-59-185 (1959).
32. M. P. Vanyukov and A. A. Mak, *Usp. Fiz. Nauk* **66**, 301 (1958); see *Soviet Phys.— Usp.* **66**, No. 1, 137 (1958).
33. F. S. Johnson, H. H. Malitson, J. D. Purcell and R. Tousey, *Astrophys. J.* **127**, 80 (1958).

3

EMISSION AND ABSORPTION
OF NEARLY MONOCHROMATIC RADIATION

The sequence of radiation processes as presented here proceeds with an increasing refinement of models and their interaction with the electromagnetic field. Thus, thermal radiation from densely packed oscillators, which is idealized in black body radiation, was treated first thermodynamically, without the benefit of a model concept, and subsequently with quantum theory on the basis of very simple mechanisms. Again, it proved not to be necessary, for the purpose of deriving the laws of heat radiation, to postulate a specific process by which vibrating charges would exchange energy with the electric and magnetic components of the radiation field. In its stead, it was possible to arrive at the end result of such interactions by purely statistical reasoning. This is reflected by the fact that the heat radiation laws do not stipulate what kind of atoms are necessary to construct a black body.

However, the specific description of radiating centers and their interactions with the electromagnetic field becomes necessary as we turn to the study of emission and absorption in isolated oscillators such as gases of low density and certain luminescent solids. The radiation from such sources is almost monochromatic, having a small, but finite spectral width for the reasons already discussed in Section 1.2.2. In other words, we are now dealing with spectral lines, or bands of lines, the wavelengths of which are so sharply determined by the emitting or absorbing atoms that the inspection of spectra represents one of the most powerful methods of analytical chemistry.

In this chapter we shall treat the question of the coupling between such sharply resonant structures and the electromagnetic field. Many aspects of these interaction problems are well understood, in fact best introduced, in terms of the classical theory. With this treatment as a starting point, we shall progress to the basic questions of atomic structure and radiation, i.e. questions which have been the major issue in the development of modern quantum theory.

3.1 Radiation Field of the Classical Oscillator

3.1.1 Hertzian Dipoles and Multipoles

In Section 1.2.2 we discussed some general properties of the harmonic oscillator and, in particular, the nature of its resonance spectrum with, or without, a damping force. This procedure leads to a mechanical model of optical resonators which, by itself, is capable of describing duration and the spectral selectivity characteristics of their interactions with radiation. However, if we are asking for the amount of energy exchange or the orientation of the oscillator with respect to the electromagnetic field, then we have to concern ourselves with the interaction between moving charges and the field. A single charge in harmonic motion $x(t)$ will produce: (1) a surrounding magnetic field proportional to the elementary current $e \cdot \dot{x}(t)$, (2) an electric field, proportional to $e \cdot \ddot{x}(t)$, induced by the magnetic field as it changes with time; and (3) a magnetic field, $\propto e\dddot{x}(t)$, which is in turn coupled to the electric field. These fields will spread as electromagnetic waves with the speed of light, the pattern markedly changing as the disturbance travels from close distances $r \ll \lambda$, the "near-zone," to the "far-zone," or $r \gg \lambda$. In the far-zone, the power radiated will be proportional to $(e\ddot{x})^2$; a charge moving at uniform velocity does not emit any radiation power.

The case of multiple charges is closely related to that just discussed. The dipole, which consists of a positive and negative charge at distance $x(t)$, is mathematically identical with the single charge in a Coulomb field; one defines for both cases a dipole moment $X = e \cdot x(t)$ with its time derivatives $e \cdot \dot{x}(t)$ and $e \cdot \ddot{x}(t)$. However, in the generalized configuration of multiple particles, one chooses a center O from which the various charges e_k have distances given by the radius vector $\mathbf{x}_k$, as is shown in Fig. 3.1. The location of the center is arbitrary and can be shifted by a constant displacement because we are interested only in the time

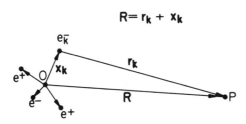

Fig. 3.1.   Geometrical relationships for the Hertzian vector.

derivatives of the x_k's. We now define a multipole moment by the Hertzian vector

$$\mathbf{X} = \sum_k e_k \cdot \mathbf{x}_k(t_k') \tag{3.1}$$

where the t_k' represent the so-called "retarded times" for each charge, the significance being the following: the radiation field measured at a distance r from a charge is due to the motion of such charge at an earlier time, namely, that preceding the present by the travel time of light r/c. Thus a point P at distance R from the center and r_k from the various charges receives at time t individual contributions which had been emitted at different retarded times

$$t_k' = t - r_k/c \tag{3.2}$$

Because of this difference in the t_k', the configuration of moving charges appears at P to be distorted from that for the multipole at rest. We assume that the particles are relatively close together and that their velocities are small compared with that of light:

$$|x_k| \ll R, \qquad |\dot{x}_k| \ll c \tag{3.3}$$

We then can develop the multipole moment, Eq. (3.1), into a converging series

$$\mathbf{X} = \sum_k e_k \mathbf{x}_k(t_0') + \sum_k e_k \dot{\mathbf{x}}_k(t_0') \frac{x_{kR}(t_0')}{c} + \cdots \tag{3.4}$$

The first member of the series makes up the largest contribution; it represents effectively a dipole system, since all charges emit at the same time—the retarded time of the center $t_0' = t - R/c$. The second term takes account of the difference between the retarded times: x_{kR} is the projection of $\mathbf{x}_k$ on $\mathbf{R}$ so that x_{kR}/c is approximately $t_k' - t_0'$. Thus, effects of higher order result due to a quadrupole moment and magnetic dipole moment.

Examples for various pole systems are, first, the models for atomic and molecular systems which may be simple dipoles. In certain molecules, such as H_2, N_2, and O_2, charges are symmetrically distributed so that the moment $\mathbf{X}$ vanishes. These molecules are, therefore, incapable of absorbing or emitting infrared radiation as a result of vibration or rotation. A dipole may, however, be induced by electric fields or the combination of different isotopes, and the missing spectrum may be obtained by these measures, although usually at relatively low intensities.

Similarly, quadrupole radiation may become important, if the dipole moment vanishes.

A most important application of the dipole concept lies in those low frequency oscillations in which actually a large number of particles participate cooperatively. Aside from certain electromagnetic radiation phenomena in plasmas, this has particular reference to dipole antennas. It was, in fact, in connection with the generation of electromagnetic waves in resonant wire structures, that Hertz[1] first founded the theory of dipole radiation (1899), the basis for modern radio.

3.1.2 Dipole Radiation Pattern

We proceed to state the values of the electric and magnetic fields at distance R from the dipole with the retarded time $t' = t - R/c$. With Maxwell's equations as a starting point, electromagnetic theory arrives at the following expression (see e.g., Becker and Sauter,[2] p. 197):

$$\mathbf{E} = \frac{1}{R^3 c^2} [\mathbf{R}[\mathbf{R}\ddot{\mathbf{X}}]] - \frac{\dot{\mathbf{X}}}{R^2 c} + \frac{3(\dot{\mathbf{X}}\mathbf{R})\mathbf{R}}{R^4 c} - \frac{\mathbf{X}}{R^3} + \frac{3(\mathbf{X}\mathbf{R})\mathbf{R}}{R^5} \tag{3.5}$$

$$\mathbf{H} = \frac{1}{R^2 c^2} [\ddot{\mathbf{X}}\mathbf{R}] + \frac{1}{R^3 c} [\dot{\mathbf{X}}\mathbf{R}] \tag{3.6}$$

It is seen that the fields contain terms the amounts of which vary as R^{-1}, R^{-2}, and R^{-3}. The last of these is, of course, most important in the near-zone, i.e. for $R \ll \lambda$. For the far-zone, however, where $R \gg \lambda$, only the first term in the field equations (3.5) and (3.6) has to be considered. We now assume that $\mathbf{X}$ represents a harmonic oscillator with the retarded time t':

$$\mathbf{X}(t') = \mathbf{X}_0 e^{i\omega t'} = \mathbf{X}_0 e^{i\omega(t - R/c)} \tag{3.7}$$

Introducing the second derivative of this equation into the first term of each of the field equations, we obtain for the far (or "wave") zone

$$\mathbf{E} = \frac{\omega^2}{R^3 c^2} e^{i\omega(t - R/c)} [\mathbf{R}[\mathbf{X}_0 \mathbf{R}]] \tag{3.8}$$

$$\mathbf{H} = \frac{\omega^2}{R^2 c^2} e^{i\omega(t - R/c)} [\mathbf{R}\mathbf{X}_0] \tag{3.9}$$

The vector product in Eq. (3.8) shows that the electric field vector $\mathbf{E}$ is (1) polarized in the plane formed by the dipole $\mathbf{X}$ and the radius vector

R, (2) normal to **R**, and (3) proportional in amount to sin θ where θ is the angle between **X** and **R**. If the dipole is oriented in the center and along the polar axis of a sphere with radius **R**, the electric field vector is tangential to the meridian at the point P as indicated in Fig. 3.2. The

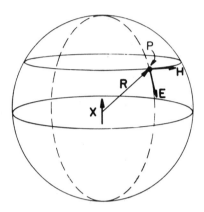

Fig. 3.2.   Electric and magnetic fields produced by oscillating dipole.

magnetic field **H**, which is described by Eq. (3.9), is normal to the radius vector **R** and the electric field **E** since (3.8) can be combined with (3.9) to yield

$$E = \frac{1}{R}[HR] \quad \text{and} \quad H = \frac{1}{R}[RE] \qquad (3.10)$$

Thus **H** is tangential to the latitude at P in Fig. 3.2.

It is also seen that the amount of the electric field is at all points of the wave zone equal to that of the magnetic field, varying radially as $R^{-1}e^{i\omega(t-R/c)}$. Thus **E** and **H**, equal and normal to each other, propagate as spherical waves from the dipole (see Fig. 3.3). Since they vary as sin θ, they vanish at the poles: the dipole does not radiate into its axial direction. Because of this fact, the terms varying as R^{-2} and R^{-3} in Eqs. (3.5) and (3.6) have to be considered near $\theta = 0$ with the result of modifying the field direction near the poles. Thus, if one plots the lines of force, one obtains the Hertzian pattern shown in part by Fig. 3.3.

3.1.3 Radiation Power of the Classical Oscillator

The power propagated by the electromagnetic field of the dipole can now be readily determined. The energy density of an electric field

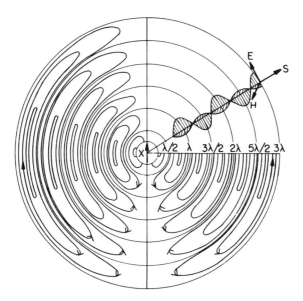

FIG. 3.3.   Hertzian pattern of spreading electromagnetic waves generated by oscillating dipole. Contours shown are instantaneous configurations of the electric field in the near- and far-zone. The complete spatial pattern is obtained as a figure of revolution around the dipole axis. Shaded curves indicate phase variation of electric and magnetic field in the direction of the Poynting vector. No alternating fields are produced in the axial direction of the dipole.

E in vacuo equals $E^2/8\pi$, a result known from electrostatics. The flux density then in uniform propagation is, according to Eq. (2.7),

$$S = \frac{c}{4\pi} E^2 \qquad (3.11)$$

a factor of 2 arising from the equal contribution of the magnetic field. More generally, one introduces the Poynting vector $\mathbf{S}$ for the power density propagated by the electromagnetic field

$$\mathbf{S} = \frac{c}{4\pi} [\mathbf{EH}] \qquad (3.12)$$

which in the far-zone of the dipole field has the direction of the radius vector (see Fig. 3.2) and the amount already stated in (3.11), since $|\mathbf{E}| = |\mathbf{H}|$. When we apply Eq. (3.8) or (3.9) to (3.11) we are only interested in the time-average of the power. The average of the time variant factor of E^2, written in real form as $\cos^2\omega(t - R/c)$, is, of course,

$\frac{1}{2}$. Thus the power density radiated at an angle θ with respect to the dipole axis is

$$S = \frac{1}{8\pi} \frac{\omega^4}{R^2 c^3} X_0^2 \sin^2\theta \qquad (3.13)$$

which contains the inverse-square law of intensity versus distance. We are, in particular, interested in the total power radiated by the dipole. The total flux P through a sphere of radius R is

$$P = \int SR^2 \, d\Omega = \frac{\omega^4 X_0^2}{4c^3} \int_0^\pi \sin^3\theta \, d\theta \qquad (3.14)$$

so that

$$P = \frac{\omega^4 X_0^2}{3c^3} = \frac{e^2}{3c^3} x_0^2 \omega^4 = \frac{2}{3c^3} \overline{|\ddot{\mathbf{X}}|^2} \qquad (3.15)$$

where we have introduced the charge e and maximum displacement x_0 of the dipole with $\mathbf{X}_0 = e\mathbf{x}_0$. Equation (3.15) is the classical expression for the radiation power from an oscillating charge in terms of amplitude and resonant frequency ω; it is seen to be proportional to the square of the displacement and the fourth power of the frequency.

3.1.3.1 *Emission of Light with Radiation Damping*

Equation (3.15) describes the radiation resulting from pure harmonic motion of dipole charges. If the motion is not sinusoidal, it can be presented as the sum of Fourier components. An important example is the damped oscillator since it represents the classical model of an atomic or molecular light source. The meaning of a damping constant γ as reciprocal oscillator lifetime (i.e. $\gamma = 1/\tau$) and its connection with spectral line broadening ($\Delta\nu_h = \gamma/2\pi$) have been discussed already in Section 1.2.2. The initial energy ϵ_0 of the dipole is, in terms of x_0 and the symbols used in Eqs. (1.5) and (1.6),

$$\epsilon_0 = \tfrac{1}{2}Gx_0^2 = \tfrac{1}{2}M(2\pi\nu_0)^2 x_0^2 \qquad (3.16)$$

This energy is dissipated during an oscillation lifetime $1/\gamma$ (see Fig. 1.3). As a result, the radiation is spread over the spectrum as described by the intensity distribution relation (1.10). Thus the energy radiated per unit frequency interval is, according to Eq. (3.15),

$$\epsilon_\nu = \frac{16\pi^4}{3c^3} e^2 x_0^2 \nu_0^4 \frac{1}{4\pi^2(\nu_0 - \nu)^2 + \gamma^2/4} \qquad (3.17)$$

The total radiated energy is obtained by integration of (3.17) over all frequencies which yields

$$\epsilon = \int_0^\infty \epsilon_\nu \, d\nu = \frac{16\pi^4}{3c^3} \frac{\nu_0^4}{\gamma} e^2 x_0^2 \qquad (3.18)$$

This energy, which is simply the radiation power P of Eq. (3.15) expended to $1/e$ over a time $\tau = 1/\gamma$, must equal the initial oscillator energy ϵ_0, if all of it has been converted into radiation. A comparison then between (3.16) and (3.18) yields a radiation damping constant γ_0,

$$\gamma_0 = \frac{8\pi^2}{3} \frac{e^2}{Mc^3} \nu_0^2 = \frac{8\pi^2}{3} \frac{e^2}{Mc} \frac{1}{\lambda_0^2} \qquad (3.19)$$

This relation, which was first derived by Planck,[3] states oscillator lifetime and natural line broadening in terms of basic constants and the wavelengths. Thus if we consider the light source as consisting of a vibrating electron with mass $m = 9.1 \times 10^{-28}$ gm and charge $e = 4.80 \times 10^{-10}$ esu, we obtain with (3.19)

$$\gamma_0 = 0.22/\lambda^2 \qquad (3.20)$$

As an example, an atom with a resonance line at 0.5μ (5×10^{-5} cm) would have a damping constant of $\gamma_0 = 0.88 \times 10^8$ sec^{-1} and a duration of oscillation $\tau = 1/\gamma = 1.1 \times 10^{-8}$ sec. This order of magnitude agrees well with observation.

There exists here also an interesting relationship to the classical electron radius r_0. Such radius results from equating the energy mc^2 of the electron at rest to its electrostatic energy e^2/r_0:

$$r_0 \approx \frac{e^2}{mc^2} \qquad (3.21)$$

Introducing (3.21) into (3.19) and expressing $\gamma_0 = 2\pi \, \Delta\nu_h$ in terms of the halfwidth $\Delta\lambda_h$, yields (cf. Heitler[62], pp. 31-34):

$$\Delta\lambda_h = \frac{\Delta\nu_h \lambda^2}{c} = \frac{4\pi}{3} \frac{e^2}{mc^2} = \frac{4\pi}{3} r_0 = 1.18 \times 10^{-4} \text{ Å} \qquad (3.22)$$

The natural line broadening remains, therefore, the same on the wavelength scale throughout the spectrum.

3.1.3.2 *Antenna Resistance for the Short Dipole Case*, $x_0 \ll \lambda$

Finally, we mention an application of Eq. (3.15) to antenna theory. If we ignore ohmic losses, the transmitter will supply to the antenna just enough power to make up for the losses to the radiation field so that there is no damping. Now the term $e\omega$ in Eq. (3.15) represents a current I_0, and the equation can be written in the form

$$P = \frac{4\pi^2}{3c} I_0{}^2 \left(\frac{x_0}{\lambda}\right)^2 = \frac{I_0{}^2 R_{\text{eff}}}{2} \tag{3.23}$$

where

$$R_{\text{eff}} = \frac{8\pi^2}{3c} \left(\frac{x_0}{\lambda}\right)^2 = 80\pi^2 \left(\frac{x_0}{\lambda}\right)^2 \quad \text{[ohms]} \tag{3.24}$$

Here R_{eff} represents, for an antenna of length x_0, an effective "radiation resistance," i.e., that resistance which with a current I_0 presents the same load as the surrounding electromagnetic field. Equations (3.23) and (3.24) are, therefore, basic to antenna design.

3.1.4 Absorption

We now turn to the case which represents the inverse of that discussed in the last two sections: we assume an existing radiation field and proceed to determine its energy transfer to a damped oscillator. The equation of motion for free oscillation (see Section 1.2.2),

$$m\ddot{x} + R\dot{x} + Gx = 0, \tag{3.25}$$

has now to be replaced by the equation containing a driving force:

$$m\ddot{x} + R\dot{x} + Gx = e\mathbf{E}_\omega e^{i\omega t} \tag{3.26}$$

We assume that the electric field E_ω has a continuous frequency distribution in the neighborhood of resonance ω_0 for the oscillator. According to the theory of differential equations, the solution of Eq. (3.26) is completely determined by the sum of the general integral of (3.25), $\mathbf{x}_{\text{tr}}$, and a particular integral of (3.26), $\mathbf{x}_{\text{st}}$. The solution for (3.25) is given by Eq. (1.5), while a particular integral of (3.26) is obtained by the following (see Section 5.2.2 for details of the solution):

$$\mathbf{x}_{\text{st}} = \mathbf{a}_\omega e^{i\omega(t-\alpha)} \tag{3.27}$$

Altogether, the solution for Eq. (3.26) is, in real quantities,

$$\mathbf{x} = \mathbf{x}_{\text{tr}} + \mathbf{x}_{\text{st}} = x_0 e^{-(\gamma/2)t} \cos(\omega_0' t - \alpha) + \frac{e\mathbf{E}_\omega}{m} \frac{\cos(\omega t - \beta)}{\sqrt{(\omega_0{}^2 - \omega^2)^2 + \gamma^2 \omega^2}} \tag{3.28}$$

In this superposition of two oscillations, x_{tr} represents the contribution of the free motion. This part is transient since it decays as $e^{-\gamma t/2}$ with a damping constant $\gamma = R/m \ll (G/m)^{1/2}$. We have to assume that the free oscillation starts randomly with a phase α at $t = 0$ and has an angular frequency of resonance

$$\omega_0' = \sqrt{\frac{G}{m} - \frac{\gamma^2}{4}} \approx \sqrt{\frac{G}{m}} = \omega_0 \qquad (3.29)$$

The steady state contribution x_{st} adopts the frequency of the driving field which it follows with a phase delay

$$\beta = \tan^{-1} \frac{\gamma\omega}{(\omega_0{}^2 - \omega^2)} \qquad (3.30)$$

It will be seen that the response vanishes for very small and very large driving frequencies, the phase difference being 0 and π, respectively. For $\omega = \omega_0$ the response is a maximum and the phase shift is $\pi/2$. As the charge e moves with the field, an energy ϵ_ω is exchanged with the field, during a time t', which is given by

$$\epsilon_\omega = e \int_{x(0)}^{x(t')} \mathbf{E}_\omega \, d\mathbf{x} = e \int_{t=0}^{t'} \mathbf{E}_\omega \dot{\mathbf{x}}(t) \, dt \qquad (3.31)$$

The contribution in $\dot{x}$ from the free oscillation depends, of course, on its phase. It is noteworthy that the net effect may consist of energy delivered to the field, resulting in negative absorption (or stimulated emission). In general, however, we will be dealing with a multitude of oscillators with random phase distribution, and thus, in the average, the energy exchanged due to $\dot{x}_{tr}$ is zero.

We encounter, in fact, a considerable simplification in computing the work done per second on the oscillator, if we apply Eq. (3.26) to Eq. (3.31) since the terms $x \cdot \dot{x}$ and $\ddot{x} \cdot \dot{x}$ are, in the average, also zero as they contain the products $\cos(\omega t - \beta) \cdot \sin(\omega t - \beta)$. Altogether, we obtain the energy absorbed by the oscillator per second per unit angular frequency with Eqs. (3.26), (3.28), and (3.31):

$$P_\omega = R\overline{\dot{x}^2} = \frac{e^2 E_\omega{}^2 \gamma}{8m} \cdot \frac{1}{(\omega_0 - \omega)^2 + \gamma^2/4} \qquad (3.32)$$

where we have replaced $(\omega_0{}^2 - \omega^2)$ by the approximation $2\omega_0(\omega_0 - \omega)$. The spectral profile of an absorption line due to a damping factor γ is seen to be the same as in emission [cf. Eq. (3.17)], and this we had to expect, at any rate, on the basis of Kirchhoff's law. Now if we assume that the radiation intensity (energy per cm²-sec)

$$I_0 \, d\omega = \frac{c}{8\pi} E_\omega^2 \, d\omega \tag{3.33}$$

is constant near resonance ω_0, we have for the total energy absorbed by the oscillator per second

$$P = \frac{e^2 E_\omega^2 \gamma}{8m\omega_0^2} \int_0^\infty \frac{d\omega}{\left(1 - \frac{\omega}{\omega_0}\right)^2 + \frac{\gamma^2}{4\omega_0^2}} = \frac{e^2 E_\omega^2 \gamma}{8m\omega_0^2} \cdot \frac{2\pi\omega_0^2}{\gamma} \tag{3.34}$$

With (3.33) we then have in terms of radiant flux density per unit angular frequency interval, I_0:

$$P = \frac{2\pi^2 e^2}{mc} I_0 \tag{3.35}$$

Thus the energy extracted by the oscillator out of the continuous radiation spectrum is independent of the line width $\gamma/2\pi$ and the resonant frequency; aside from the fundamental constants e, m, and c, it is given by the product of field intensity and time (see Fig. 3.4).

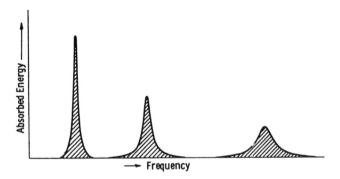

Fig. 3.4. Spectral profile of classical oscillator absorption. The area under the curve is independent of the broadening.

3.2 Discrete Energy States of the Atom

The results of the classical treatment presented so far in this chapter are valid to the extent to which the model of the pure or damped harmonic oscillator is applicable. This is the case for the features of lifetime and line broadening, the ν^4-dependence of emitted intensity, the formula for the energy absorbed, and the properties of the electromagnetic field in the neighborhood of the radiating dipole. The classical method is particularly compatible for phenomena observed in the low-frequency region of the spectrum where the number of quanta per unit energy is large so that the discreteness of the radiation flux is little noticeable. However, when processes are involved in which the discrete energy structure of matter is relevant, the methods of electromagnetic theory will be insufficient. Historically, this showed itself first in the inability, in effect, to explain the spectral position of oscillator resonances. As a result, just as the spectral profile of heat radiation could be explained only with the assumptions of quantized energy states in an oscillator continuum, so did the appearance of gaseous line spectra require the postulate of quantized energy states in individual atoms.

3.2.1 EXPERIMENTAL BACKGROUND

Emission lines can be produced in gases and, where applicable, in solids by electric excitation such as arcs, sparks, or electron beam bombardment, but also by thermal and optical energies, i.e. by thermoluminescence and fluorescence, respectively. Now the observations which were inexplicable by classical theory concerned the fundamental relationships that existed for the position and spacing of lines, or groups of lines, which could be preferentially excited. The most impressive orderliness of this kind occurs in hydrogen and hydrogenlike atoms. The emission spectra of these gases contained certain sequences of lines which obeyed the following empirical relationship (Rydberg-Ritz):

$$\nu_{n_1 n_2} = R_z\left(\frac{1}{n_1{}^2} - \frac{1}{n_2{}^2}\right) \tag{3.36}$$

where n_1 and n_2 represent integer numbers. At fixed n_1, the sequence $n_2 = n_1 + 1$, $n_1 + 2$, $n_1 + 3$, ..., generates the frequencies of the observed line series according to (3.36). R_z is a constant which, while different for different isotopes, is correlated to a fundamental "Rydberg constant" R_∞ with an accuracy that served as a yardstick of reliability for the unfolding atomic theory.

The discovery of the various line series occured in stepwise advances as successive regions of the optical spectrum were made accessible to spectroscopic observation. First to be recognized was a visible series in hydrogen for $n_1 = 2$, $n_2 = 3, 4, 5, ...$, which was named the Balmer

series after its discover in 1885. Subsequently Lyman found the series
based on $n_1 = 1$ in the extreme ultraviolet, while Paschen, Brackett,
and Pfund found those for $n_1 = 3$, 4, and 5, respectively, in the infrared
spectrum of hydrogen. Furthermore, a principle of combining frequencies from different series was recognized by Ritz in the fact that the
sum of $\nu_{n_1 n_2}$ and $\nu_{n_2 n_3}$ is equal to $\nu_{n_1 n_3}$, etc., if the corresponding lines
exist.

3.2.2 BOHR ATOM

A variety of attempts notwithstanding, it was found impossible to
explain the line series of hydrogen and other spectra by purely classical
models which would, e.g., interpret Eq. (3.36) as overtones of a fundamental vibration. It was against this background of uncertainty about the
origin of line spectra that Bohr[4] (1913) set forth his theory of the hydrogen atom (see Fig. 3.5). Here it was possible to derive, in a single step,
all salient features of the hydrogen series, to give a quantitative account
of Eq. (3.36), and to lay the foundation for the modern theory of the
atom.

One aspect of Bohr's contribution lies in the application of quantum
concepts to the atom model which had been previously proposed by
Rutherford[5] (1911). Rutherford's theory visualized an atom of order
number Z_0 as consisting of Z_0 electrons which describe planetary motions
around a nucleus of charge eZ_0. This planetary system with a diameter
in the order of 10^{-8} cm surrounds a central nucleus which has linear
dimensions of less than 10^{-12} cm. A hydrogenlike atom is characterized
by a configuration in which a single electron revolves around a charge
eZ (e.g., hydrogen H, singly ionized helium He$^+$, doubly ionized Lithium
Li^{++}, etc.). If the orbits are circles of radius r, the central Coulomb force

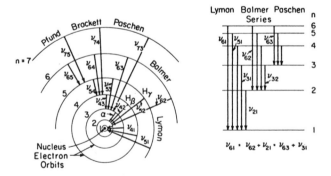

FIG. 3.5.   Hydrogen series and Ritz combination principle in Bohr Atom.

e^2Z/r^2 is just equal and opposite to the centrifugal force $mr\omega^2$ which is due to the circular velocity at angular frequency ω. Setting the potential energy of the electron zero at infinite distance from the nucleus, one has for the total energy (potential and kinetic)

$$\epsilon = -\frac{e^2Z}{r} + \frac{m\omega^2r^2}{2} = -\frac{e^2Z}{2r} = -\frac{m}{2}r^2\omega^2 \tag{3.37}$$

Bohr now made the assumption that only certain orbital states could exist, namely only those for which the angular momentum was a multiple of $\hbar$ ($\hbar = h/2\pi$):

$$mr^2\omega_n = n\hbar \qquad n = 1, 2, 3 \ldots \tag{3.38}$$

By the simple operation of eliminating ω_n and r in Eqs. (3.37) and (3.38), one obtains for the discrete energy states of the allowed orbits

$$-\epsilon_n = \frac{2\pi^2me^4}{h^2}\frac{Z^2}{n^2} \tag{3.39}$$

The final assumption in Bohr's theory was that the transition of an electron from an orbit with quantum number n_1 to another state with number n_2 was accompanied by the emission or absorption of a light quantum $h\nu_{n_1n_2}$ of energy $\epsilon_{n_1} - \epsilon_{n_2}$. We divide Eq. (3.39) by hc to obtain the frequencies, as customary, in wave numbers (i.e., number of wavelengths per centimeter, rather than cycles per second). Thus,

$$\tilde{\nu}_{n_1n_2} = R_0Z^2\left(\frac{1}{n_1^2} - \frac{1}{n_2^2}\right) = R_z\left(\frac{1}{n_1^2} - \frac{1}{n_2^2}\right) \tag{3.40}$$

where R_0 is the Rydberg constant

$$R_0 = \frac{2\pi^2me^4}{ch^3} \tag{3.41}$$

Equation (3.40) is, of course, the theoretical counterpart of the empirically found line series law, Eq. (3.36). Figure 3.5 shows the mechanism by which transitions between the various stationary orbits give rise to the spectral series of hydrogen and to the combination principle stated by Ritz.

Because of the accuracy in wavelength assignment and abundance of suitable spectra which are both available to the measurement of the Rydberg constant, the latter represents an important input value for the determination of the fundamental constants.[6] It must be pointed out, in this connection, that in the derivation of Eqs. (3.40) and (3.41) we ignored the relatively very slight motion of the nucleus around a center of mass in common with the orbiting electron. If the finite mass of the nucleus is taken into consideration, the results must be corrected by replacing the electron mass m with a *reduced* mass

$$m_r = \frac{mM}{m + M} = \frac{m}{1 + m/M} \tag{3.42}$$

This leads to small, but perceptible differences between the R_0-values of different atoms and even the isotopes of the same element so that the spectral lines of such an element may show a resulting hyperfine structure. Birge[7], in an analysis of experimental data, presented the following values for the Rydberg constant of various isotopes:

$$R_H = 109{,}677.581 \pm 0.007_5 \quad \text{cm}^{-1}$$
$$R_D = 109{,}707.419 \pm 0.007_5$$
$$R_{He} = 109{,}722.263 \pm 0.007$$
$$R_\infty = 109{,}737.303 + 0.017$$

where we have replaced the subscript 0 by the isotope symbol and R_∞ represents the extrapolation to infinite nuclear mass.

It remains to be discussed that there are energy states which are not quantized at all. When an ionized atom recaptures an electron into the nth orbit, the energy available for reradiation is the sum of that for the series limit and the kinetic energy before recombination:

$$E = \frac{R_0 hcZ}{n^2} + \frac{mv^2}{2} \tag{3.43}$$

The last term is not quantized. Therefore, the recombination radiation consists of a continuum extending from each series limit towards shorter wavelengths. Such spectra are indeed observed experimentally.

3.2.3 THE SCHRÖDINGER WAVE EQUATION

In the preceding section we have discussed the origin of discrete energy states of individual atoms. Bohr and others developed this theory beyond the stage of describing hydrogenlike atom configurations and their spectra. In fact, an important application of the original concepts was made by Bohr when he developed the theory of the structure of the periodic system. Nevertheless, the method of introducing integer numbers n to arrive at discrete states, which the original quantum theory employs so successfully, is in itself arbitrary and implausible.

This need for a more cogent derivation of quantized energy levels and modes of motion was filled by the quantum mechanics of Schrödinger[8] and Heisenberg.[9] We shall briefly introduce this subject matter to the extent to which it is pertinent to our purposes.

3.2.3.1 *Standing Waves in Electron Orbits*

The starting point of quantum mechanics is the de Broglie[10, 11] relation (see Section 1.3.3) which assigns to an electron of momentum $p = mv$ a wave character such that the wavelength is

$$\lambda = h/p \tag{3.44}$$

We now recall a property of standing waves in elastic bodies such as strings, namely, of existing only in modes for which the wavelengths assume a set of discrete values (see Section 1.2.3). Clearly, an obvious analogy arises at this point: if the wave equation of the electron, too, has solutions only for certain proper values of the wavelength, then the introduction of integer quantum numbers in the Bohr atom is at once made plausible.

Thus let us assume tentatively that an electron orbit $2\pi r$ must consist of an integer number of wavelengths (see Fig. 3.6) since the value of the electron wave function must be everywhere unique and continuous:

$$2\pi r = n\lambda \tag{3.45}$$

Then, if we use the relation (3.44), we indeed obtain from (3.45)

$$pr = m\omega r^2 = nh/2\pi \tag{3.46}$$

which is the Bohr quantum condition [cf. Eq. (3.38)].

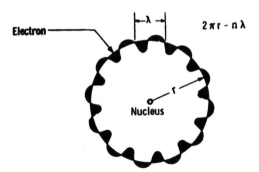

Electron

$2\pi r - n\lambda$

λ

r

Nucleus

FIG. 3.6. Interpretation of electronic wave function as elastic standing wave in Bohr orbit.

3.2.3.2 *Wave Equations*

Electron wave functions, therefore, when applied to the atom, give rise to the Bohr quantum conditions just as the standing waves of a displacement or density variable give rise to proper modes in an elastic body. It has, of course, yet to be determined what form the wave functions will assume under various conditions of potential energy and how one has to interpret the physical significance of the results. In analogy to the wave equation in elastic media, we can postulate quite generally the following expression in terms of the Laplacian operator ∇:

$$\nabla^2\Psi = \frac{\partial^2\Psi}{\partial x^2} + \frac{\partial^2\Psi}{\partial y^2} + \frac{\partial^2\Psi}{\partial z^2} = \frac{1}{c^2}\frac{\partial^2\Psi}{\partial t^2} \tag{3.47}$$

where in the spirit of the original de Broglie theory, $\Psi = \Psi(x, y, z, t)$ signifies a measure of electron density or probability of being present at the point (x, y, z) at the time t. It is known from the theory of differential equations that the solution of (3.47) can be presented as the product of functions with separate variables:

$$\Psi(x, y, z, t) = \psi(x, y, z) \cdot \varphi(t) \tag{3.48}$$

The function containing the time will be simply periodic so that we can write

$$\Psi = \psi e^{i\omega t} \tag{3.49}$$

If this expression is introduced into Eq. (3.47), one obtains

$$\nabla^2\psi + \frac{\omega^2}{c^2}\psi = 0 \tag{3.50}$$

Replacing ω by $2\pi\nu = 2\pi c/\lambda$ and λ by the De Broglie relation, Eq. (3.44), we write

$$\nabla^2\psi + \frac{4\pi^2}{\lambda^2}\psi = \nabla^2\psi + 4\pi^2\frac{p^2}{h^2}\psi = 0 \tag{3.51}$$

This equation will represent standing waves for certain discrete values of the momentum p. We are interested, however, in the total energy W of the electron because it is that which appears in spectral terms and other data directly determined by measurement. The relationship between momentum, kinetic energy T, potential energy $V(x, y, z)$, and the total energy is this:

$$p^2 = m^2v^2 = 2mT = 2m(W - V) \tag{3.52}$$

If we introduce this expression into Eq. (3.51), we obtain

$$\nabla^2\psi + \frac{8\pi^2 m}{h^2}\,[W - V(x, y, z)]\psi = 0 \qquad (3.53)$$

This is Schrödinger's wave equation. More precisely, it is the amplitude equation for the complete function which contains the time according to Eq. (3.49). Schrödinger derived (3.53) more generally by means of the Hamiltonian function for the total energy, assigning formally the significance of a mathematical operator to p and W. For instance, in the Hamiltonian for the one-dimensional case

$$H(p_x, x) = \frac{p_x^2}{2m} + V(x) = W \qquad (3.54)$$

p_x is replaced by the operator $(h/2\pi i)(\partial/\partial x)$ and W by $(- h/2\pi i)(\partial/\partial t)$. If these operators are applied to the Hamiltonian (3.54), one has

$$H\Psi = -\frac{h^2}{8\pi^2 m}\frac{\partial^2\Psi}{\partial x^2} + V\Psi = W\Psi = -\frac{h}{2\pi i}\frac{\partial\Psi}{\partial t} \qquad (3.55)$$

from which it is possible to derive Eq. (3.53) for the x-coordinate:

$$\frac{\partial^2\psi}{\partial x^2} + \frac{8\pi^2 m}{h^2}\,[W - V(x)]\psi = 0 \qquad (3.56)$$

3.2.3.3 *Proper Energy Values and Continua*

A large number of conclusions can be drawn from the wave equation on the basis of the requirement that the function must be single-valued, continuous, and finite at all points. It is, for example, because of these restrictions that the total energy W can assume only discrete values in certain regions if the Schrödinger equation (3.53) or (3.56) is to be fulfilled. This can be conveniently proven by graphically constructing the solutions of Eq. (3.56), which latter we now write in the form

$$\frac{\partial^2\psi}{\partial x^2} = \frac{8\pi^2 m}{h^2}\,[V(x) - W]\psi \qquad (3.57)$$

In its upper part, Fig. 3.7 exhibits an energy diagram plotted against the x-coordinate. The potential energy $V(x)$ is assumed to have a bell-shaped profile which is symmetrical with respect to $x = 0$, and which approaches zero for very large and very small x. We thus start the construction of $\psi(x)$ with an arbitrary position of $\psi(0)$ at $x = 0$ and with the slope $(d\psi/dx)_{x=0} \equiv \psi'(0) = 0$.

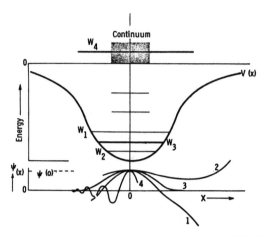

Fig. 3.7.   Construction of proper wave function on the basis of Schrödinger equation. Among the alternatives shown, only curves 3 and 4 corresponding to the discrete energy value W_3 and the continuum region, respectively, fulfill the boundary conditions.

If we now select, by trial and error, a value W, the course of $\psi(x)$ is completely determined by Eq. (3.57). The second derivative $\psi''(x)$ is related to the curvature by the geometric relation

$$\kappa = \frac{\psi''}{(1 + \psi'^2)^{3/2}} \tag{3.58}$$

Thus the curvature has the same sign as $\psi(x)$ for $V(x) > W$ and the opposite sign for $V(x) < W$. For example, if we choose the value W_1 indicated in Fig. 3.7, the curvature is negative for small $|x|$, and changes sign twice to become negative also for large $|x|$ since both $(V - W)$ and ψ change sign. As a result, the tentative curve $\psi_1(x)$ approaches minus infinity for large $|x|$ and has to be disregarded. Similarly $\psi_2''(x)$ is positive for large $|x|$ and increases with $|x|$; the resulting curve $\psi_2(x)$ thus approaches plus infinity. Clearly, there can only be one discrete value W_3 between W_1 and W_2 such that the resulting $\psi'(x)$ and $\psi(x)$ go to zero together, and $\psi(x)$ approaches zero asymptotically. In this manner, the necessity for proper energy values is made evident.

One can also show the existence of energy continua by this method. For values of W which are everywhere larger than V, such as W_4 in Fig. 3.7, $\psi''(x)$ has always a sign opposite to that of $\psi(x)$. Therefore the function turns towards the x-axis at all points for increasing x and oscillates around zero with decreasing amplitude. Hence all values

$W > V$ provide acceptable solutions in agreement with the Bohr theory in which this condition corresponds to the energy continuum of the recombination region (see Section 3.2.2).

3.2.3.4 *Physical Significance*

Although the Schrödinger equation can be derived by formal postulates, as in Eq. (3.55), or by an analogy to elastic waves, as in Eq. (3.47), its actual justification lies in the agreement of its predictions with the facts of observation. The process of deriving proper energy levels (or *Eigenwerte*), which was performed in the previous section qualitatively by graphical methods, can be achieved quantitatively by the analytical means known from the theory of differential equations. From the starting point of merely rendering plausible the quantum conditions of the Bohr atom, quantum mechanics progressed to the construction of atomic energy level schemes with a universality and refinement of which the old theory was quite incapable. Along with this went a deeper insight into the nature of matter on the atomic level and into the processes in which elementary particles are involved. The Schrödinger equation assumes, therefore, a role similar to that of Maxwell's equations or the second law of thermodynamics and other indirectly stated laws which have in common that, while they have only a tenuous relationship to experimental facts, they themselves represent concise statements of first principles.

Aside from the general conclusions, the Schrödinger equation makes definite statements about the function ψ. The physical significance of the wave function is that its square $| \psi(x, y, z) |^2$ represents the probability of an electron to be at the point $P(x, y, z)$. This is, of course, equivalent to its interpretation as electron density distribution which is closest to the mechanical analog of the elastic density wave Eq. (3.47). Finally, the statistical interpretation is not limited to position x, y, z of a particle, but yields the average of its time and space dependent functions.

3.3 Transition Probabilities and Intensities

It has been seen at the beginning of this chapter that emission and absorption of electromagnetic radiation are quantitatively accounted for in the classical theory by the deceleration and acceleration, respectively, of electric charges. As an example, the process of spontaneous emission is accompanied by the gradual dissipation of the initial oscillator energy over a lifetime $\tau = 1/\gamma$ as illustrated in Fig. 1.3. Thus the radiated power is distributed in time according to an exponential attenuation law and it is also spread in space in spherical waves, as shown in Fig. 3.3. Quantum theory, however, as was already brought out in Section 1.3.5, treats the processes of emission and absorption as more or less sudden, or at least as sharply confined, events which have to be dealt with on the basis of statistical concepts. Such a statistical treatment was given by Einstein[12,13] in an investigation which by way of a remarkably short derivation of Planck's radiation law succeeded in correlating the equilibria of radiation, molecular motion, and quantized states.

3.3.1 EINSTEIN TRANSITION PROBABILITIES

For the following it will be sufficient to consider a very simple case although the conclusions have general applicability. We are concerned with the radiation equilibrium of a gas the atoms of which can exist in a number of discrete energy states such as ϵ_1 and ϵ_2 ($> \epsilon_1$) (see Fig. 3.8). The probability that an atom is in the energy level ϵ_1 is given by the Boltzmann factor $g_1 e^{-\epsilon_1/kT}$. Here g_1 is the statistical weight, i.e. the number of different states or sets of quantum numbers for the particular energy level (as discussed further in Chapter 4, the g-values are usually unity, each atomic state having a distinct value of energy, except for so-called "degenerate" levels). Similarly the probability for an atom to be in the state ϵ_2 is $g_2 e^{-\epsilon_2/kT}$. It follows that the numbers of atoms n_1 and n_2 in the energy states ϵ_1 and ϵ_2, respectively, are related by the equation

$$n_1 = \frac{g_1}{g_2} e^{-(\epsilon_1-\epsilon_2)/kT} \cdot n_2 = \frac{g_1}{g_2} e^{h\nu/kT} \cdot n_2 \qquad (3.59)$$

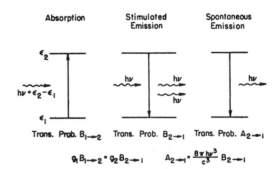

FIG. 3.8.   Einstein transition probabilities.

We now assume that an equilibrium between energy states and radiation exists such that per unit time as many atoms are raised from state ϵ_1 to ϵ_2 with the absorption of a photon $h\nu = \epsilon_2 - \epsilon_1$, as undergo a transition from state ϵ_2 to ϵ_1 with the emission of such quantum. As shown in Fig. 3.8, there exist three types of transition processes by which the equilibrium establishes itself:

(1) Spontaneous emission from the higher to the lower level which Einstein postulated in analogy to radioactive decay. Thus the speed of this process is determined by a transition probability $A_{2\to1}$ per unit time. In the absence of competing processes, this is equal to the γ-value (see Section 3.1.3.1) or equal to the reciprocal of the lifetime τ_2.

(2) Stimulated emission from the higher to the lower level in response, and proportional, to the radiation density $u(\nu)$. The stimulated transfer of energy from the oscillator to the electromagnetic field has been predicted already by classical theory (see Section 3.1.3); but it occurs in quantum theory according to a transition probability $B_{2\to1}$ per unit time and energy density.

(3) Absorption of a photon with subsequent transition of the atom from the lower to the higher level. This process is governed by a transition probability $B_{1\to2}$ per unit time and energy density.

According to the equilibrium condition stated earlier, as many individual transitions of the type (3) must occur in any time interval as of type (1) and (2) added together. Thus, the equilibrium condition has the form

$$n_1 B_{1\to2} u(\nu)\, dt = n_2\, dt(B_{2\to1} u(\nu) + A_{2\to1}) \tag{3.60}$$

Introduction of the relationship (3.59) between the populations leads to an expression for the energy density of radiation

$$u(\nu) = \frac{A_{2\to1}}{B_{2\to1}} \frac{1}{\dfrac{g_1}{g_2} \dfrac{B_{1\to2}}{B_{2\to1}} e^{h\nu/kT} - 1} \tag{3.61}$$

which must, of course, agree with Planck's radiation formula

$$u(\nu) = \frac{8\pi h \nu^3}{c^3} \frac{1}{e^{h\nu/kT} - 1} \tag{3.62}$$

If this is to be the case, then there must exist the following relationships between the Einstein transition probabilities:

$$g_1 B_{1\to2} = g_2 B_{2\to1} \tag{3.63}$$

or for the case of nondegeneracy, $g_1 = g_2 = 1$, with only one atomic state per energy level,

$$B_{1\to2} = B_{2\to1} \qquad (3.64)$$

Again by comparison of Eqs. (3.61) and (3.62), we find

$$A_{2\to1} = \frac{8\pi h \nu^3}{c^3} B_{2\to1} \qquad (3.65)$$

These equations are basic for the elementary radiation processes. First, (3.63) and (3.64) assert that, *cet. par.*, positive and negative absorption have equal probabilities; that is, a photon encountering a resonant upper state is as likely to cause stimulated emission, as it is to be absorbed when encountering a lower state. Of course, in thermal equilibrium the higher energy states are less populated than the lower states. Normally, therefore, events of absorption are far more frequent than those of stimulated emission, the difference being made up by spontaneous transitions. The probability for the latter is the subject of the second statement made by the Einstein equations. According to Eq. (3.65), the occurrence of spontaneous emission increases with ν^3; and this behavior reflects the fact that excited atoms can spontaneously decay into any of $8\pi\nu^2 \, d\nu/c^3$ modes per unit volume, whereas stimulated emission is prescribed in the single mode of the incident photon. For we can write Eq. (3.65) in the form

$$A_{2\to1} = \underbrace{\frac{8\pi\nu^2}{c^3}}_{\substack{\text{no. modes} \\ \overline{\text{unit freq. interv.} \times \text{unit vol.}}}} \times \underbrace{h\nu}_{\text{1 quantum}} \times \underbrace{B_{2\to1},}_{\substack{\text{stim. em. rate} \times \text{unit freq. interv.} \\ \overline{\text{energy density}}}}$$

$$(3.65a)$$

which states that the spontaneous transition probability equals the product of the number of modes and the absorption rate for one quantum per mode. Thus the existence of spontaneous emission can be interpreted as stimulated by *virtual photons* in the various cells of phase space. To express Eq. (3.65a) in other terms, the probability of spontaneous emission per mode equals that of negative or positive absorption in the presence of one quantum in that mode.

3.3.1.1 *Directionality of Emission Processes*

The preceding considerations of the atomic transition probabilities have only dealt with the radiation equilibrium of energy levels; so far nothing was discussed regarding the accompanying impulse exchange. That momentum is at all exchanged in these processes, is in itself, of course, a significant statement of quantum theory (cf. Section 1.3.2); the emission of spherical waves according to the electromagnetic model of a light source does not impart a momentum to the radiating dipole. Averaged over a large number of events, spontaneous transitions will, in fact, behave according to the classical picture and leave the momentum of the atom unaltered. For single events, however, an impulse $h\nu/c$ is exchanged between the atom and photon with sharp directionality. The transfer of quantized energies and momenta, nevertheless, must leave the Maxwell-Boltzmann velocity distribution of the gas undisturbed (or must indeed establish it, if atom collisions are negligible in number compared to the photon exchanges in thermal equilibrium). Einstein showed in the works cited before[12, 13] (see also Dirac[14]) that this is the case provided that the three types of transitions occur according to the following momentum rules:

(1) In absorbing a quantum $h\nu$, an atom receives the momentum $h\nu/c$ in the direction of the incident photon.

(2) In stimulated emission a light quantum is emitted with the same direction and frequency as the incident radiation while the atom moves in the opposite direction.

(3) In spontaneous emission, a photon propagates in a single random direction.

This confirms that the processes of positive and negative absorption occur in definite modes prescribed by the incident photons, while spontaneous emission can occur in any direction.

3.3.2 Transition Probabilities in Quantum Mechanics

We have seen in Section 3.2.3 that wave mechanics represented the role of mediator between classical physics and the radical quantum conditions which it evolved plausibly on the basis of an elastic wave equation. Similarly, wave mechanics is capable of explaining a number of features pertaining to radiative transitions, namely, aspects interpreted in the old quantum theory by postulates which in themselves need further clarification. Examples for such postulates are the Bohr transition conditions which demand that an orbiting electron does not radiate unless it goes over to another orbit. The circular motion of an electron in the x-y plane, however, appears to an observer in that plane as a harmonic oscillation; thus it should give rise to dipole radiation according to the classical rules discussed in Section 3.1.2. Again, the Einstein transition probabilities are determined only in relation to each other and not in terms of other atomic constants. This was the situation that became clarified by Schrödinger's wave mechanics[8] and Heisenberg's matrix mechanics,[9] the general term "quantum mechanics" applying to both theories together. The common name is justified by the mathematical equivalence of the two theories which was shown to exist by Schrödinger[8] (p. 734) and Eckart.[15] The treatment of radiative transitions is considerably simpler in wave mechanics, however, particularly with the use of the correspondence principle and Born's[16, 17] statistical interpretation of the wave functions.

3.3.2.1 *Matrix Element of Dipole Moment*

The key concept of the quantum mechanical radiation theory is represented, as we shall presently explain, by integrals of the type

$$\mathbf{X}_{nm} = \int \Psi_n{}^* \, \mathbf{X}(t) \Psi_m \, dv \tag{3.66}$$

in which ψ_n and ψ_m are the wave functions of an upper state n and a lower state m, respectively, and $\mathbf{X}(t)$ is here the dipole moment

$$\mathbf{X} = e\mathbf{x}$$

while the asterisk denotes the conjugate complex wave function (in which i is replaced by $-i$). $\mathbf{X}_{nm}$ is called the matrix element of the dipole moment $\mathbf{X}$, the significance of Eq. (3.66) being the following.

The physical meaning of the wave function Ψ_n as discussed in Section 3.2.3.4, is that

$$|\Psi_n|^2 \, dv = \Psi_n{}^* \Psi_n \, dv \tag{3.68}$$

represents the probability for an electron in the state n to be in the volume element dv. Altogether, of course, the probability is unity that the particle is somewhere, and we must adjust the Ψ's such that

$$\int \Psi_n{}^* \Psi_n \, dv = 1 \tag{3.69}$$

It follows that (3.68) also represents a weighting factor for functions of positions, say $F(x, t)$. A particularly simple case is that of the dipole moment $e \cdot x$, with x being the distance of the electron from the nucleus. The total moment of state n (after integration over all volume elements with their respective probabilities of being occupied by the electron) is thus

$$\mathbf{X}_{nn} = e \int \Psi_n{}^* \mathbf{x} \Psi_n \, dv \qquad (3.70)$$

which represents a special case of Eq. (3.66). Now, it follows from symmetry considerations that

$$\mathbf{X}_{nn} = 0, \qquad (3.71)$$

i.e., that the moment is zero. More generally, if we write the wave functions with Eqs. (3.49) and (3.55) in the form

$$\Psi_n{}^* = \psi_n{}^* \exp\left[\frac{-2\pi i W_n}{h} t\right] = \psi_n{}^* \exp[-2\pi i \nu_n t] \qquad (3.72)$$

$$\Psi_m = \psi_m \exp\left[\frac{2\pi i W_m}{h} t\right] = \psi_m \exp[2\pi i \nu_m t] \qquad (3.73)$$

it follows from (3.70) that $\mathbf{X}_{nn}$ is independent of time, regardless of symmetry properties. Here then is the explanation of the fact that the electron in the stationary energy state W_n does not radiate: the power emitted by a dipole is according to Eq. (3.15) proportional to the square of its second derivative:

$$P \propto |\ddot{\mathbf{X}}|^2 \qquad (3.74)$$

This is zero in the stationary case.

The situation is quite different for the time during which an atom undergoes a transition from state n to m. To be consistent, we must now write for the dipole moment associated with such transition

$$\mathbf{X}_{nm} = e \int \Psi_n{}^* \mathbf{x} \Psi_m \, dv \qquad (3.75)$$

If we now introduce Eqs. (3.72) and (3.73), we have

$$\mathbf{X}_{nm} = e \int \psi_n{}^* \mathbf{x} \psi_m \, dv \cdot \exp\left[-2\pi i \left(\frac{W_n - W_m}{h}\right) t\right] = e \mathbf{x}_{nm} \exp[-2\pi i \nu_{nm} t] \quad (3.76)$$

where

$$\mathbf{x}_{nm} = \int \psi_n{}^* \mathbf{x} \psi_m \, dv \tag{3.77}$$

represents the matrix element of the position coordinate x, and where

$$\nu_{nm} = \frac{W_n - W_m}{h} \tag{3.78}$$

confirms the Bohr transition rule. In fact, the latter is made physically plausible in this picture since

$$\nu_{nm} = \nu_n - \nu_m \tag{3.79}$$

now appears as a beat frequency between the vibration of the upper and the lower state.

Of particular interest also is the amount of power radiated out by the atom. For this purpose, we first compute $|\ddot{\mathbf{X}}_{nm}|^2$ from (3.76),

$$|\ddot{\mathbf{X}}_{nm}|^2 = 16\pi^4 \nu_{nm}^4 e^2 |\mathbf{x}_{nm}|^2 = e^2 \omega_{nm}^4 |\mathbf{x}_{nm}|^2 \tag{3.80}$$

Furthermore, we assume that the principle of correspondence between classical and quantized states exists. In this case, the amplitude of the classical dipole oscillation is related to the matrix as follows:

$$\mathbf{x}_0{}^2 = 2\overline{\mathbf{x}^2} = 4\,|\mathbf{x}_{nm}|^2 \tag{3.81}$$

Introducing Eq. (3.81) into (3.15) we obtain for the power radiated by the atom

$$P = \frac{4}{3} \frac{e^2}{c^3}\,\omega_{nm}^4 |\mathbf{x}_{nm}|^2 \tag{3.82}$$

3.3.2.2 Transition Probabilities

In terms of the Einstein transition probabilities, the power radiated by an atom is

$$P = h\nu A_{n \to m} \tag{3.83}$$

A comparison between (3.82) and (3.83) yields at once a value for $A_{n \to m}$ in terms of atomic constants:

$$A_{n \to m} = \frac{64\pi^4}{3} \frac{e^2}{hc^3}\,\nu_{nm}^3 |\mathbf{x}_{nm}|^2 \tag{3.84}$$

It is seen that the spontaneous transition probability increases with the third power of frequency for a *given oscillator amplitude*. However, since, other things being equal, x_{nm} will tend to be inversely proportional to the binding energy of the electron and hence to ν (because of smaller orbits for tighter coupling), the magnitude of A_{nm}—in the absence of others factors—will go as ν through the electromagnetic spectrum.

The transition probabilities for negative and positive absorption are, according to Eqs. (3.63) and (3.65),

$$B_{n \to m} = \frac{8\pi^3}{3} \frac{e^2}{h^2} \mid x_{nm} \mid^2 \tag{3.85}$$

and

$$B_{m \to n} = \frac{g_n}{g_m} B_{n \to m} = \frac{8\pi^3}{3} \frac{g_n}{g_m} \frac{e^2}{h^2} \mid x_{nm} \mid^2 \tag{3.86}$$

3.3.2.3 *Magnetic Dipole and Electric Quadrupole Transitions*

The physical meaning of the term $e \mid x_{nm} \mid$ is that of an electric dipole moment, and its product with ω^2 represents its second derivative with respect to time as a result of its oscillation with an angular frequency ω. When the subject of dipoles was introduced on a classical basis by way of the Hertzian vector (section 3.1.1), the generalized model of multiple charges made its appearance. The vibration of such a system consisting of more than two point charges—or a more complex charge distribution—may give rise, in second or yet higher order, to multipole radiation as a result of time retardation effects. However, it follows already from classical arguments that this higher order radiation will be much less intense than that normal for electric dipole oscillations. This can be seen by rewriting Eq. (3.4) in the following form:

$$X = \sum_k e_k x_k + \sum_k e_k \dot{x}_k \frac{x_{kR}}{c} + \cdots = \sum_k e_k x_k \left(1 + i\omega \frac{x_{kR}}{c} + \cdots \right) \tag{3.4a}$$

It follows from this equation that the second order moment is smaller than the first for each component charge k by an amount

$$\frac{\omega x_{kR}}{c} = 2\pi \frac{x_{kR}}{\lambda} \tag{3.4b}$$

Since intensity is proportional to the square of the oscillation amplitude, the second order multipole radiation is weaker than the first by a factor

$$\frac{A''}{A'} \cong \left(\frac{2\pi a}{\lambda}\right)^2 \tag{3.4c}$$

where we have replaced x_{kR} by a mean atom radius a. It follows from Eq. (3.4c) that in the visible region ($\lambda \approx 5000$ Å) magnetic dipole and electric quadrupole radiation is by a factor of 10^{-6} less intense than the normally occurring electric dipole radiation. The same factor reduces, in general, the intensity of the $(n + 1)$th order with respect to the nth term. Higher order radiation is significant mainly, if first order radiation cannot occur, i.e. if x_{nm} vanishes.

The quantum mechanical treatment of magnetic dipole and electric quadrupole transitions is analogous to that of the electric dipole case, except for the fact that the matrix element x_{nm} must be replaced by a higher moment. This involves the more detailed atom model discussed in the next chapter where we shall resume the subject of multipole radiation (Section 4.4).

3.3.3 Theory of Line Intensities[*]

In gases of low concentrations, observed in optically *thin* layers, the contributions of the oscillators are simply additive. Thus we have for the total power radiated from N_n atoms per unit volume in the nth level into 4π steradian:

$$I_{\text{em}} = N_n h\nu_{nm} A_{n\to m} = \frac{64\pi^4}{3} \frac{e^2}{c^3} \nu_{nm}^4 \,|\, x_{nm}\,|^2 N_n \tag{3.87}$$

Furthermore, by virtue of the definition implicit in Eq. (3.60), we can compute the power I_{ab} absorbed by N_m atoms in the mth level in a radiation field of energy density u_ν (assumed to be constant in the neighborhood of the line ν_{nm}):

$$I_{\text{ab}} = N_m h\nu_{nm} u_\nu B_{m\to n} = \frac{8\pi^3}{3} \frac{g_n}{g_m} \frac{e^2}{h} \,|\, x_{nm}\,|^2 \nu_{nm} u_\nu N_m \tag{3.88}$$

Under ordinary conditions, for $N_n \ll N_m$, the contribution I_{st} of stimulated emission is negligible. However, the populations in the upper and lower states tend to become more nearly equal for small

---

[*] As a choice between conflicting designations elsewhere in the literature, we adopt here the following notation: the initial and final states in a transition are indicated by the order of the subscripts, hence $A_{nm} \equiv A_{n\to m}$ and $f_{mn} \equiv f_{m\to n}$, although $\nu_{mn} \equiv \nu_{nm}$ and $|\, x_{nm}\,|^2 \equiv |\, x_{mn}\,|^2$. Furthermore, increasing energy states follow, in general, numerical or alphabetical order, hence $E_2 > E_1$, $E_n > E_m$.

$h\nu/kT$ or in the presence of strong "pumping" light (cf. also Chapter 6). In such a case, the total effect is usually that of a reduced *net absorption* I'_{ab}, although the "twinning" process inherent in stimulated emission alters the fluctuation and noise characteristic of the photon stream (see Chapter 7). We have for the net absorption, according to Eqs. (3.86) and (3.88),

$$I'_{ab} = I_{ab} - I_{st} = N_m h\nu_{nm} u_\nu B_{m\to n}\left(1 - \frac{g_m}{g_n}\frac{N_n}{N_m}\right) \tag{3.89}$$

where the term $g_m N_n/g_n N_m$ is given by the Boltzmann factor $\exp(-h\nu/kT)$ in case of thermal equilibrium [cf. Eq. (3.59)].

From Eqs. (2.9) and (3.89), we can derive* (cf. Unsöld,[18] p. 279) the net absorption coefficient k_ν' per centimeter length which we have already defined in Section 2.4.3.2:

$$\int k_\nu' \, d\nu = \frac{I'_{ab}}{cu_\nu} = \frac{N_m h\nu_{nm} B_{m\to n}}{c}\cdot\left(1 - \frac{g_m N_n}{g_n N_m}\right) \tag{3.90}$$

3.3.3.1 Oscillator Strengths

Many aspects of the classical radiation theory are reaffirmed by the quantum mechanical treatment. Agreement exists, for instance, on the form of the emission law [cf. Eqs. (3.15) and (3.87)], especially the ν^4-dependence for a given amplitude, and also on the invariance of absorption with respect to line width [cf. Eqs. (3.35) and (3.86)]. It is therefore meaningful to define a so-called *oscillator strength* f_{mn} (Ladenburg[19]) as the number of classical oscillators which are as effective in terms of power absorbed or emitted as a single atom in the transition $m \to n$. Given a radiant flux density $I_{0\nu}$ per unit frequency interval $d\nu$ (rather than angular frequency interval $d\omega$), Eq. (3.35) yields for the power absorbed by $\mathcal{N}$ classical oscillators

$$I_{ab} = \frac{\pi e^2}{m_e c} I_{0\nu}\mathcal{N} = \frac{\pi e^2}{m_e} u_\nu \mathcal{N} \tag{3.91}$$

where m_e the electron mass and, by definition,[†]

$$\mathcal{N} = N_m f_{mn} \tag{3.92}$$

---

* For isotropic radiation, of which the power P_ν is assumed constant near the spectral line, the absorption is given by $dP_\nu = k_\nu P_\nu \, d\nu \, d\Omega$. Integrating over angle $d\Omega$ and frequency $d\nu$, one obtains for the total absorbed power $P_{ab} = 4\pi P_\nu \int k_\nu \, d\nu = cu_\nu \int k_\nu \, d\nu$.

† The original definition[19] of $\mathcal{N}$ included the negative contribution by stimulated emission from the upper level:

$$\mathcal{N} = N_m f_{mn}\left(1 - \frac{N_n}{N_m}\frac{g_m}{g_n}\right)$$

If this is applied to Eq. (3.94), the right-hand side has to be replaced by that of Eq. (3.89) so that f_{mn} results as in Eq. (3.95).

The corresponding classical expression for the (positive) absorption coefficient is thus[19]

$$\int k_\nu \, d\nu = \frac{I_{ab}}{cu_\nu} = \frac{\pi e^2}{m_e c} N_m f_{mn} \tag{3.93}$$

The theoretical f-values are at once obtained by setting the classical absorption Eq. (3.91) equal to the quantum-mechanical value, Eq. (3.88):

$$\frac{\pi e^2}{m_e} u_\nu N_m f_{mn} = h\nu_{nm} B_{mn} u_\nu N_m \tag{3.94}$$

Hence, the oscillator strengths for absorption and stimulated emission are, respectively,

$$f_{mn} = \frac{h\nu_{mn} m_e}{\pi e^2} B_{mn} = \frac{8\pi^2}{3} \frac{g_n}{g_m} \frac{m_e \nu}{h} |\, \mathbf{x}_{nm} \,|^2 \tag{3.95}$$

$$f_{nm} = \frac{g_m}{g_n} f_{mn} = \frac{8\pi^2}{3} \frac{m_e \nu}{h} |\, \mathbf{x}_{nm} \,|^2 \tag{3.96}$$

It should be noted that from the classical viewpoint an atom in the level m represents a single oscillator, whereas the quantum-theoretical picture is that of an atom which is capable of an unlimited number of transitions from a level m, each with an oscillator strength f_{mn}. If the two models are to be commensurate, then it can be argued that the classical amount of absorptivity at a single frequency has been spread over a quantized distribution of resonances in addition to the continuum, the total amount of response being equal to the classical case. This, indeed, is the statement of the f-sum rule by Kuhn,[20] Thomas, and Reiche[21, 22]:

$$\sum_n f_{mn} - \sum_k f_{mk} = 1 \tag{3.97}$$

where the second term represents the contribution of negative absorption owing to stimulated transitions from the levels m to lower levels k, according to Eq. (3.96). More generally, if one deals with more than one classical oscillator, e.g., z valence electrons, the right hand side of Eq. (3.97) is not unity, but z. For Eq. (3.97) to be valid, all transitions have to be considered for which the f-values are not zero, even those which the electron cannot actually undergo because the lower state is filled (Pauli principle, Chapter 4).

The computation of the oscillator strengths and transition probabilities requires the evaluation of the matrix elements $\mathbf{x}_{nm}$ on which they are based through the relations (3.95), (3.96), and (3.84)–(3.86), respectively.

Hydrogen or hydrogenlike configurations, consisting of a single electron in the field of a positive charge, represent the least complexity, and for these conditions it was possible to carry through completely the computation of the oscillator strengths parameters. Table 3.1 lists the transition probabilities A_{nm} and corresponding oscillator strengths f_{mn} for the Lyman series ($m = 1$) of hydrogen (Bethe[23, 24], Unsöld[18]).

TABLE 3.1

<small>Transition Probabilities and Oscillator Strengths of the Lyman Series[a]</small>

Wavelength λ (Å)	Transition	Transition probability $A_{n \to m}$ [sec^{-1}]	Oscillator strengths $f_{m \to n}$
1216	$m = 1 \quad n = 2$	6.25×10^8	0.4162
1026	3	1.64	0.0791
973	4	0.68	0.0290
950	5	0.34	0.0139
938	6	0.195	0.0078
931	7	0.122	0.0048
926	8	0.082	0.0032
$\leqslant$923	n		$1.6n^{-3}$
$\leqslant$912	Continuum		0.4359
			$\sum\limits_{n} f_{mn} = 1$

[a] H. A. Bethe, *in* "Handbuch der Physik" (H. Geiger and K. Scheel, eds.), 2nd ed. Vol. 24, Part I, Chapter 3. Springer, Berlin, 1933; H. A. Bethe and E. E. Salpeter, *in* "Handbuch der Physik" (S. Flügge, ed.), Vol. 35, p. 88. Springer, Berlin, 1957.

It is seen that the oscillator strengths and transition probabilities decrease rapidly for lines of increasing upper quantum number n. In fact, the oscillator strength of the first line in the Lyman series amounts to 41.6 % of the total sum, including the continuum. Under ordinary conditions, the number N_1 of atoms in the unexcited state $m = 1$ is overwhelmingly large compared to that of the excited levels. Therefore, absorption of lines starting at the ground state is particularly strong [cf. Eq. (3.88)]. It follows that the most intense and most easily excited transition is the first of the series with $m = 1$. Such transitions are therefore called *resonance lines*. Since the lower level of the Lyman series is the ground state, there are no negative contributions to the f-sum. In the other hydrogen series, the fine structure due to the orbital quantum numbers has to be considered as discussed in Chapter 4. We mention here merely that the sum of the spontaneous transition probabilities A_{nm} from a level n represents its total natural decay rate, hence the reciprocal natural lifetime τ of the level:

$$\sum_m A_{nm} = \frac{1}{\tau} \tag{3.98}$$

The lifetimes of the lower hydrogen levels are mostly in the order of 10^{-9} to 10^{-7} sec.

3.3.3.2 *Interrelationships between Line Strength Parameters*

In addition to the formulas given in the preceding sections, the connections between transition probabilities, oscillator strengths, matrices, and line intensities can be expressed in yet other ways. We list these alternative forms to the extent to which they contribute to a further insight into their meaning or to which they are practically useful.

From Eqs. (3.84) and (3.95) follows Ladenburg's relationship between the oscillator strength f_{mn} and the corresponding probability A_{nm} of transition from level n to m:

$$f_{mn} = \frac{m_e c^3}{8\pi^2 e^2 \nu^2} \frac{g_n}{g_m} A_{nm} = \frac{g_n}{g_m} \frac{A_{nm}}{3\gamma_0} \tag{3.99}$$

On the right-hand side of this equation, we have introduced the classical damping constant γ_0 of Eq. (3.19) so that the oscillator strength reveals its meaning in terms of the ratio between the quantum-theoretical transition probability A_{nm} and the classical decay rate, $\gamma_0 = 1/\tau_0$, of the electromagnetic radiator. Similarly we have for the·oscillator strength of stimulated emission :

$$f_{nm} = \frac{g_m}{g_n} f_{mn} = \frac{A_{nm}}{3\gamma_0} \tag{3.100}$$

If Eq. (3.99) is introduced into Eq. (3.93), one obtains an important result for the absorption coefficient:

$$\int k_\nu \, d\nu = \frac{\pi e^2}{m_e c} N_m f_{mn} = \frac{1}{8\pi} \frac{g_n}{g_m} N_m \lambda^2 A_{nm} \tag{3.101a}$$

Thus when integrated over the spectral environment of a transition ν_{nm}, the absorption coefficient does not depend on type and extent of the line broadening. It follows that the theory may be extended to dense gases and even solids. More generally, because individual lines may merge into a continuum we have to sum over all transitions from the state m:

$$\int_0^\infty k_\nu \, d\nu = \sum_{n,k} \int k_\nu \, d\nu = \frac{\pi e^2}{m_e c} N_m z \tag{3.101b}$$

Here we have applied the f-sum rule to z electrons per absorbing center for transitions from a state m to upper states n and lower states k. Therefore, the integrated absorption coefficient depends only on the density of the centers and the number of their participating electrons and is otherwise independent of the atom species and the physical state of matter. In the important case that there is only one absorbing electron per atom, $z = 1$, we obtain for the maximum value k_0 (after substituting the profile of $k(\nu)$ by a rectangle of height k_0 and a representative width $\nu_2 - \nu_1$):

$$k_0 = \frac{1}{\nu_2 - \nu_1} \cdot \int_0^\infty k_\nu \, d\nu = 1.1 \times 10^{-16} \frac{N_m}{\nu_2 - \nu_1} \qquad (3.101\text{c})$$

where k_0 is given in cm^{-1} and the bandwidth in the equivalent number of electron-volts. Equations (3.101a)–(3.101c) have been found valid not only for gases, but they are approximately correct also for intrinsic and extrinsic semiconductors and insulators. The general validity is emphasized in Eq. (3.101c) which contains only N_m and $\nu_2 - \nu_1$ as variables.

For the case of a simple resonance line, in which $A_{nm} = \gamma = 1/\tau$, Eq. (3.101a) can be written in the form

$$k_0 = \frac{1}{\gamma} \int k_\nu \, d\nu = \frac{1}{8\pi} \frac{g_n}{g_m} \cdot N_m \lambda^2 \qquad (3.102)$$

For natural broadening, Eq. (3.102) is capable of an illustrative interpretation: the left-hand side, as before, represents the spectral peak of the absorption coefficient, and this equals the product of the number N_m of absorbing atoms and an *absorption cross section* which is in the order of λ^2.

Through the substitution of the transition probabilities, line intensities can be given in terms of oscillator strengths. The expression for the (spontaneously emitted) light power of a line results, for instance, from Eqs. (3.87) and (3.99) as

$$I_{em} = \frac{8\pi^2 h e^2 \nu^3}{m_e c^3} \frac{g_m}{g_n} f_{mn} N_n \qquad (3.103)$$

The number N_n of excited states in thermal equilibrium with the population N_1 of the ground state, is known from the Boltzmann relation

$$N_n = \frac{g_n}{g_1} N_1 e^{-(E_n - E_1)/kT} \qquad (3.104)$$

so that Eq. (3.103) becomes

$$I_{em} = \frac{8\pi^2 he^2\nu^3}{m_e c^3} \frac{g_m}{g_1} f_{mn} N_1 e^{-(E_n-E_1)/kT} \qquad (3.105)$$

In the case of hydrogen, it will be shown in Chapter 4 that an electron orbit of quantum number n can have $2n^2$ different configurations of identical energy; hence $g_m = 2m^2$, $g_1 = 2$.

In conclusion we mention the concept of *line strength* introduced by Condon and Shortley.[25] Consider two energy levels A and B—we adopt here the designations of these authors—such that the upper level, A, have g_A fold degeneracy, consisting of g_A states $a = 1, 2, ..., g_A$, with similar degeneracy g_B of b-states in the lower level B. The line strength $S(A, B)$ is then defined on the basis of individual transition matrix elements between the a- and b-states [where we write $(a \mid \mathbf{P} \mid b)$ instead of $e \cdot \mid \mathbf{x}_{ab} \mid$]:

$$S(A, B) = \sum_{a,b} (a \mid \mathbf{P} \mid b)^2 \qquad (3.106)$$

We obtain for the transition probability $A(A, B)$ and the oscillator strength $f(B, A)$ in absorption and $f(A, B)$ in stimulated emission:

$$A(A, B) = \frac{64\pi^4\nu^3}{3hc^3} \frac{1}{g_A} S(A, B) \qquad (3.107)$$

$$f(B, A) = \frac{8\pi^2 m_e}{3he^2} \nu \frac{1}{g_B} S(A, B) \qquad (3.108)$$

$$f(A, B) = \frac{8\pi^2 m_e}{3he^2} \nu \frac{1}{g_A} S(A, B) \qquad (3.109)$$

It is seen that the asymmetry between Eqs. (3.95) and (3.96) has disappeared in the corresponding expressions Eqs. (3.108) and (3.109).

3.3.4 Experimental Determination of the Intensity Parameters

Although it has been the practice, since the early days of spectroscopy, to assign to newly found lines approximate intensity values on an arbitrary scale, this information is of value only for purposes of identification. For it is clear from the preceding discussions that even the relative line intensities must vary in first order with changes in type or degree of excitation, or with different conditions of density, temperature, and radiation path. In fact, the only invariant parameters of intensity are the atomic constants, i.e., the transition probabilities or oscillator strengths and the statistical weights. For this reason, a large amount of effort has been expended on the experimental determination of the atomic parameters, especially for that majority of elements in which the electronic configurations are too complex for a theoretical treatment. It is, of course, important to compare, wherever possible, the results of theory and experiment as a test for the validity of the atomic model used. However, the region of overlap between the two approaches is still relatively narrow, and even then the difficulties facing either the computation or the experiment are often such that agreement can be expected only by order of magnitude. Aside from this aspect, a knowledge of oscillator strengths makes it possible, in principle, to predict intensities of emission or absorption for given laboratory conditions. Conversely, the measurement of intensities provides an input value for the determination of density or temperature, if other indicators are not available.

3.3.4.1 *Self-absorption and Resonance Imprisonment*

The relationship for line intensities, Eqs. (3.87)–(3.89), and those derived from them, are valid only for optically thin media, or more accurately, they are valid only for the case that the number N of atoms in the radiation field is not large enough for a mutual interference of the emission or absorption processes. The opposite limit, namely, that of densities so large that a thermal equilibrium results, has been discussed in connection with plasma radiation (Section 2.4.3.2). The measurement of line intensity parameters, of course, will be performed in general under conditions which minimize self-absorption and re-emission (or in which the factors causing them are varied so as to extrapolate to the zero point of interaction). If these precautions are not taken or if, alternatively, no analytical corrections for self-absorption and re-emission are made, then the intensity measurements do not permit a conclusion regarding the atomic constants. The modifications which a beam of light experiences in its passage through a medium will be treated in Chapter 5 on the phenomena of propagation. We shall indicate here briefly, however, the corrections necessary for self-absorption in the case of emission, and for re-emission in the case of absorption.

(a) *Self-Absorption.* For a homogeneously radiating slab of thickness L, Ladenburg[26] found, by computing the absorption of the line for various intensity profiles and applying Kirchhoff's law, that the intensity formula, Eq. (3.87), had to be modified by a factor S:

$$I = N_n A_{nm} h\nu_{nm} \cdot SL \qquad (3.110)$$

where N_n represents here the number of upper states per unit volume and where for the case of Doppler broadening (see Section 3.4.2)

$$S_{\text{Doppler}} = 1 - \frac{C}{2\sqrt{2}} + \frac{C^2}{6\sqrt{3}} - \cdots + (-1)^n \frac{C^n}{(n+1)!\sqrt{n+1}} \quad (3.111)$$

with

$$C = \frac{2\mathcal{N}e^2L}{\nu_0 m_e \bar{v}_x}$$

$\mathcal{N}$ is the number of classical oscillators per unit volume and $\bar{v}_x$ the mean velocity of the atoms. Similar relationships follow for the S-factors based on other line profiles, although these are of less practical importance for the type of intensity measurement under discussion.

The effect of self-absorption can be surprisingly large: for atoms at room temperature, radiation in the visible at $\lambda = 5000$ Å over a path length of $d = 1$ cm, and $\mathcal{N} = 10^{10}$ [e.g., $N_m = 10^{11}$, $f_{mn} = 0.1$, see Eq. (3.92)], it follows from Eq. (3.111) that $S \cong 0.60$. Thus even at that low concentration almost half of the emitted photons remain locked in. These considerations are particularly important for transitions to the ground state. For example, if in the preceding example the path length is 0.1 mm and the pressure 1 mm Hg (so that $N_m \cong 3 \times 10^{16}$ cm^{-3}), with all other things being equal, then $S \cong 0.0006$; hence, less than 0.1 % of the original radiation emerges.

(b) *Resonance Imprisonment.* Every normal atom can participate in the self-absorption of resonance radiation. Therefore such absorption, as shown in the example of the previous paragraph, represents in order of magnitude a class by itself. In fact, there results, to an extent depending on atom density and geometric configuration of the radiation space, a so-called *resonance imprisonment* because the excitation energy of the resonance level is essentially trapped within a group of atoms by a large number of emission and absorption processes before a photon can finally escape. Thus the decay of the excited levels follows a law of the form $e^{-g\gamma t}$ in which the decay rate γ of an individual atom is reduced by an *escape factor g*. If events of stimulated emission are negligible, g will assume values between zero and unity, and it will be inversely proportional to the number of emissions and absorptions provided the number is large enough.

The first observation—and correct interpretation—of resonance imprisonment was reported by Wood[27] (1912) who noticed a diffusive spread of radiation in mercury vapor from one excited portion to another, even if the two regions were separated by a quartz window. The decay of the 2537 Å radiation in mercury after pulsed excitation was measured later with increasing refinement.[28, 29]

A full account of the quantitative behavior observed in resonance imprisonment has been given by the theory of Holstein[30, 31] (see also Bieberman[32]). The theory stresses a previously not recognized feature of photon transfer through resonant media, namely, that such transfer cannot be described in terms of a mean free path and a diffusion equation. There rather exists a distribution of escape probabilities, the photons of a v farther away from resonance having a relatively large range while those near resonance are quickly absorbed. The calculation is thus based on the probability $T(\rho)$ that quanta emitted with a distribution $P(v)$ can reach a distance ρ without being absorbed by the medium with an absorption coefficient $k(v)$:

$$T(\rho) = \int P(v)e^{-k(v)\rho}\,dv \tag{3.112}$$

$P(v)$ can be shown to be proportional to $k(v)$ for Doppler and pressure broadening, even though the existence of a thermal equilibrium (as the condition for this form of Kirchhoff's law) is not immediately obvious. We cite in Table 3.2 the results of the Holstein theory for the cases of Doppler- and impact-broadening in terms of the escape factors:

TABLE 3.2

ESCAPE FACTORS g [Holstein (*24*)][a]

	Doppler line shape	Impact broadening
Cylindrical[b] case (radius R)	$\dfrac{1.60}{k_0R\,\sqrt{\pi\log_e k_0R}}$	$\dfrac{1.115}{\sqrt{\pi k_p R}}$
Slab[c] (thickness L)	$\dfrac{1.875}{k_0L\,\sqrt{\pi\log_e k_0L}}$	$\dfrac{1.150}{\sqrt{\pi k_p L}}$

[a] T. Holstein, *Phys. Rev.* **72**, 1212 (1947); **73**, 1159 (1951).

[b] $k_0 = \dfrac{\lambda^3 N_1}{8\pi^{3/2}}\dfrac{g_2}{g_1}\dfrac{\gamma}{v_0}$.

[c] $k_v = \dfrac{\lambda^2 N_1}{2\pi}\dfrac{g_2}{g_1}\dfrac{\gamma}{\gamma_p}$.

v_0 mean velocity of atom.
g_2, g_1 statistical weights of resonance level and ground state.
$\gamma = 2\pi \Delta v_h$ natural halfwidth in angular frequency.
$\gamma_p = 2\pi \Delta v_{h,p}$ impact halfwidth in angular frequency.

The reciprocal of the escape factor, $1/g$, represents the ratio of the radiation persistence time to the lifetime of the individual excited state.

It is seen from Table 3.2 that, e.g., for $k_0R = 10^2$, there results an increase of the apparent lifetime by a factor of several hundred. The effect of resonance imprisonment has therefore to be taken into account in all measurements in which the lifetime or the excitation efficiency of any level combining with the ground state have to be known.[33, 34, 35]

3.3.4.2 Experimental Methods for A- and f- Determinations

The theory of line intensities deals with a considerable number of physical concepts and quantities—to wit (in the order in which they have been introduced in the preceding paragraphs): the various transition probabilities, the matrix elements, the intensities derived from them, oscillator strengths, lifetimes, and line strengths. If no confidence at all were to be placed in the theoretical approach, the task of determining these magnitudes by experiment would be all but impossible because, in general, measurement has access to one or the other, but not to all, of the parameters. Fortunately, as we have seen, the various concepts and magnitudes are tightly linked to each other by interrelationships the validity of which cannot be in question; for these relationships are based on first principles such as the laws of thermodynamics.

Thus it is sufficient to determine a minimum number of parameters—the transition probability A_{nm}, for instance, if also g_m and g_n are known. While such minimum experimental effort is sufficient, it is also often necessary because in many cases the computation of the transition matrix is valid only as an order of magnitude approximation. The measurement, then, will be concerned with transitions either from a lower or an upper state; but the experimental possibilities are by no means limited to intensity determinations in absorption or emission. Other physical properties depend on the transition probabilities as well, and we shall list in the following the more important among them along with the direct methods.

(a) *Absorption Measurements.* The most obvious method of determining f_{mn}- and B_{mn}-values from the relationships given in the preceding pages consists of absorption measurements. If radiation from a continuum source passes through a layer of thickness Δx containing a known number N_m of absorbing atoms, the intensity I suffers a relative reduction [cf. Eq. (3.93)]

$$\frac{\Delta I}{I} = -\Delta x \int k_\nu \, d\nu = -\frac{\pi e^2}{m_e c} N_m f_{mn} \Delta x \qquad (3.113)$$

This relation presumes that stimulated emission from the level n can be ignored. The assumption is practically always justified, if absorption from the ground level is observed in the absence of other excitation.

A more difficult problem is the measurement of absorption and determination of the f_{mn}-values from excited levels. The effect of negative absorption, although often negligible, enters Eq. (3.113) as indicated in Eq. (3.89). At any rate, the population densities of excited levels now appear in the equations. These densities are related to the normal atom concentration through the Boltzmann factor which is known only if the effective temperature can be determined by some means. Alternatively, if all transitions from a given level are measured out, it may be possible to use the f-sum rule for normalization and absolute determination of the oscillator strengths (see Section 3.3.4.3). These difficulties do not exist, of course, if one is content with relative values: in the intensity ratios of lines starting from the same level, the population density cancels out.

A further problem in the measurement of absorption from excited levels is spontaneous emission of the line since in the usual discharges all levels are more or less populated. It is thus necessary to discriminate in some manner against the emission from the absorption medium. The simplest method for this purpose consists of pulsing the primary radiation,[36, 37] e.g., by a chopping wheel between emitter and absorber, at a frequency for which the detector amplifier is selective.

Equation (3.113) is correct only for the limit of small optical thicknesses $k(\nu)\,\Delta x \ll 1$ and an incident intensity I_0 which is independent of ν. If these conditions are not fulfilled, especially if the same line is used for emission and absorption, one has to consider the intensity reduction separately for each frequency interval. Thus one has

$$I_{\nu ab}(\nu, x)\,d\nu = I_{\nu 0}(\nu)(1 - e^{-k(\nu)x})\,d\nu \qquad (3.114)$$

The total fraction absorbed in the environment of the line is thus

$$\frac{I_{ab}}{I_0} = \frac{\int I_{\nu 0}(\nu)(1 - e^{-k(\nu)x})\,d\nu}{\int I_{\nu 0}(\nu)\,d\nu} \qquad (3.115)$$

For the limits cited in the beginning, this goes over into Eq. (3.113). In general, however, approximation methods have to be used (see e.g., Kopfermann and Wessel[38] for absorption and absolute f-values in a beam of iron atoms).

(b) *Anomalous Dispersion Measurements* (*Hook Method*) In Chapter 5 it will be brought out that the propagation velocity of light through a a medium, measured by the index of refraction, n, undergoes a more or less sharp variation in magnitude and slope with wavelength at regions of resonance. Such behavior is called *anomalous dispersion* since dis-person, $dn/d\lambda$, normally represents a small, negative change of n for increasing wavelength. The resonance regions of the medium are characterized also by their absorption, and thus it is not unexpected that anomalous dispersion is a direct measure of the corresponding oscillator strengths. In particular, we have in case of low absorber densities such as gases in the neighborhood of an isolated line ν_{mn}:

$$n - 1 = \frac{N_m e^2}{2\pi m_e} \frac{f_{mn}}{(\nu_{mn}^2 - \nu^2)} \left(1 - \frac{g_m}{g_n} \frac{N_n}{N_m} \right) \tag{3.116}$$

The effect at low concentrations is small (i.e., $n - 1 \ll 1$); nevertheless it is possible to measure the dispersion quite accurately by interferometric means as a variation of optical path length $n \cdot d$. The use of the dispersion method for f-value determinations has certain advantages over absorption measurements. The restriction to optically thin layers is not necessary. Moreover, the method is applicable to excited media so that the f_{mn}-values from levels m other than the ground state can be determined. This approach was used by Ladenburg and co-workers[39, 40, 41] with the arrangement shown in Fig. 3.9. The light from a carbon arc source, which contains a spectal continuum, traverses a Jamin interferometer where it is split into two parallel beams. One of the beams passes through a length d of the excited medium (e.g., the positive colum of a neon discharge); the other through an equal length of the normal, unexcited medium and a compensation plate K. After the two beams are superimposed again they are analyzed in a spectrograph. The pattern of most of the resulting spectrum consists of interference

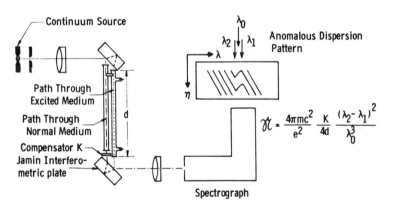

Fig. 3.9. Anomalous dispersion (Hook) method. Interference pattern of spectrum is shown in upper right.

fringes which have a slope in direct proportion to the dispersion of the compensator plate. In the neighborhood of the excited lines, however, the bright fringes flare out into hyperbolic "hooks" (see Fig. 3.9), in direct relation to the anomalous dispersion (Roschdestwensky[42]). In particular, the wavelength distance of the two dispersion peaks can be accurately measured out to yield the effective number,

$$\mathcal{N}_e = N_m f_{mn}\left(1 - \frac{g_m}{g_n}\frac{N_n}{N_m}\right)$$

of "classical dispersion electrons" which contains the negative contribution of emissions stimulated by the arc light with an effective oscillator strength f_{nm}.

Again, if the occupation numbers N are not known, and if the negative dispersion can be ignored (low excitation), it is still possible to evaluate relative oscillator strengths and transition probabilities.

(c) *Magnetic Rotation Method* Weingeroff[43] has described a procedure which combines the measurement of absorption with that of magnetic rotation. When plane polarized light from a continuum source traverses a medium which is subjected to a magnetic field parallel to the beam, optical activity is observed in the region of the absorption lines. Specifically, the plane of polarization is rotated by an angle which is proportional to the field and to the number $f \cdot N$ of classical oscillators. Thus, if the medium is placed between crossed polarizers, the absorption lines will appear bright against a dark background. For other angles of the polarizers, the lines will appear dark against a bright background. Moreover, at certain critical angles the lines will disappear in the continuum, and at this point the effects of absorption and magnetorotation on the spectral contrast have cancelled each other. Since essentially phenomena are observed in which, as the detailed analysis shows,[44] N and f do not enter solely as a product, the oscillator strength can be determined without exact knowledge of the density.

(d) *Emission Measurements* We now turn to methods which, in essence, probe spontaneous transitions from an upper level n with probabilities A_{nm}. The most direct approach, of course, is a measurement of line intensities. The upper level, by definition, is always an excited state of which the occupation density can be determined only with limited accuracy. Furthermore, as has been emphasized in Section 3.3.4.1(a), self-absorption has to be considered—always, if the lower level is the ground state, and in many cases, even if the lower level is also an excited state. Taking all this into account, the highest accuracy is possible only for the determination of relative probabilities A_{n1}, A_{n2}, etc., of spontaneous transitions from a common level n, particularly, if the Ladenburg factors S have been determined by some means. One has then for the ratio of the transition probabilities [cf. Eq. (3.108)]:

$$\frac{A_{n1}}{A_{n2}} = \frac{I_{n1}}{I_{n2}} \frac{\lambda_{n1}}{\lambda_{n2}} \frac{S_{n1}}{S_{n2}} \qquad (3.117)$$

so that the measurement is reduced to determining the ratio of intensities by a calibrated detector (see, e.g., Krebs[37] and Garbuny[45, 46]).

(d) *Lifetime Measurements.* The sum of the spontaneous transition probabilities of a level is, according to Eq. (3.98), the reciprocal of its lifetime, provided the latter is not foreshortened by collisions or other perturbations. In particular, if only one transition has to be considered, either because others do not exist or because they have negligible intensities, the lifetime yields an unambiguous and absolute A-value. The same end is served by suppressing a competing transition. This was accomplished, in effect, by resonance imprisonment in helium in the method of Heron, McWhirter, and Rhoderick[34]: the resultant rate of decay due to a transition at 537 Å to the ground state and another line at 5016 Å is given by $\gamma = 1/\tau = A_{5016} + gA_{537}$; hence the decay rate decreases linearly with decreasing escape factor g (increasing density, cf. Table 3.2), and extrapolation to $g = 0$ yields the absolute transition probability for 5016 Å. It is also possible to combine lifetime measurements with relative transition probabilities to obtain absolute values as we shall discuss later.

An astounding variety of experimental methods exists for measuring lifetimes either directly or by way of a time-dependent parameter. Below, a few of the more characteristic examples are cited.

The *direct* methods measure basically the time interval τ between the excitation of atoms to a given energy level and the radiation decay of the excited population to $1/e$. Excitation is possible by a variety of energy sources, but the most important are those of light and of electron collisions. The former involves *fluorescence*, viz., the excitation of a level by light and the subsequent spontaneous re-emission of quanta of the same transition or of photons resulting from a return to the ground state by a different route. Fluorescence has the practical advantage that levels can be excited selectively since intense monochromatic sources with matching photon energy are usually available. Strong selective excitation is more difficult to achieve by electron impact; on the other hand, electric pulses can be produced with shorter duration and better definition than those of light.

The simplest direct procedure [Fig. 3.10(a)] is to illuminate periodically a sample with a pulse of light and observe the decay of the fluorescence with a photomultiplier the output being displayed on a high-speed oscilloscope. This method can accomodate lifetimes of as low as 10^{-6} or at most 10^{-7} seconds, the limit being set by the definition of the optical pulse at the needed intensity. Since most atomic lifetimes range between 10^{-7} and 10^{-9} sec, this simple technique is not sufficient in general, although such means as Kerr cell shuttering or multiple reflections from rotating mirrors may increase the speed under favorable conditions.

The method of delayed coincidence, first developed for nuclear experimentation, was found suitable also in the optical region because it

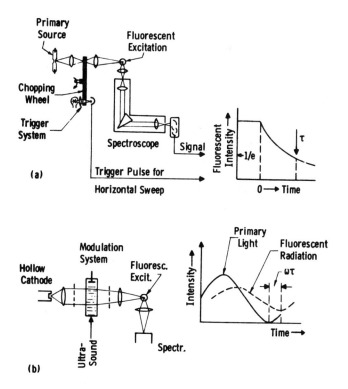

FIG. 3.10. Two direct methods for lifetime determination of excited states. (a) Observation of fluorescence decay with time [D. Alpert, A. O. McCoubry, and T. Holstein, *Phys. Rev.* **76**, 1257 (1949)]; (b) Determination of phase difference between modulated primary light and the fluorescent emission [W. Demtröder, *Z. Physik* **166**, 46 (1962); K. Ziock, *ibid.* **147**, 99 (1957)].

could measure time constants of the required magnitude. In this application,[34] a pulse is used for the electron impact excitation of atomic levels and brought to delayed coincidence with the photomultiplier output resulting from the desired transition. By varying the delay time, the decay curve can be traced. In a different application, coincidences could be determined for "cascade" decay through an intermediate level of which the lifetime was measured as the delay between the emission of the first and second photon.[47]

Another direct method utilizes excitation by sinusoidally modulated light[48, 49] [see Fig. 3.10(b)]. It can be shown that the fluorescent radiation follows the same periodicity, but that it is delayed in phase with respect to the primary light. The phase angle equals $\tan^{-1} \omega\tau$ (ω = the angular frequency of modulation) and remains unchanged, even if ω is converted to much lower values by a heterodyne technique. The modulation of the light may be produced by ultrasound[48] ($\omega \approx 10^7$–10^8 sec^{-1}) generated in a suitable medium (e.g., organic liquids). If the medium is placed into the light path between two gratings in registry with respect to each other, the time-variant diffraction by the ultrasonic standing waves causes a sinusoidal variation of the emergent beam intensity.

The *indirect* methods determine first, not the lifetime τ, but parameters which are known functions of τ. Thus when an excited atom or ion is observed in a predictable trajectory, the time of its light emission reveals itself indirectly by the position or velocity reached at such time.

The *canal ray* method is an important example (Wien[50, 53]). Canal rays consist of positive ions which have been accelerated in the cathode drop region of a glow discharge and which are allowed to pass through a channel in the cathode [see Fig. 3.11(a)]. Because of charge or energy exchange with other particles in the channel, however, excited neutral atoms may continue the journey with a velocity v and emerge into the space behind the cathode. If this latter region is higly evacuated, the radiation from the various levels n will decay at mean distances $v \cdot \tau_n$. In other words, the intensities of different spectral lines drop off to $1/e$ at distances behind the channel which correspond to the lifetime of their origin. Thus to determine τ, one has to measure the glow distance and the velocity v—the latter usually by means of the Doppler effect (see Section 3.4.2).

Somewhat similar is the method of Maxwell.[54, 55] A focused electron beam produces and excites ions of mass M in a narrow and well defined region between two electrode plates [Fig. 3.11(b)]. At sufficiently low pressures the ions will follow the field E and reach distances $eE\tau_n^2/2M$ before their radiation has fallen off to $1/e$ for various spectral lines n, provided τ_n is small compared to the collision time. In this manner,

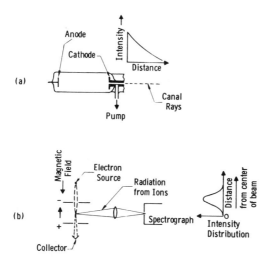

FIG. 3.11. Two indirect methods for lifetime measurements of excited states. (a) Canal ray method utilizes the fact that particles passing through a perforated cathode remain in an excited state over a distance given by their velocity and lifetime. (b) The determination of luminescence range is also the basis of Maxwell's method in which ions are produced and excited by electron bombardment in a small, well-defined region from which they are accelerated by an electrostatic field. Decay of each spectral line can be observed or photographed with spectrograph in the direction of the slit.

Maxwell showed the increase of τ for the helium lines originating at levels with increasing n.

As an alternative to measuring the distance the ion has reached in τ, one may simply determine[56] the final velocity $\bar{v} = eE\tau_n/M$ by means of the Doppler effect. In contrast to the older methods mentioned in the preceding, the starting point of the ion trajectory—a source of uncertainty—does not have to be known; conditions can be so arranged that at time $t = 0$, $\bar{v} = 0$.

3.3.4.3 *Methods of Evaluation*

The experimental results of line intensity and lifetime measurements do not usually lend themselves to a direct evaluation of all the intensity parameters because of the complexity which in general the energy levels and their transitions exhibit. We shall see in the next chapter that there exists for a given quantum number n, not always just a single energy value E_n, but a finer structure of *energy terms* which result in the simple case of Russell-Saunders coupling from an electronic orbital momentum L and a total spin momentum S. Thus each set of values $(n; L; S)$ gives rise to a different energy term; but each term can still have finer *energy levels* owing to various orientations of the orbital momentum with respect to the spin. These fine structure levels are characterized by a quantum number J. Finally, for a given level J, there are $g = 2J + 1$ degenerate *states* which are different, not in energy, but in orientation of the whole atom and which must be counted in statistical distributions with the weight g. Experimentally all this poses a problem in that the different levels of a term—and even the different terms belonging to the same n—often cannot be separately observed. This situation can arise when wavelength resolution has to be sacrified for intensity so that lines belonging to different terms or different levels are measured together.

There are, however, simplifications. The various states of a level and the various levels of a Russell-Saunders term (see Chapter 4) have the same lifetime as shown by Condon and Shortley.[25] From this follow certain conclusions with the aid of Eqs. (3.87), (3.98), and (3.104). If the states are in thermal equilibrium, the intensities of lines having a common upper level add up to a sum which is proportional to the statistical weight $2J + 1$ of the level; similarly, the sum of line intensities for the same lower level is proportional to the statistical weight of the latter. In this so-called *Burger-Dorgelo*[57] *sum rule* the small differences between energies and frequencies can be ignored since we compare transition from, and to, closely spaced levels. The ratio of the intensity sum for lines from level J_1 to that of level J_2 is thus

$$\frac{\sum I_1}{\sum I_2} = \frac{2J_1 + 1}{2J_2 + 1} \qquad (3.118)$$

The *average* spontaneous probability A_{nm} of transition from various levels J of a term n to various levels J' of a term m, therefore, is given by

$$\bar{A}_{nm} = \frac{\sum\limits_{JJ'} g_j A_{nJ \to mJ'}}{\sum\limits_{J} g_J} = \frac{\sum\limits_{JJ'} (2J + 1) A_{nJ \to mJ'}}{\sum\limits_{J} (2J + 1)} \qquad (3.119)$$

where, consistent with the statement about the equality of lifetimes, we have

$$\sum\limits_{J'} A_{nJ_1 \to mJ'} = \sum\limits_{J'} A_{nJ_2 \to mJ'} \qquad (3.120)$$

The average oscillator strength of transitions from various levels J' of a term m to various levels J of a term n, is correspondingly

$$f_{mn} = \frac{\sum\limits_{J'J} g_{J'} f_{mJ' \to nJ}}{\sum\limits_{J'} g_{J'}} = \frac{\sum\limits_{J'J} (2J' + 1) f_{mJ' \to nJ}}{\sum\limits_{J'} g_{J'}} \qquad (3.121)$$

It will be recognized that the preceding relationships and, in particular, Eqs. (3.118)–(3.121), allow in simple cases the determination of individual line parameters from a minimum of experimental information, e.g., from a measurement of the lifetime $\tau = 1/\bar{A}_{nm}$. This possibility is of importance, of course, for the evaluation of spectra that cannot be resolved; but it is of value also in other cases because it simplifies the task of measurement.

Thus the last four equations, and others similar to them, relieve the complexity resulting from the fine structure of transitions; but their validity is restricted to the interrelationship within such fine structure. In the more general case, particularly, if we deal with transitions from a level n to various levels of quantum number m, a combination of two or more measurements may be necessary.

Absolute A-values, for instance, can be obtained from a combination of relative transition probabilities and the lifetime of the upper state. This can be seen if the equation

$$\frac{1}{\tau} = \sum_m A_m$$

is divided by A_s which we have selected among the various A_m. We obtain then an expression for the absolute value of A_s,

$$A_s = \frac{1}{\sum\limits_m A_m/A_s} \cdot \frac{1}{\tau} \tag{3.122}$$

in which the relative values A_m/A_s and τ represent the input data. A corresponding alternative exists for the determination of absolute f-values from relative measurements of the oscillator strengths combined with the f-sum rule (cf. Section 3.3.3.1):

$$f_s = \frac{z}{\sum\limits_n f_n/f_s} \tag{3.123}$$

where z is the number of electrons in the level n. A necessary condition for the validity of these two methods is that *all* transitions are included. The link between A-values and oscillator strengths is given by Eq. (3.99) which, after the introduction of numerical values for the constants and conversion to wavelengths, has the form

$$f_{mn} = 1.499\lambda^2 \frac{g_n}{g_m} A_{nm} \tag{3.124}$$

λ being measured in centimeters.

The other alternatives for the experimental determination of absolute transition probabilities or oscillator strengths require the knowledge of the atom distribution over the excited levels, except, of course, for transitions from the ground state. As has been pointed out before, this amounts to a determination of the excitation temperature (see e.g., Corliss[58]). Such a measurement may not be valid for all the terms at once; moreover, it is less direct and, therefore, usually less accurate than, e.g., lifetime measurements.

3.3.5 Test of the Physical Model

A comparison between experimental and theoretical line intensity parameters sheds light on the reliability of either approach; but it is of particular significance as a measure of the reality which attaches to the physical picture underlying the calculations. What is, in fact, tested here, is the combination of classical and quantum-theoretical concepts. The quantitative results for the emission and absorption power of the classical oscillator, as given by electromagnetic theory, appear again in quantum theory, although with modified interpretation. Thus the absorption intensity of an individual electron transition is measured as fractional strength of a classical oscillator and the sum of all these fractions must be unity for the electron. The f-sum rule has indeed been confirmed experimentally, although of necessity only in those few cases in which measurements were possible and unambiguous (see, e.g., Table 3.4).

A comparison between theory and experiment is made in Table 3.3 for the lifetimes of some excited states in hydrogen, neutral and ionized helium, and lithium. The data computed or measured by the various workers (see footnotes to the table) refer to excited levels or terms as a whole, although they have unresolved fine structure, and this along with the designations is further explained in Chapter 4. The theory can treat completely and strictly only the case of hydrogen and hydrogenlike spectra. In the case of hydrogen, we can cite only early and relatively uncertain experimental results based mainly on the canal ray method. Agreement is here by order of magnitude only, although the predicted trend of lifetimes towards larger values for higher quantum numbers is confirmed.

The calculations on hydrogen are immediately applicable to He^+, Li^{++}, and generally hydrogen-like ions with charge Z, since the transition probabilities are proportional to Z^4 so that the lifetimes change as $1/Z^4$. This can be seen from the following proportionality considerations: $A_{nm} \propto \nu^3 \mid x_{nm} \mid^2$, Eq. (3.84); $\nu \propto Z^2$, Eq. (3.40); $\mid x_{nm} \mid \propto r \propto Z/\epsilon \propto 1/Z$,

TABLE 3.3

EXPERIMENTAL AND THEORETICAL LIFETIMES OF SIMPLE ATOMIC LEVELS

Particle	Energy level	Lifetime measurement		Lifetime calculation (sec)
		Result (sec)	Method	
H		$\times\ 10^{-9}$		$\times\ 10^{-9}$
	$n = 2\ (s, p)$	$\{$ 6.7[a]	Canal	2.1[b]
		$\{$ 12[c]	Direct pulse	
	$n = 3\ (s, p, d)$	$\{$ 6.7[a]	Canal	10.2[b]
		$\{$ 18.5[d]	Canal	
	$n = 4\ (s, p, d, f)$	18.5[d]	Canal	33.5[b]
	$n = 5\ (s - g)$			88[b]
	$n = 6\ (s - h)$			196[b]
He$^+$	$n = 4\ (s, p, d, f)$	0.9[e]	Indir., vel.	2.1 (from hydrogen)
	$n = 6\ (s - h)$	11 ± 2^f	Indir., pos.	12 (from hydrogen)
He	3^1P		Direct,	$\{$ 74.7[g]
	$(3 \rightarrow 2$ only; $3 \rightarrow 1$	74 ± 1^h	Del. coinc.	$\{$ 72.0[i]
	suppressed)			$\{$ 75.5[j]
	4^3S	67.5 ± 1^h	Del. coinc.	$\{$ 64[g]
				$\{$ 102[j]
	3^3P	115 ± 5^h	Del. coinc.	$\{$ 97[g]
				$\{$ 83[i]
				$\{$ 118[j]
	4^3P	153 ± 2^h	Del. coinc.	$\{$ 138[g]
				$\{$ 127[j]
	3^3D	10 ± 5^h	Del. coinc.	$\{$ 13.9[g]
				$\{$ 15.2[i]
				$\{$ 12.8[j]
Li	2^2P	28 ± 1^k	Absorption	27.2[g,l]

[a] W. Wien, *Ann. Physik* [4] **83**, 1 (1927).

[b] H. A. Bethe, *in* "Handbuch der Physik" (H. Geiger and K. Scheel, eds.), 2nd ed., Vol. 24, Part I, Chapter 3. Springer, Berlin, 1933.

[c] F. G. Slack, *Phys. Rev.* **28**, 1 (1926).

[d] H. Kerschbaum, *Ann. Physik* [4] **83**, 294 (1927).

[e] M. Kagan and Y. P. Koritskii, *Opt. i Spektroskopiya* 11, 308 (1961); see *Opt. Spectry.* (*USSR*) (*English Transl.*) 11, 166 (1961).

[f] L. Maxwell, *Phys. Rev.* **38**, 1664 (1931).

[g] D. R. Bates and A. Damgaard, *Phil. Trans. Roy. Soc.* **A242**, 101 (1949).

[h] S. Heron, R. W. P. McWhirter, and E. H. Rhoderick, *Proc. Roy. Soc.* **A234**, 565 (1956).

[i] L. Goldberg, *Astrophys. J.* **90**, 414 (1939).

[j] E. A. Hylleraas, *Z. Physik* **106**, 395 (1937).

[k] G. Stephenson, *Nature* **167**, 156 (1951).

[l] O. S. Heavens, *J. Opt. Soc. Am.* **51**, 1058 (1961).

Eqs. (3.37) and (3.39); hence, $A_{nm} \propto Z^4$ and $\tau \propto 1/Z^4$, *q.e.d.* Furthermore, $f_{mn} \propto A_{nm}/\nu^2$, according to Eq. (3.99), and thus is independent of Z—a plausible result in view of the f-sum rule for a single electron, $z = 1$. The theoretical values for He^+ are simply $\frac{1}{16}$ of the lifetimes calculated for hydrogen. The two available measurements agree with the theory to the extent of their accuracy limits.

As soon as the theory proceeds beyond the configurations containing a single electron, a greater complexity results. The presence of the additional electrons then must be accounted for by some assumed shape of the potential field $V(r)$ which affects the Schrödinger equation as shown in Eq. (3.53) and Fig. 3.7. The computation of the matrix elements follows thus from approximation methods. For the simpler atoms the so-called *Coulomb-approximation* method of Bates and Damgaard[59] has proven particularly successful. This theory assumes the validity of the Coulomb potential in the important range of r, the main difference to hydrogen being modified—so-called *effective*—quantum numbers n which adjust the Born-Rydberg equation (3.40) for hydrogen to the observed spectra of the more complicated elements. The Coulomb approximation, as well as theoretical approaches preceding it, yield values for helium and lithium which are generally in close agreement with rather accurate experimental results as shown in Table 3.3.

A comparison between calculated and measured oscillator strengths is made in Table 3.4 for a few heavier elements. The f-values refer to transitions between individual levels, or as indicated in the second column, to transitions between unresolved terms and are either quoted directly from the references or converted from lifetimes and transition probabilities. For the alkali metals the theory of Bates and Damgaard accounts very well for the measured values. It is also seen that the sum of oscillator strengths converges very fast towards unity. Nevertheless, the negative values of $f_{n \to n-1}$, etc., must also be included in the sum, except for lithium (cf. Section 3.3.3.1); although they are small, they will in part compensate the contribution of the transition to the continuum.

The theory is, of course, less reliable for the more complex atoms, and this is shown for the example of gallium in Table 3.4.

A rather complete bibliography of measurements and calculations on lifetimes, transition probabilities, and oscillator strengths has been published by Glennon and Wiese.[60] Approximate transition probabilities for the lines of 70 elements were determined by Corliss and Bozman[61] who evaluated the relative intensities of 39,000 spectral lines. The intensity measurements were available as the result of a previous program in which the spectra were taken of 70 elements added as impurities to a copper arc of known temperature. However, the automatic processing which had to be applied in an undertaking of this magnitude did not have the flexibility needed for the measurement of self-absorption and other corrections.

TABLE 3.4

EXPERIMENTAL AND THEORETICAL f-VALUES FOR SELECTED TRANSITIONS

Atom	Transition	λ (Å)	Measurement		Calculation
			Method	f-value	f-value
Li	$2^2S_{1/2} - 2^2P_{1/2,3/2}$	6708	Absorption	0.72 ± 0.03^a	0.74 (total)^{c,d}
			Emission	0.71^b	
Na	$3^2S_{1/2} - 3^2P_{1/2}$	5896	τ-Phase shift	0.328 ± 0.003^e	$0.313^{a,e}$
			Magneto-rot.	0.323 ± 0.012^a	
	$3^2S_{1/2} - 3^2P_{3/2}$	5890	τ-Phase shift	0.654 ± 0.007^e	$0.626^{a,c}$
			Magneto-rot.	0.648 ± 0.024^a	
	$3^2S_{1/2} - 4P_{1/2,3/2}$	3303		$0.014(?)^f$	$0.014^{a,d}$
	$3^2S_{1/2} - 5P_{1/2,3/2}$	2853			$0.0024^{a,d}$
	$3^2S_{1/2} - 6P_{1/2,3/2}$	2680			0.0007^f
				$\sum\limits_{n=3}^{4} f_{3n} = 0.991$	$\sum\limits_{n=3}^{6} f_{3n} = 0.956$
K	$4^2S_{1/2} - 4^2P_{1/2}$	7699	Magneto-rot.	0.328 ± 0.011^a	$0.331^{a,c}$
	$4^2S_{1/2} - 4^2P_{3/2}$	7665	Magneto-rot.	0.652 ± 0.022^a	$0.654^{a,c}$
Rb	$5^2S_{1/2} - 5^2P_{1/2}$	7947	Magneto-rot.	0.333 ± 0.010^a	$0.350^{a,c}$
	$5^2S_{1/2} - 5^2P_{3/2}$	7800	Magneto-rot.	0.635 ± 0.020^a	$0.652^{a,c}$
Ga	$4^2P_{1/2,3/2} - 5S_{1/2}$	4033	τ-Phase shift	0.086 ± 0.001^e	$0.058^{a,e}$
		4172			
	$4^2P_{1/2,3/2} - 6S_{1/2}$	2660	Absorption	0.016^e	$0.007^{a,e}$
		2720			
	$4^2P_{1/2,3/2} - 4^2D_{1/2,3/2}$	2874	Absorption	0.36^e	$0.37^{a,e}$
		2944			
	$4^2P_{1/2,3/2} - 5^2D_{1/2,3/2}$	2450	Absorption	0.10^e	$0.09^{a,e}$
		2500			

[a] G. Stephenson, *Nature* **167**, 156 (1951).
[b] E. Hinnov and H. Kohn, *J. Opt. Soc. Am.* **47**, 156 (1957).
[c] D. R. Bates and A. Damgaard, *Phil. Trans. Roy. Soc.* **A242**, 101 (1949).
[d] O. S. Heavens, *J. Opt. Soc. Am.* **51**, 1058 (1961).
[e] W. Demtröder, *Z. Physik* **166**, 46 (1962).
[f] L. Bierman, *in* "Zahlenwerte und Funktionen aus Physik, Chemie, etc." (Landolt-Börnstein), 6th ed., Vol. I, Part 1, p. 260. Springer, Berlin, 1950.

3.4 Shape and Broadening of Spectral Lines

Most aspects of monochromatic radiation involve just those processes which bring about a broadening of the ideal, infinitely narrow frequency response of the emitting or absorbing oscillator. In fact, we shall see that spectral lines, defined as transitions between quantized energy levels, may have values[*] of $Q' = \nu/\Delta\nu$ which vary from 10^2 to at least 10^{12}. On a logarithmic scale of frequency, therefore, lines show larger differences among each other than they do with respect to the continua. The fact that such a great variety of spectral widths exists at all, must be attributed to a corresponding variety of physical causes. It is, of course, not always true that any particular broadening effect is limited to a typical width. Nevertheless, it is possible to classify line shapes according to their origin and assign to them such magnitudes as are characteristic for the practical conditions under which they usually occur (see Fig. 3.12).

Spectral lines and line shapes can be observed in emission and absorption; in solids, liquids, and gases; and in the presence or absence of a continuum. Most of the work on line broadening deals with gases;

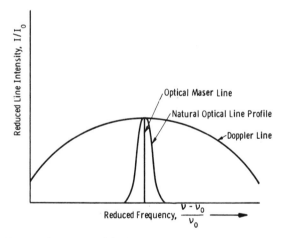

FIG. 3.12. Characteristic line widths for various broadening conditions. The sharpest lines in the optical region are obtained with laser beams. Before the arrival of maser technology the natural line width represented an unattainable lower limit of broadening. In normal, incoherent light sources Doppler broadening is always present; only the top of the spectral profile can be shown in comparison with the others.

---

[*] For convenience, the value $\nu/\Delta\nu$ is indicated here by the symbol Q'. Its meaning is identical with that of Q only if the broadening is determined by the decay time of the oscillation.

because of tradition and convenience, the isolated oscillator has usually been studied as the emitting or absorbing atom of a gas or plasma. But crystal field spectroscopy, too, has become a subject of increasing importance. Many factors causing line broadening in solids are, at least qualitatively, the same as in gases. Other phenomena, however, occur in solids only and thus have to be treated in their own right.

Similarly, it is only for specific effects necessary to discuss emission and absorption separately. It is a consequence of Kirchhoff's law that the spectral profile of an individual oscillator must be the same in emission as in absorption. If this profile is observed through thick layers of the medium, it may become distorted. For instance, the light emitted from the central regions of a glow discharge tube will be reabsorbed at a nonlinear rate as it propagates to the walls. The result of such *self-absorption* is that emission and absorption at the spectral peak of the line approaches an equilibrium while the intensity at the wings continues to grow so that the line as a whole tends towards a square shape. Such a mechanism has been discussed in Section 2.4.3.2 and others will be dealt with in Chapter 5. Here we shall assume that the optical thickness of the medium is small enough so as to have no effect on the line shape.

3.4.1 Natural Line Width

We have seen (Sections 1.2.2 and 1.3.5) that a vanishing spectral width corresponds to a vibration of unlimited duration or, in terms of propagation, to a wavetrain of infinite length. Conversely, a finite lifetime of the vibration gives rise to a broadened response (cf. Section 1.2.2) of the form

$$I(\omega) = I_0 \frac{\gamma}{2\pi} \cdot \frac{1}{(\omega - \omega_0)^2 + \gamma^2/4} \qquad (3.125)$$

The factor $\gamma/2\pi$ on the right-hand side of Eq. (3.125) results from the requirement that the intensity distributed over the spectrum must total I_0, i.e. the intensity obtained without considerations of line broadening as in Eq. (3.87). Equation (3.125) defines a "natural" halfwidth $\Delta\omega_h$ related to the time constant τ and the damping term γ by

$$\Delta\omega_h = \gamma = 1/\tau \qquad (3.126)$$

In the quantum theory this last relationship results from Heisenberg's uncertainty principle

$$\Delta E \, \Delta t = \hbar \, \Delta\omega\tau = \hbar \qquad (3.127)$$

as applied to the transition from an excited level E_2 to the ground state E_1. In this case an uncertainty of energy ΔE, hence of the angular frequency, $\Delta\omega$, is associated with a finite lifetime τ of the upper level. This lifetime is the reciprocal of the transition probability $A_{21} = \gamma$, provided the ground state is the only level to which a jump is possible [see Fig. 3.13(a)]. This is the simplest possible case; the one usually

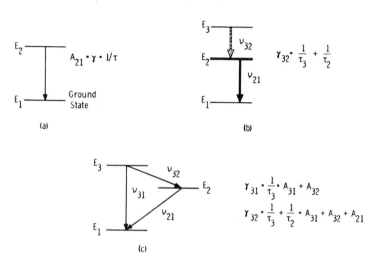

(a)   (b)

(c)

Fig. 3.13. Natural line widths γ and their dependence on transition probabilities and term scheme combinations. (a) Case in which transition to only one lower level is possible; natural width is then exactly equal to spontaneous transition probability. (b) Transition to intermediate level has line width which depends on lifetime of upper and lower state and may be broad but weak (contrary to classical oscillator behavior). (c) Line widths for transitions among more complex schemes.

considered. However, other alternatives exist. In the case, shown in Fig. 3.13(b), that the transition from a higher level with lifetime τ_3 occurs to a lower level which is not a ground state but has a lifetime τ_2, the resulting uncertainties add to $\Delta E = \Delta E_3 + \Delta E_2$. Thus one has for the resulting halfwidth and transition probabilities the expression (if E_1 is the ground state):

$$\Delta\omega_{32} = \gamma_{32} = \frac{1}{\tau_3} + \frac{1}{\tau_2} = A_{32} + A_{31} + A_{21} \qquad (3.128)$$

The natural width of the line ν_{32}, therefore, is affected by the lifetime of the upper as well as the lower state. Thus although the transition probability A_{32} may be small and the line weak, the latter may yet be broad (Heitler,[62] Weisskopf and Wigner[63]). Even more complicated relationships between the halfwidths and the transition probabilities exist when several branching decay routes exist as shown in Fig. 3.13(c). In practical cases, of course, it is usually only one lifetime τ, the shortest, which is dominant in equations such as (3.128). In any case, it is sufficient for most purposes to consider only a single effective γ or τ, although these values may be the resultant of several contributions. It is possible by artificial means to prolong in effect the lifetime τ, namely, by means of stimulated emission of coherent radiation, as explained in Chapter 6. This leads to an observable narrowing of the spectral frequency interval below the natural limit as indicated schematically in Fig. 3.12. On the other hand, for ordinary sources of atomic spectra such as flames and gas discharges, the line breadth will exceed the natural width by orders of magnitude. This additional broadening results ultimately from the fact that optical radiation is emitted, not by a single oscillator, but at random by a large number of individual sources at various velocities with respect to the observer and under conditions of mutual perturbation. The practical importance of line broadening lies, first, in the limitations it places on the usefulness of monochromatic light sources. Furthermore, just as the position and number of the spectral lines reveal the chemical nature and atomic structure of the source (see Chapter 4), the width is indicative, albeit to a much more limited extent, of the temperature and pressure of the source.

The main difficulty facing the diagnostic evaluation of line shapes lies in the simultaneous presence of several broadening effects. Of course, in many situations observed in the laboratory only one of these may be dominant, and the determination of physical parameters such as the temperature is then simple. In other cases, several causes contribute to the line shape in complex fashion, although conclusions can be drawn from various parts of the profile, the wings as contrasted with the peak, for instance. Some broadening effects which may appear at first to be different phenomena are, in fact, basically the same, although it is difficult to treat them from a unified point of view.

3.4.2 DOPPLER BROADENING

One ever-present process affecting line shapes of ordinary light sources is based on the Doppler effect, i.e. the frequency shift in the radiation received from a moving source. When an oscillator at frequency ν_0 speeds with a velocity v_x towards an observer, the emitted wavetrains arrive at a higher frequency ν. This can be seen simply on the basis of the classical argument that wavecrests, generated with period T_0 at a moving source, follow each other in arrival intervals shorter by $T_0 v_x/c$. Hence the fractional change in frequency is

$$\frac{\nu - \nu_0}{\nu_0} = \frac{v_x}{c} \tag{3.129}$$

where c is the velocity of wave propagation. If v_x is negative, a lower frequency or red-shift is observed (as for instance in the spectra of distant galaxies which thus reveal themselves as part of an expanding universe).

The process of line broadening results from the statistical distribution of the radiating atoms over a continuous range of velocities. Such a distribution establishes itself in thermal equilibrium and for gases is given by the Maxwell-Boltzmann relation. Only the velocity component in the viewing direction contributes to the Doppler effect. The Maxwell-Boltzmann distribution then has the form

$$dn = n\sqrt{\frac{M}{2\pi kT}} \exp\left(- \frac{Mv_x^2}{2kT}\right) dv_x \tag{3.130}$$

where dn represents the number of atoms of mass M which, among a total of n, have a velocity component between v_x and $v_x + dv_x$. To each velocity corresponds a certain frequency shift $\nu - \nu_0$, according to Eq. (3.129), and the introduction of this relation into (3.130) yields the Doppler line shape

$$I_\nu d\nu = I_0 \exp\left[- \frac{Mc^2}{2kT}\left(\frac{\nu - \nu_0}{\nu_0}\right)^2\right] d\nu \tag{3.131}$$

The Doppler profile of line intensity follows, therefore, a Gaussian distribution of frequencies of the form $\exp[-\alpha(\nu - \nu_0)^2]$. The frequencies for which $I = I_0/2$ in Eq. (3.131) determine the *Doppler width*

$$\Delta\nu_h = 2\nu_0\sqrt{(2\ln 2)\frac{kT}{Mc^2}} \tag{3.132}$$

If we introduce numerical values for the atomic constants in Eq. (3.132) and replace the mass M of the atom by the atomic (or molecular) weight M', we obtain

$$\nu_h = 7.17 \times 10^{-7}\nu_0\sqrt{\frac{T}{M'}} \tag{3.133}$$

or, in terms of wavelengths,

$$\frac{\Delta\lambda_h}{\lambda} = 7.17 \times 10^{-7}\sqrt{\frac{T}{M'}} \tag{3.134}$$

For example, the lines near 6000 Å of neon ($M' = 20$) at $T = 300°\text{K}$ have a Doppler width of about 1.66×10^{-2} Å. This value exceeds natural broadening by about two orders of magnitude. More generally, the larger the mass and the lower the temperature, the narrower is the Doppler width and vice versa.

3.4.2.1 *Recoil Shift and Mössbauer Effect*

A nonthermal cause of Doppler shift is due to recoil of the atom as it emits or absorbs a photon. The momentum of a photon equals $h\nu/c$ (see Section 1.3.2), and thus the effect is quite pronounced for high frequencies, in the region of gamma rays. A momentum $Mv = h\nu/c$ is transferred to the radiation source, such as an isolated nucleus of mass M. Because the source recedes with a velocity v the photon emerges from it with a relative red shift in the order of

$$\frac{\Delta\nu}{\nu} \approx \frac{v}{c} = \frac{h\nu}{Mc^2} \tag{3.135}$$

A more accurate description is the statement that a recoil energy

$$E_R = \frac{Mv^2}{2} = \frac{(h\nu)^2}{2Mc^2} \tag{3.136}$$

is transferred to the atom from the photon which, because it loses this amount, experiences a corresponding red shift. In the optical spectrum the shift is small compared with the natural line width. Even in the ultraviolet, for $h\nu = 10$ ev, the relative shift $\Delta\nu/\nu$ for an atomic mass $M' = 100$ ($Mc^2 = 10^{11}$ ev) is only 10^{-10} according to Eq. (3.135). However, if a nucleus of such mass emits a gamma quantum of $h\nu = 100$ kev, the relative shift amounts to 10^{-6}, corresponding to an absolute value of 0.1 ev, i.e., in the order of thermal Doppler shifts.

This "detuning" effect is of importance for the resonance absorption of radiation between like oscillators. Since a recoil energy E_R has to be communicated to the source, as well as to the absorber, the quantum falls short of achieving resonance by an energy amount of $2E_R$ (Fig. 3.14). In the optical region, there exists no difficulty in detecting resonance between atoms of the same kind by fluorescence or stimulated emission since, as we have shown, the recoil shift is negligible. Gamma rays,

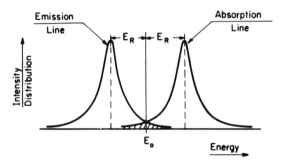

FIG. 3.14.   Recoil shift of high energy line for emission and absorption. Resonance between emitter and absorber, even though both oscillate at the identical frequency, is limited to the shaded area, hence a statistically rare event for gamma rays. In the optical region, emission and absorption lines virtually overlap (except for a range less than 10^{-10} of their width).

however, correspond to high energy transitions between excited nuclear states which may have considerable lifetimes, hence, sharply defined energy levels. As an example, the nucleus of Fe^{57} has an excited level at 14.4 kev above the ground state with a lifetime of 1×10^{-7} sec. Therefore, it has, according to Eq. (3.127), a natural width of only 4.6×10^{-9} ev so that $\Delta\nu/\nu = 3.2 \times 10^{-13}$. This high frequency fidelity, corresponding to a Q'-value of 3×10^{12}, should lend itself, in principle, to the design of experiments involving the measurement of time with an unheard-of accuracy. Nevertheless, the recoil energy at the emitting as well as the absorbing nucleus amounts to 4.0×10^{-3} ev so that the gap between the incident and the resonant frequency is a million times larger than the natural line width itself. It was possible to compensate for the frequency gap by superimposing, in effect, an opposing Doppler shift, e.g., by orbiting the emitter towards the absorber on a centrifuge (with a tangential speed of $v = 2\,\Delta\nu c/\nu = 1.6 \times 10^4$ cm/sec) or by heating the samples so that their thermal line widths overlapped. However, the high precision inherent to the high Q'-values of the nuclear oscillators is essentially lost by such compensation.

In 1958, Mössbauer[64, 65, 66] reported a nuclear resonance absorption experiment in which the recoil energy loss was avoided rather than compensated. The method employed was the inverse of a previous procedure: Ir^{191} crystallites as absorbers and emitters, were cooled to $88°K$ (rather than heated). The atoms in solids, in contrast to those in gases, are bound to certain positions of equilibrium about which they vibrate at rms-velocities of the same order as those in gases (at least, if compared at sufficiently high temperatures). Nevertheless, the motions of all the individual atoms are interdependent or coupled; indeed, they arrange themselves cooperatively in certain distinct modes which correspond to the elastic waves of the Debye model (see Section 1.2.3). The lattice energy in these modes is quantized (like photons in cavities), consisting of phonons $h\nu$. The mechanism of transferring an impulse to the lattice is based on the excitation of the proper phonons, i.e. on transitions between allowed energy states (Lamb[67]). There exists an upper limit ν_m for the spectrum of lattice vibrations which is given by Eq. (1.21) and which defines an upper phonon energy $h\nu_m = k\theta$ (where θ represents the "Debye temperature" for the material). If the recoil energy supplied by the gamma quantum is large compared to $h\nu_m$, an excitation of proper lattice vibrations is always possible; the momentum then is taken up by the individual nucleus of mass M. However, if the recoil energy is smaller than $h\nu_m$, a probability exists—which increases with decreasing operating temperatures—that the momentum cannot be transferred to the quantized vibrational states.

We wish to explain this fact in simple terms. Although momentum must be conserved, the recoil energy can assume a variety of values since various configurations of atoms with various effective mass can participate in elastic interaction. There are thus a large number of transitions possible (see, e.g., Lipkin[68] and Lustig[69]) between initial states $E(n_i)$ and final states $E(n_f)$ of vibrational energy. Each such transition has a probability $P(n_i, n_f)$ which shall be normalized to unity. We can now make the plausible statement that the sum of the weighted energy level transitions must be equal to the recoil energy for a single atom:

$$\sum_{i,f} [E(n_f) - E(n_i)]P(n_f, n_i) = \frac{(h\nu)^2}{2Mc^2} = E_r \qquad (3.137)$$

If E_r is smaller than the Debye limit $k\theta$, then there are many transitions possible for which

$$| E(n_f) - E(n_i) | > E_r \qquad (3.138)$$

The occurence of these high energy exchanges is, however, limited by Eq. (3.137). Thus there will be a sizable probability for the case

$$E(n_i) - E(n_f) = 0 \qquad (3.139)$$

in which no phonon energy is exchanged at all.

In this case, then, the impulse is directly communicated to the bulk of the crystallite, and M in Eqs. (3.135) and (3.136) is now the mass of the crystal, virtually infinite in comparison to the nuclear mass. This means that the recoil energy lost to the photon is practically zero and that the red shift of the nuclear line is negligible compared with its width. The probability of such "recoilless" gamma emissions can be quite substantial; it is for high θ and low temperature often larger than the probability of emissions which excite lattice vibrations. One observes under these conditions strong lines in nuclear resonance absorption, if emitter and absorber are at rest with respect to each other. The extraordinary narrow line shape is, in fact, explored by slowly moving the source at varying speeds with respect to the absorber so that $\Delta v/v = v/c$. In case of Fe^{57}, for instance, a velocity interval of 0.05 cm/sec is found to cover approximately the line width. In all, it is often possible to approach the theoretical limit by careful experimentation. Therein lies the great value of the Mössbauer effect as applied to measurements such as of hyperfine structure of energy levels, internal fields in solids, and the gravitational red shift.

3.4.3 Pressure Broadening

Interactions between randomly emitting atoms are another major cause of line broadening. These perturbations are collectively named pressure broadening; for although one may differentiate in detail between various mechanisms, these effects have all in common that they increase with pressure, that they are, often to an indeterminable degree, simultaneously active, and that they ultimately are based on the same forces such as Coulomb fields. However, two limiting cases are readily distinguished. The first of these is collision broadening, which is the premature foreshortening of the oscillator lifetime by a collision between the light emitting center and another atom, the "perturber." The second case is broadening by coupling of the oscillators, the latter being considered at rest.

3.4.3.1 *Collision Broadening*

If the rate at which excited states decay under light emission is increased beyond the spontaneous transition probability $\gamma_{tr} = A_{21} = 1/\tau_{tr}$ by a collision rate γ_{coll}, the resulting damping parameters and associated lifetimes are given by Eq. (3.140) and (3.141), respectively*:

$$\Gamma = \gamma_{tr} + \gamma_{coll} \tag{3.140}$$

$$\frac{1}{\tau} = \frac{1}{\tau_{tr}} + \frac{1}{\tau_{coll}} \tag{3.141}$$

The spectral line shape in collision broadening is thus simply given by the relation for radiation damping, except for the increased factor Γ, viz.

$$I(\omega) = I_0 \frac{\Gamma}{2\pi} \cdot \frac{1}{(\omega - \omega_0)^2 + \Gamma^2/4} \qquad (3.142)$$

In Fig. (3.15) the intensity profile due to Doppler (Gaussian curve) broadening is compared with that resulting from collisions (Lorentz-shape). The width due to collision broadening can be easily computed, if collision cross sections and velocities of the interacting centers are known.* The collision frequency of nitrogen at $300°K$ and 10 mm Hg pressure is, for instance, about 10^8 sec^{-1} so that in this case collisions and radiation damping make nearly equal contributions to the line widths. However, in hot, dense plasmas the pressure effects are sizable.

It is noteworthy that Eq. (3.142) is valid regardless of the actual mechanism of the collisions. Several types of interaction are, in fact, possible. The original theory, proposed by Lorentz[71] assumed that the radiating oscillators experienced inelastic collisions with other atoms such that the emission process was interrupted and the remaining energy was taken up in kinetic form. This kind of process will reduce the total intensity of the line besides broadening its shape.

A somewhat different encounter between the radiating center and a perturber proceeds with little or no net exchange of kinetic energy.

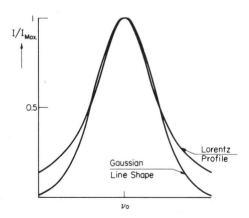

FIG. 3.15. Comparison of the two most important line shapes. Profiles are normalized to the same maximum and halfwidth.

---

* $\tau_{coll} = 1/\gamma_{coll}$ equals here half the mean free time $\bar{l}/\bar{v}$, where $\bar{v} = (8kT/\pi M)^{1/2}$ is the mean velocity and $\bar{l} = kT/(4\sqrt{2}\pi\rho^2 p)$ the mean free path for a cross section $\pi\rho^2$ and pressure p. Hence, $\gamma_{coll} = 32\sqrt{\pi}\rho^2 p/(MkT)^{1/2}$ (cf. Born,[70] p. 438).

In this case the frequency of the oscillator is temporarily shifted or "detuned" as the two particles pass through the point of closest approach. As a result, the wavetrain, although it resumes its original frequency, proceeds with dislocated phase, and this amounts to an interruption of coherence.

Other types of collision can be postulated. It is conceivable that the light from a center is eclipsed or scattered by a perturber during the emission process so that effectively again an interruption of the wavetrain results. Experimentally it is found that if conditions favor pressure broadening over Doppler effects, such as in the case of heavy gases at low temperatures and high densities, the profile of emission and absorption lines agrees usually quite well with that predicted by Eq. (3.142). Furthermore the line width in general shows, as it should, a linear increase with the density of added foreign gases (so chosen as to avoid resonance with the observed line). Already the earliest observations, however, also indicated that there were other aspects to pressure broadening than the shortening of oscillator lifetimes by collisions. Aside from the occurences of line shift and asymmetry, the most compelling sympton was that the line profiles, although usually following the Lorentz shape, had about 2 to 3 times the theoretically expected width.

3.4.3.2 *Stark Broadening*

Since the pressure-dependent line width cannot be solely explained on the basis of collisions between fast moving luminescent centers and perturbers, additional broadening mechanisms must be sought in the detuning of neighboring oscillators. The first suggestion of this possibility was made by Stark[72] who demonstrated the splitting of spectral lines in strong electric fields and who pointed out that this effect must contribute to the broadening of spectral lines in that the radiating atoms and their neighbors exert a distribution of electric perturbations on each other. On the basis of the Stark effect, Holtsmark[73, 76] developed a theory of line broadening which, in combination with the other mechanisms of broadening, yielded the correct order of magnitude for the widths occuring in the Balmer series of hydrogen as well as in other spectra. The Holtsmark theory assumes that the radiating atoms are in the electric fields of either ions, dipoles, or quadrupoles. Under the influence of a given field a spectral line is perturbed in a predictable manner, i.e. it may be split (linear Stark effect) or shifted (as in the quadratic Stark effect). The radiating center is surrounded by a great number of neighbors which, as a result of their thermal motion, are randomly distributed. Therefore the net field, although it is here considered to be constant in time for each center, is statistically distributed over all the emitting atoms. Thus the broadening results as the probability distribution of the various frequency displacements. The shape is found to follow that given by Eq. (3.125) in some cases. An important example is that of dipole action on atoms exhibiting a *linear Stark effect*. The latter occurs in hydrogenlike spectra and is characterized by the fact that a line under the effect of an electric field F splits into two components with a frequency interval

$$\Delta \nu_{st} = aF \tag{3.143}$$

where a represents the split width per unit field strength. The radiation observed from a large number of atoms, which experience in the average a field F_D from surrounding dipoles, consists of a broadened line where the halfwidth is given by

$$\Delta \nu_h = \frac{\gamma}{2\pi} = aF_D \tag{3.144}$$

and where, according to the Holtsmark theory, the line shape is the same as in Lorentz-type broadening:

$$I(\nu) = \frac{\gamma I_0}{(\omega - \omega_0)^2 + \gamma^2/4} = \frac{2I_0}{\pi} \frac{aF_D}{4(\nu - \nu_0)^2 + (aF_D)^2} \tag{3.145}$$

The field F_D depends on the concentration n of the dipoles and is proportional to the moment X such that the halfwidth is

$$\Delta \nu_h = aF_D = 4.54 \, anX \tag{3.146}$$

The line shapes for the linear effect in the field of ions and quadrupoles were determined graphically by Holtsmark. The result is shown in Fig. 3.16 for the three cases of ions, dipoles and quadrupoles, the three curves normalized for equal halfwidths. It is seen that dipole and quadrupole

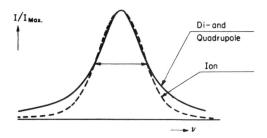

FIG. 3.16.   Profile for field broadening by various perturbers.

fields produce practically the same Lorentz-type intensity distribution. However, the graphically determined halfwidths for ions and quadrupoles differ from that given in Eq. (3.146) by a small factor and also depend on the concentration n in higher than first order, so that for ions of charge e,

$$\Delta v_h = 1.25aF_i = 3.25aen^{2/3} \qquad (3.147)$$

while for quadrupoles of moment A,

$$\Delta v_h = 0.67aF_q = 5.53a \, A \, n^{4/3} \qquad (3.148)$$

We mention briefly the quadratic Stark effect which may appear in nonhydrogenic spectra and which consists, not of a split, but of a shift of the line:

$$\delta v = v - v_0 = b \cdot F^2 \qquad (3.149)$$

The statistical distribution of a large number of field components which is observed in emitting or absorbing atoms results then in a broadened and shifted line.

3.4.3.3 *Statistical and Impact Theories of Pressure Broadening*

The two models of collision and coupling effects discussed in the preceding paragraphs represent the limiting cases of a more general and complex physical situation. One has to consider, under this more generalized viewpoint, that the radiating centers and their perturbers are in each others' fields, these fields being functions of the interatomic distances which in turn vary more or less rapidly with time. The manner in which such time-variant configuration gives rise to the two extreme mechanisms will be understood on the basis of Fig. 3.17. The two curves shown represent two electronic energy states of a radiating center at various distances r from just one other atom which exerts a perturbation. At large distances the two energy levels are undisturbed, and a transition between them corresponds to a frequency normal for the radiating atom. At very close proximity the two atoms exert a strongly repelling force on each other, as seen from the first derivative of the energies, dE/dr. In general the two energy levels have minima $(dE/dr = 0)$ which represent the positions of equilibrium. However, the energy difference between the two states is in most cases not constant so that the emitted frequency depends on the instantaneous position r of the radiating atom. Broadening by coupling results, therefore, from the statistical distribution of distances which various neighbors may assume with respect to the radiating atom during times presumed to be long in

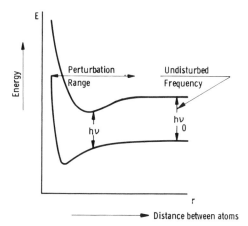

FIG. 3.17. Frank-Condon potential curves. Two energy levels of an atom are shown as function of distance from another atom. The two levels in general do not remain parallel so that the original frequency ν_o of transition is detuned as a perturber approaches.

comparison with that of the emission process. Refinements of the calculations based on this viewpoint are collectively called the *statistical* theory (Margenau[77-81]).

If, conversely, the velocities of the perturbing particles are so high that the times of interaction are small compared to the duration of the emission process, a type of collision process will result. We can in this case—which is the starting point of the so-called *impact* theory—proceed on the basis of a model which is due to Lenz (1924)[82] and was quantitatively elaborated by Weisskopf (1932).[83, 84] As the perturbing particle approaches the radiating atom with uniform velocity v [see Fig. 3.18(a)], the normally emitted frequency ω_0 is detuned [Fig. 3.18(b)] to an extent $\Delta\omega$ which follows from the energy curves of the type shown in Fig. 3.17. As mentioned before, this transient frequency deviation, Fig. 3.18(c), results in a net phase change η, to wit

$$\eta = \int_{-\infty}^{+\infty} \Delta\omega(r)\,dt = \int_{-\infty}^{+\infty} \Delta\omega(\sqrt{v^2t^2 + \rho^2})\,dt \qquad (3.150)$$

where $\Delta\omega$ depends in some manner on the perturber distance r and where r, in turn, is expressed in terms of the closest approach parameter ρ. The function $\Delta\omega(r)$ will have the form

$$\Delta\omega = \frac{c}{r^a} \qquad (3.151)$$

in which α is characteristic of the forces involved at the pertinent distances.

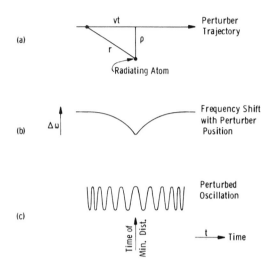

FIG. 3.18. Phase change of oscillation in transient interaction between radiating atom and passing perturber. Total phase shift is given by the difference in radians between the perturbed (c) and unperturbed oscillation within a given time interval.

Thus, the closer the approach of the two particles, the larger will be, in general, the integrand and the phase change in Eq. (3.150). One can, for any velocity, determine a closest distance ρ, such that the resulting phase change is in the order of unity. Somewhat arbitrarily, this may be considered as the criterion for the destruction of coherence, and thus $\pi\rho^2$ assumes the role of an *optical cross section* for collisions. The latter will always be larger than the gas-kinetic cross section so that the previously mentioned failure of the Lorentz theory to yield large enough halfwidths is partially removed (cf. footnote to Section 3.4.3.1). With these assumptions, for instance, Weisskopf derived for the case of the sodium D-lines (see Chapter 4), perturbed by foreign atoms ($\alpha = 6$), the value $\rho = 6.8$ Å. However, the interaction between like sodium atoms ($\alpha = 3$) results in a value $\rho = 44$ Å. The difference in the order of magnitude between pressure broadening in pure and admixed gases is thus explained. The criterion that a phase shift of $\eta \approx 1$ is just sufficient for incoherent interruption is, of course, at best a very rough approximation. A refinement of the model (Lenz,[85] Lindholm[86-89]) considers the effect of a large number of small phase perturbations, and in this manner line shifts are explained which could not be obtained from the simpler theory.

In their simplest forms, the statistical and impact theories can explain quantitatively many detailed phenomena of pressure broadening such as the larger pressure effects of the same atom species. However, much larger demands are made on the theory in the case of dense plasmas in which there exist large and equal concentrations of positive and negative charge carriers, i.e., ions and electrons. We have seen in Section 2.4.3 that there exists in plasmas of increasing densities a smooth transition from line to continuum spectra. It is in this transition region with Q'-values in the order of 100 or less, that complex problems arise. Experimental work was performed mainly on hydrogen and helium, with electron and ion concentrations typically of 10^{15} to 10^{17} cm^{-3} and electron temperatures of 10^4 to 10^5 deg, although practical instances are encountered in which these limits are exceeded. Impact and quasi-static considerations apply here simultaneously: the light emitting atoms and ions are in motion; they experience strong Stark splitting of their levels, hence broadening of the overall resulting radiation; and additional broadening results from encounters and collisions, particularly with the swiftly moving electrons. In such situations, the main features of Stark broadening turn out to be dominant. Quantum mechanical calculations of impact broadening combined with the theory of quasi-static broadening were applied to hydrogen-type lines by Griem and others.[90, 91, 92] The results yield line shapes which are in excellent agreement with experimental observations; but they represent often only a small improvement over the greater part of the pure Holtsmark profile (see, e.g., Griem and Shen[92]). The detailed features of the line broadening theory in dense plasmas were evolved by Margenau,[77–81] Griem,[90] Baranger,[93] and others. We refer here to a number of review articles which present comprehensive discussions of this growing field.[94, 95, 96]

3.4.3.4 *Field Broadening of Crystal Spectra*

It has been brought out before that the emission and absorption of optical radiation in solids belongs, in general, to the class of continua. This is merely a consequence of the fact that strong coupling exists between the atoms which are in solids ten times nearer to each other than they are in gases under normal pressure and temperature.

Nevertheless, there exist even in solids instances in which radiation interacts with more or less isolated oscillators. This is the case of foreign atoms, added as impurities to a host lattice. Indeed, here we may encounter line spectra again, albeit, usually of considerable broadening. To a certain extent, the width can be reduced by cooling; broadening increases as $T^{\frac{1}{2}}$ at higher temperatures. It must be pointed out that the Doppler effect is not the cause of this relationship. The decay time of an excited state is always long compared to the period of vibration which the radiating center undergoes as part of the infrared lattice oscillation spectrum. In fact, the emission process may last in excess of 10^{-3} sec, during which time the center will perform vibrations of small amplitude at the rate of 10^{13} to 10^{14} sec^{-1}. The frequency shift due to the velocity—as it varies periodically during each cycle—is, however, much smaller so that the net Doppler effect is practically zero. Therefore the origin of the line-width dependence on temperature in solids must be sought in the randomness with which the surrounding ions perturb the radiating state and which can be expected to increase with some power of T (see also Section 4.6.2).

A special situation exists for ions of the rare earths embedded in the lattice of dielectric materials such as the oxides of silicon or aluminum or even in glass. The rare earths have this outstanding characteristic: they contain unfilled inner shells (see Chapter 4) so that electron transitions can take place within them which are shielded from the crystal fields by the outer electrons, as if they were inside a microscopic Faraday cage. Hence, such line spectra are relatively sharp at sufficiently low temperatures, exhibiting Q'-values in the order of 10^5. We shall have occasion to discuss details later in connection with maser materials.

3.4.4 Comparison of Broadening Effects and Their Magnitudes

We have seen that the deviations from ideal monochromaticity can be assigned, according to type and cause, to various physical processes. Most important among these processes are (a) the termination of lifetimes, either by "natural" or external causes; (b) velocity distribution of the light emitters with resulting Doppler effects and (c) detuning of the oscillator frequency. The variety of the detail mechanisms is very large, and even more impressive is the range of magnitudes characterizing the line widths which can vary from almost arbitrarily small values, in the maser process, to the continuum characteristic of heat radiation.

A few examples for the comparative study of the most typical broadening effects and magnitudes are cited in Table 3.5. Monochromatic radiation sources are listed in the order of increasing relative width $\Delta\nu/\nu$ (or decreasing Q' in column II). The halfwidths $\Delta\nu_h$ are presented in terms of frequency, i.e. cycles per second. The correlation to the halfwidths in other customary units, such as to $\Delta\tilde{\nu}$ in cm^{-1}, $\Delta\lambda$ in angstroms, ΔE in electron volts, is given by

$$\Delta\nu = c\,\Delta\tilde{\nu} = \frac{c\,\Delta\lambda}{\lambda^2} \times 10^8 = 2.418 \times 10^{14}\,\Delta E \qquad (3.152)$$

As explained before, the line width in each case may be the result of one, or more, dominant mechanisms. Each mechanism of broadening

TABLE 3.5

<small>Comparison of Line Widths Resulting from Various Causes</small>

I	II	III	IV
Radiation source	$Q' = \nu/\Delta\nu_h$	Halfwidth $\Delta\nu_h$ (sec^{-1})	Responsible mechanism and theoretically expected limits (in sec^{-1}) of $\Delta\nu_h$
Mössbauer 0.86 Å from Fe^{57}	1.5×10^{12}	2.3×10^6 [a]	2.2×10^6 ($= 2\gamma$), from 1.0×10^{-7} sec lifetime of Fe^{57} state
0.096 Å from Ir^{191}	1.4×10^{10}	2.2×10^9 [b]	2.2×10^9, from 1.4×10^{-10} sec lifetime of Ir^{191}
Optical maser 11,530 Å-Ne: He at 0.1:1.0 mm Hg	$>3 \times 10^{13}$	$<10^c$	< 10 for coherent radiation with power $P = 10^{-2}$ watt in single mode and $\nu_{doppler} = 2.5 \times 10^8$ sec^{-1}
Glow discharge 5791 Å in Hg (at 140°C)	1.7×10^6	3.0×10^8 [d]	2.7×10^8 for pure Doppler broadening
6563 Å (H$_\alpha$) in H at 50°C	1.4×10^5	3.3×10^9	3.3×10^9 for pure Doppler broadening
Shock-heated plasmas of H 6563 Å (Hα)	850	5.4×10^{11} [e]	5.6×10^{11} for $T = 25 \times 10^3$ °K
4861 Å (Hβ)	120	5.2×10^{12}	5.2×10^{12} for $T = 40.7 \times 10^3$ °K
4340 Å (Hγ)	89	7.8×10^{12}	8.0×10^{12} for $T = 49.2 \times 10^3$ °K from pressure broadening theory of Griem^g
Hydrogen arc 4861 Å, 15 amp	300	2.0×10^{12} [f]	Pressure broadening
4861 Å, 250 amp	76	7.9×10^{12}	Approximate Holtsmark contour

[a] G. K. Wertheim, *Phys. Rev.* **124**, 764 (1961).
[b] R. L. Mössbauer, *Z. Physik* **151**, 124 (1958); *Naturwissenschaften* **45**, 538 (1958); *Naturforsch.* **14a**, 211 (1959).
[c] A. Javan, W. R. Bennett, Jr., and D. R. Herriott, *Phys. Rev. Letters* **6**, 106 (1961).
[d] A. A. Michelson, *Phil. Mag.* [5] **34**, 280 (1892).
[e] H. F. Berg, A. W. Ali, R. Lincke, and H. R. Griem, *Phys. Rev.* **125**, 199 (1962).
[f] P. Bogen, *Z. Physik* **149**, 62 (1957).
[g] H. R. Griem, A. C. Kolb, and K. Y. Shen, *Phys. Rev.* **116**, 4 (1959).

may be viewed as a process of disorder; hence the theoretical values, as shown in Colum IV of Table 3.5, represent a lower limit to the widths which are practically achievable under the stated conditions.

The most ideal monochromatic behavior is exhibited by those processes in which the line width is determined by an oscillator lifetime of considerable natural or artificial length, as is the case in nuclear resonance and optical masers, respectively. In the Mössbauer effect, line sharpness can indeed reach the theoretically predicted limit as quoted for the cases of iron 57 and iridium 191. A prerequisite for agreement with theory is careful experimentation (use of very thin foils for emitter and absorber, for instance). There are, of course, also cases in which the Mössbauer lines are in the order of 10 times broader than the natural width owing to unresolved hyperfine splitting and other causes. In terms of absolute widths, optical masers promise considerably smaller values than the Mössbauer lines. Furthermore, as will be discussed in Chapter 6, line sharpening by orders of magnitude is in the offing for these processes so that masers with Q'-values in excess of 10^{14} may ultimately be attainable.

It is noteworthy that, until the arrival of the high-Q' oscillations just discussed, the natural line width of 10^{-4} Å corresponding to a Q' of about 10^8 in the visible region was considered as the upper—and practically unattainable—limit for the definition of optical frequencies. In fact, the sharpest lines which are spontaneously emitted from a gas discharge exhibit a Doppler broadening of at least 10 times the natural width, even under the highest, practically achievable ratios of mass and temperature. Doppler broadening usually presents the dominant limitation for gases in the pressure region of a few mm Hg or below, with Q'-values of the order 10^5–10^6 throughout the optical spectrum. A small contribution to the width from pressure effects is noticeable in the example cited for mercury vapor in Table 3.5. In the regime of pressure broadening the evolution of theory has arrived at an agreement with observation which is remarkable in view of the complexity facing calculations of line widths beyond 1 Å. An example is shown in Fig. 3.19 (Berg et al.[97]) in which the doubly peaked Stark profile of the H_β line, emitted from a shock-heated plasma, is compared with the theory.[90, 91, 92] Although the spectral contour with a halfwidth of about 50 Å has several detail features, the theory is seen to be accurate within the order of 10 %.

The test of the theory, at any rate, is indirect in the case of dense plasmas. Temperature and density of the electrons have to be determined, the temperature being determined by absolute or relative line intensities (see Section 3.3.3.2) and the density from the Saha equation which

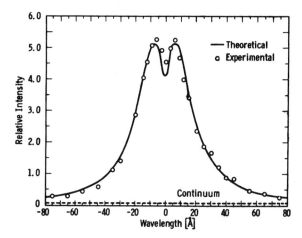

FIG. 3.19. Profile of H_β line emitted from shock heated plasma [H. F. Berg, A. W. Ali, R. Lincke, and H. R. Griem, *Phys. Rev.* **125**, 199 (1962)]. Theoretical details [H. R. Griem, A. C. Kolb, and K. Y. Shen, *Phys. Rev.* **116**, 4 (1959); H. R. Griem, *Astrophys. J.* **132**, 883 (1960); H. R. Griem and K. Y. Shen, *Phys. Rev.* **122**, 1490 (1961)] are well confirmed by experiment.

describes the equilibrium between electrons and ions as a function of temperature. In addition, the fields and concentrations of the ions and neutral atoms have to be known as discussed before. The real significance of the agreement (or limitations of such agreement) between theory and observation is twofold: the comparison establishes to what extent line shapes reveal the physical condition of the source; and our understanding of the laws governing emission and absorption processes experiences further tests and confirmation.

REFERENCES

1. H. HERTZ, *Ann. Physik* [3]ʼ **36**, 1 1(889).
2. R. BECKER and F. SAUTER, "Theorie der Elektrizität," Vol. 1. Teubner, Stuttgart, 1959.
3. M. PLANCK, *Ann. Physik* [3] **60**, 577 (1897).
4. N. BOHR, *Phil. Mag.* [6] **26**, 1 and 476 (1913).
5. L. M. RUTHERFORD, *Phil. Mag.* [6] **21**, 669 (1911).
6. J. W. M. DuMOND and E. R. COHEN, *Rev. Mod. Phys.* **20**, 82 (1948).
7. R. T. BIRGE, *Phys. Rev.* **60**, 766 (1941).
8. E. SCHRÖDINGER, *Ann. Physik* [4] **79**, 361, 489, and 734 (1926).
9. W. HEISENBERG, *Z. Physik* **33**, 879 (1925).
10. L. DE BROGLIE, *Ann. Phys.* [10] **3**, 22 (1925).
11. L. DE BROGLIE, Thesis, Paris (1924).
12. A. EINSTEIN, *Verhandl. Deut. Phys. Ges.* **18**, 318 (1916).

13. A. EINSTEIN, *Physik. Z.* **18**, 121 (1917).
14. P. A. M. DIRAC, *Proc. Roy. Soc.* **A106**, 581 (1924).
15. C. ECKART, *Phys. Rev.* **28**, 711 (1926).
16. M. BORN, "The Mechanics of the Atom." Bell, London, 1927.
17. M. BORN, "Atomic Physics," 5th ed. Hafner, New York, 1951 (7th ed., 1962).
18. A. UNSÖLD, "Physik der Sternatmosphären," 2nd ed. Springer, Berlin, 1955.
19. R. LADENBURG, *Z. Physik* **4**, 451 (1921).
20. W. KUHN, *Z. Physik* **34**, 408 (1925).
21. W. THOMAS, *Naturwissenschaften* **13**, 627 (1925).
22. F. REICHE and W. THOMAS, *Z. Physik* **34**, 510 (1925).
23. H. A. BETHE, *in* "Handbuch der Physik" (H. Geiger and K. Scheel, eds.), 2nd ed., Vol. 24, Part I, Chapter 3. Springer, Berlin, 1933.
24. H. A. BETHE and E. E. SALPETER, *in* "Handbuch der Physik" (S. Flügge, ed.), Vol. 35, p. 88. Springer, Berlin, 1957.
25. E. U. CONDON and G. H. SHORTLEY, "The Theory of Atomic Spectra." Cambridge Univ. Press, London and New York, 1935.
26. R. LADENBURG and S. LEVY, *Z. Physik* **65**, 189 (1930).
27. R. W. WOOD, *Phil. Mag.* [6] **23**, 689 (1912).
28. M. W. ZEMANSKY, *Phys. Rev.* **29**, 513 (1927).
29. D. ALPERT, A. O. MCCOUBRY, and T. HOLSTEIN, *Phys. Rev.* **76**, 1257 (1949).
30. T. HOLSTEIN, *Phys. Rev.* **72**, 1212 (1947).
31. T. HOLSTEIN, *Phys. Rev.* **83**, 1159 (1951).
32. L. M. BIEBERMAN, *J. Exptl. Theor. Phys. USSR* **17**, 416 (1947).
33. J. H. LEES and H. W. B. SKINNER, *Proc. Roy. Soc.* **A137**, 186 (1932).
34. S. HERON, R. W. P. MCWHIRTER, and E. H. RHODERICK, *Proc. Roy. Soc.* **A234**, 565 (1956).
35. A. V. PHELPS, *Phys. Rev.* **110**, 1362 (1958).
36. F. G. HOUTERMANS, *Z. Physik* **83**, 19 (1933).
37. K. KREBS, *Z. Physik* **101**, 604 (1936).
38. H. KOPFERMANN and G. WESSEL, *Z. Physik* **130**, 100 (1951).
39. R. LADENBURG and S. LORIA, *Verhandl. Deut. Phys. Ges.* **10**, 858 (1908).
40. R. LADENBURG, *Ann. Physik* [4] **38**, 249 (1913).
41. H. KOPFERMANN and R. LADENBURG, *Z. Physik* **48**, 26 (1926).
42. D. ROSCHDESTWENSKY, *Ann. Phys.* (*Paris*) **39**, 307 (1912).
43. M. WEINGEROFF, *Z. Physik* **67**, 679 (1931).
44. G. STEPHENSON, *Proc. Phys. Soc.* (*London*) **A64**, 458 (1951).
45. M. GARBUNY, *Z. Physik* **107**, 362 (1937).
46. M. GARBUNY, T. P. VOGL, and J. R. HANSEN, *Rev. Sci. Instr.* **28**, 826 (1957).
47. E. BRANNEN, F. R. HUNT, R. H. ADLINGTON, and R. W. NICHOLLS, *Nature* **175**, 810 (1955).
48. W. DEMTRÖDER, *Z. Physik* **166**, 42 (1962).
49. K. ZIOCK, *Z. Physik* **147**, 99 (1957).
50. W. WIEN, *Ann. Physik* [4] **60**, 597 (1919).
51. W. WIEN, *Ann. Physik* [4] **66**, 229 (1921).
52. W. WIEN, *Ann. Physik* [4] **73**, 483 (1924).
53. W. WIEN, *Ann. Physik* [4] **83**, 1 (1927).
54. L. MAXWELL, *Phys. Rev.* **32**, 721 (1928).
55. L. MAXWELL, *Phys. Rev.* **38**, 1664 (1931).
56. M. KAGAN and Y. P. KORITSKII, *Opt. i Spektroskopiya* **11**, 308 (1961); see *Opt. Spectry.* (*USSR*) (*English Transl.*) **11**, 166 (1961).

57. H. B. DORGELO and H. C. BURGER, *Z. Physik* **23**, 258 (1924).
58. C. H. CORLISS, *J. Res. Natl. Bur. Std.* **A66**, 5 (1962).
59. D. R. BATES and A. DAMGAARD, *Phil. Trans. Roy. Soc.* **A242**, 101 (1949).
60. B. M. GLENNON and W. L. WIESE, *Natl. Bur. Std.* (*U.S.*) *Monograph* **50** (1962).
61. C. H. CORLISS and W. R. BOZMAN, *Natl. Bur. Std.* (*U.S.*) *Monograph* **53** (1962).
62. W. HEITLER, "The Quantum Theory of Radiation," 3rd ed. Oxford Univ. Press, (Clarendon), London and New York, 1954.
63. V. WEISSKOPF and E. WIGNER, *Z. Physik* **63**, 54 (1930); **65**, 18 (1930).
64. R. L. MÖSSBAUER, *Z. Physik* **151**, 124 (1958).
65. R. L. MÖSSBAUER, *Naturwissenschaften* **45**, 538 (1958).
66. R. L. MÖSSBAUER, *Z. Naturforsch.* **14a**, 211 (1959).
67. W. E. LAMB, *Phys. Rev.* **55**, 190 (1939).
68. H. J. LIPKIN, *Ann. Phys.* (*N.Y.*) **9**, 332 (1960).
69. H. LUSTIG, *Am. J. Phys.* **29**, 1 (1961).
70. M. BORN, "Optik." Springer, Berlin, 1933.
71. H. A. LORENTZ, *Proc. Acad. Sci. Amsterdam* **18**, 134 (1915).
72. J. STARK, "Elektrische Spektralanalyse chemischer Atome." Hirzel, Leipzig, 1914.
73. J. HOLTSMARK, *Ann. Physik* [4] **58**, 577 (1919).
74. J. HOLTSMARK, *Physik. Z.* **20**, 162 (1919).
75. J. HOLTSMARK, *Physik. Z.* **25**, 73 (1924).
76. J. HOLTSMARK, *Z. Physik* **34**, 722 (1925).
77. H. MARGENAU, *Phys. Rev.* **40**, 387 (1932).
78. H. MARGENAU, *Phys. Rev.* **43**, 129 (1933).
79. H. MARGENAU, *Phys. Rev.* **44**, 931 (1933).
80. H. MARGENAU, *Phys. Rev.* **48**, 755 (1935).
81. H. MARGENAU, *Phys. Rev.* **82**, 156 (1951).
82. W. LENZ, *Z. Physik* **25**, 299 (1924).
83. V. F. WEISSKOPF, *Z. Physik* **75**, 287(1932).
84. V. F. WEISSKOPF, *Z. Physik* **77**, 398 (1932).
85. W. LENZ, *Z. Physik* **80**, 423 (1933).
86. E. LINDHOLM, *Z. Physik* **109**, 223 (1938).
87. E. LINDHOLM, *Z. Physik* **113**, 596 (1939).
88. E. LINDHOLM, *Arkiv Mat. Astron. Fysik* **28A**, No. 3 (1942).
89. E. LINDHOLM, *Arkiv Mat. Astron. Fysik* **32A**, No. 17 (1946).
90. H. R. GRIEM, A. C. KOLB, and K. Y. SHEN, *Phys. Rev.* **116**, 4 (1959).
91. H. R. GRIEM, *Astrophys. J.* **132**, 883 (1960).
92. H. R. GRIEM and K. Y. SHEN, *Phys. Rev.* **122**, 1490 (1961).
93. M. BARANGER, *Phys. Rev.* **111**, 494 (1958).
94. R. G. BREENE, *Rev. Mod. Phys.* **29**, 94 (1957).
95. S. CH'EN and M. TAKEO, *Rev. Mod. Phys.* **29**, 20 (1957).
96. H. MARGENAU and M. LEWIS, *Rev. Mod. Phys.* **31**, 569 (1959).
97. H. F. BERG, A. W. ALI, R. LINCKE, and H. R. GRIEM, *Phys. Rev.* **125**, 199 (1962).

4

SPECTRA OF ATOMS, MOLECULES, AND SOLIDS

The preceding three chapters dealt with the basic emission and absorption processes between light and matter. In this treatment we considered primarily the mechanism by which oscillators exchanged energy with the radiation field: either individually, giving rise to nearly monochromatic radiation; or as strongly coupled and "detuned" oscillators producing a spectral continuum. The light sources with which we are concerned, in the range between the far infrared and the ultraviolet, are for many purposes adequately described in terms of these elementary processes. Thus heat radiation, which for practical use and in natural occurrence is the most important source of the infrared and the visible spectrum, is generated by incandescent surfaces or thermally excited atmospheres, and these behave more or less like black bodies, the spectra of which have been discussed in Chapter 2. There are also cases of monochromatic radiation in which we have to know little more than that we are dealing with a transition of a certain energy and probability. On the other hand, such sources as represented by most free atoms, which may be excited by an electrical discharge in gases, emit more complex line spectra, and these have yet to be discussed. It has already been indicated in the preceding chapter that the complexity of spectra reflects a complexity of atomic structure. Indeed, spectroscopy offers the most important experimental channel for the exploration of nuclear and atomic properties. Moreover, a knowledge of spectroscopy is of practical importance in many disciplines: in chemistry, for the identification of composition and structural bondings; in astrophysics, to determine the physical and chemical condition of stellar atmospheres; and in applied physics, for the choice or design of spectrally selective light sources or for specific interactions, such as maser technology. It is, in particular, these latter kinds of applications which concern us in some of the following chapters.

4.1 Atomic Spectra for One Valence Electron

4.1.1 Spectra and Degrees of Freedom

In the narrower sense of its meaning, the science of spectroscopy is concerned with the emission or absorption of light from atoms and molecules and the interpretation of the spectral lines and bands, observed in emission or absorption, in terms of atomic and molecular structure. The simplest of the spectra has already been discussed in the preceding chapter: the line series of hydrogenlike atoms, i.e. atoms which consist of a single electron orbiting around a nucleus of charge Z. However, even this one-electron configuration embodies a number of structural details and subtleties which have only gradually been brought to light. In the treatment of the Bohr atom, we discussed principally only one variable of position, namely, the distance r of the electron from the nucleus. Because energy, along with orbital momentum, is subject to quantization rules and because both depend functionally on r, the latter, too, must assume certain discrete values which are characterized by an orbital quantum number n. Actually, the electron is free to assume three coordinates of position; we may also say it has in this respect three degrees of freedom. Following the experimental results which were achieved in spectroscopy with improving methods and instruments, a steady refinement of the atomic model took place. The electron was found to be a spinning top; it has a quantized spin (Uhlenbeck and Goudsmit,[1, 2] 1925, 1926). The same was also found to be true for the proton and for nuclei in general (Pauli,[3] 1924; Russel,[4] 1927; Dennison,[5] 1927; see in this connection also Goudsmit,[6] 1961). This means that there exist additional degrees of freedom: for each electron orbit, there are various alternatives for the orientation of nuclear and electron spins.

What we have discussed so far applies to the hydrogen atom and similarly constructed ions such as He^+, Li^{++}, Be^{3+}, and so on. As we progress along a row of elements in the periodic table, adding one electron from column to column, the degrees of freedom increase correspondingly. Thus we reach a maximum of complexity in the possible configurations towards the middle of a row of elements. Then, however, the trend is reversed because of a rule, which is most important for the structure of the periodic system, namely, the Pauli exclusion principle.[7] This, it will be seen, restricts the choice of orbital and spin states (no two electrons can have the same set of quantum numbers). Thus as we move towards the end of a row, electrons are forced into fewer available states, and greater simplicity results again.

There is yet another set of degrees of freedom to be mentioned which gives rise to an enormous multiplicity of configurations. This refers to the vibrations and rotations which atoms can perform with respect to each other, either individually or grouped in radicals. This, of course, is the subject of molecular spectroscopy. We are justified in continuing this line of reasoning for just one more step. As we increase the number of closely coupled particles, we progress from more or less resolvable patterns to the continua of the radiation from solids (see Section 1.2.3), most perfectly represented by black body emission.

4.1.1.1 Energy Terms and Spectral Lines

What significance the number of degrees of freedom has for the theory of radiation, was already discussed in Chapter 1. The equipartition law of statistical mechanics allots a mean energy of $kT/2$ to each degree of freedom. This law is modified, of course, in quantum theory which assigns quantized states to the various degrees of freedom, including the state of zero excitation. This means, e.g., that hydrogenlike atoms have states determined by 5 quantum numbers: 3 for electron position and 2 for the two spins. Usually a discrete value of energy belongs to each set of quantum numbers, just as in Bohr's theory a particular value ϵ_n belonged to the single quantum number n in Eq. (3.39). However, these proper energy values are not necessarily all different for different sets of quantum numbers. There are conditions under which the energy of the atom remains unchanged with respect to the particular value which one or more quantum numbers may assume. In other words, there result *degenerate* states, characterized by different sets of quantum numbers but the same energy such as we have discussed before in Chapter 3 and analogous to the G_j modes at a given energy ϵ_j in the oscillator continuum (cf. Section 2.3.2). In the case of atoms, the degeneracy can be removed in general by such means as the application of electric or magnetic fields, because the various degenerate configurations then unfold to assume different energies. However, even if a multiplicity of states remains latent, it still exerts an important influence on the statistical distribution of populations over the energy levels, in that each of the degenerate states has still an equal chance of being occupied. We thus assign to a degenerate level a statistical weight g which is the number of states with the same energy (cf. Section 3.3). For atomic systems, the g-values are, if not unity, relatively small integer numbers. On the other hand, the possible modes within a given frequency or energy interval for an oscillator continuum are represented by a very large number commensurate with the number of atoms counted in the solid.

The manner in which the various degrees of freedom contribute to the total energy of the atomic system is schematically shown in Fig. 4.1, in which the atomic levels or energy terms are indicated on the right. Three successive stages of model refinement are considered: (1) the gross structure, as given by Eq. (3.39) in which only one degree of freedom and thus only one quantum number, the principal number n,

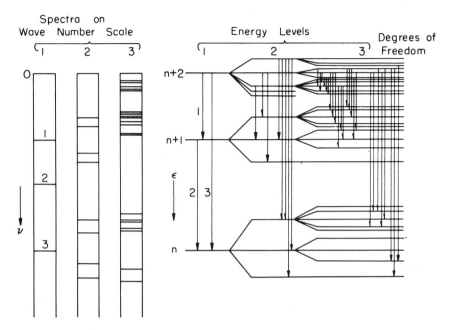

Fig. 4.1. Scheme of energy levels and spectral lines. For a hypothetical case, the increase of complexity is shown in three stages resulting from various degrees of freedom of the oscillating structure.

appears, three terms being shown (n, $n + 1$, $n + 2$); (2) gross and fine structure combined so that each of the first levels splits into several new levels which are energetically not too far apart; and (3) a yet finer structure of levels into which some, or all, of the terms under (2) have split due to the interaction with a third degree of freedom. It should be pointed out that an arbitrary scheme has been assumed here for the purpose of generality; the number and magnitude of the term splittings are not meant for any particular case.

However, the three types of energy contribution could, for instance, correspond to three degrees of freedom for the position of a single electron in the Coulomb field of a nucleus and a magnetic field which would remove degeneracies. This case applies to the hydrogen atom which requires a refinement of the original Bohr model to explain the observed fine structure of its spectrum. The first step in this direction was undertaken by Sommerfeld[8] who postulated *elliptic* electron orbits around the nucleus. Such orbits are fixed by three parameters as shown in Fig. 4.2. Therefore, three numbers are necessary for their complete quantization: (1) a *radial* quantum number which determines the average

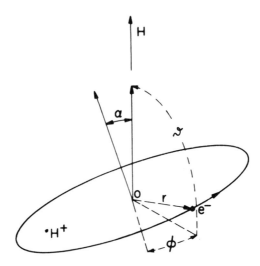

FIG. 4.2.   Spherical coordinates of electron in elliptic orbit around proton.

electron-nuclear distance and hence the size of the ellipse; (2) an *azimuthal* quantum number which defines the form or ellipticity; and (3) also a *magnetic* quantum number which indicates discrete angular orientations of the orbit plane with respect to a favored direction such as that established by a magnetic field.

Again, schematically, Fig. 4.1 could stand for a two-electron configuration, in which (1) one of the electrons assumes three successive principal numbers; (2) the effect of orbital interaction is added; and (3) the resultant effects of the spins yield the final structure.

Once a term scheme has been established, it is a relatively simple task to read off the spectral lines which correspond to the various possible transitions between energy levels. A direct method is that of plotting spectral lines on a scale of frequencies v or wave numbers $\tilde{v} = v/c$. Since $v_{mn} = (\epsilon_m - \epsilon_n)/h$, the wave number of a line plotted from some zero point is proportional to the vertical distance of the corresponding energy terms. This has been done on the left of Fig. 4.1. The first of these spectra corresponds to the first of the level diagrams; it contains the three transitions possible between n, $n + 1$, and $n + 2$. The second and third spectrum shows increasing complexity of the line patterns, although only a few of the possible transitions were drawn. At any rate, not all transitions can occur. As we shall see, the number of possible combinations between all the levels is sharply limited by so-called *selection rules*.

4.1.2 THE SPECTRA OF HYDROGENLIKE ATOMS

The preceding paragraphs have described qualitatively the general mechanism by which a multiplicity of atomic orbit and spin configurations comes into existence and gives rise to the observed complex patterns of energy terms and spectral lines. In the following, the quantitative aspects of energy level schemes will be discussed in broad outline. The first step in this treatment is the refinement of the hydrogen atom model on the basis of wave mechanics as the most valid approach.

4.1.2.1 *Wave Functions of the Hydrogen Atom*

The three-dimensional position problem of the single electron with reduced mass m_r [see Eq. (3.42)] in the Coulomb field of the nucleus is solved with the Schrödinger equation (3.53) in polar coordinates[9]:

$$\frac{1}{r^2}\frac{\partial}{\partial r}\left(r^2\frac{\partial\psi}{\partial r}\right) + \frac{1}{r^2\sin^2\vartheta}\frac{\partial^2\psi}{\partial\varphi^2} + \frac{1}{r^2\sin\vartheta}\frac{\partial}{\partial\vartheta}\left(\sin\vartheta\frac{\partial\psi}{\partial\vartheta}\right) + \frac{8\pi^2 m_r}{h^2}(W-V)\psi = 0$$

$$(4.1)$$

where r is the distance from the nucleus and where φ and ϑ are the angles of azimuth and inclination, respectively (cf. Fig. 4.2 for the Bohr-Sommerfeld atom). As discussed in Section 3.2.3.4, the meaning of the electron wave function $\psi(\varphi, \vartheta, r)$ follows from the fact that $|\psi|^2$, the square of its absolute value or the equivalent expression, $\psi\psi^*$, the product of the absolute value and its complex conjugate value, represents the *probability distribution* of the electron. In other words, $\psi\psi^*$ is the probability that the electron finds itself in a given volume element at the point $\{\varphi, \vartheta, r\}$. It must be pointed out, however, that no assumptions regarding the electron motion—such as elliptic orbits—underlie the Schrödinger equation.

A differential equation of the type (4.1) can be treated by the separation of variables. That is to say, we assume for the solution the form

$$\psi = \phi(\varphi) \cdot \theta(\vartheta) \cdot R(r) \qquad (4.2)$$

Each of the three functions, ϕ, θ, and R must fulfill the requirement stated before in Section 3.2.3.3 of being everywhere finite, single-valued, and continuous. The introduction of (4.2) into (4.1) yields after dividing by ψ

$$\frac{1}{R}\frac{d}{dr}\left(r^2\frac{dR}{dr}\right) + \frac{1}{\phi\sin^2\vartheta}\frac{d^2\phi}{d\varphi^2} + \frac{1}{\theta\sin\vartheta}\frac{d}{d\vartheta}\left(\sin\vartheta\frac{d\theta}{d\vartheta}\right) + \frac{8\pi^2 m_r}{h^2}(W-V)r^2 = 0$$

$$(4.3)$$

Equation (4.3) can be separated into three differential equations, each containing one of the three position coordinates as a single independent variable. As a first step multiplication by $\sin^2\vartheta$ in Eq. (4.3) isolates the term $(1/\phi)\,d^2\phi/d\varphi^2$ from all those not containing φ. It follows that the term must be a constant, which we set equal to $-m^2$ to make the solution more convenient:

$$\frac{d^2\phi}{d\varphi^2} = -m^2\phi \tag{4.4}$$

The normalized solution of Eq. (4.4) is the periodic function

$$\phi_m(\varphi) = \frac{1}{\sqrt{2\pi}}\,e^{im\varphi} \tag{4.5}$$

The factor $1/\sqrt{2\pi}$ has been chosen so as to fulfill the normalization requirement

$$\int_0^{2\pi} \phi^*\phi\,d\varphi = \frac{1}{2\pi}\int_0^{2\pi} e^{im\varphi}e^{-im\varphi}\,d\varphi = 1 \tag{4.6}$$

Furthermore, to be single-valued, $\phi(\varphi + 2\pi) = \phi(\varphi)$, so that m, *the magnetic quantum number*, can assume only integer values $0, +1, -1, +2, -2, \ldots$, and it may be restricted still further to just a few values among these.

The derivation of the θ- and R-functions is more involved. We present here merely the results whithout going into the details of their solution.

When Eq. (4.4) is introduced into Eq. (4.3), φ is eliminated as a variable, and the remaining terms can be separated into two parts which contain either only ϑ or only r. Each part can therefore be equated to a constant so that there result two equations for $\theta(\vartheta)$ and $R(r)$. For the first of these we have

$$\frac{1}{\theta\sin\vartheta}\frac{d}{d\vartheta}\left(\sin\vartheta\,\frac{d\theta}{d\vartheta}\right) - \frac{m^2}{\sin^2\theta} = C \tag{4.7}$$

The normalized solution of this equation (see, e.g., White[8a] or Pauling and Wilson[9]) appears in terms of *associated Legendre polynomials* of $\cos\vartheta$:

$$\theta_{l,m}(\vartheta) = \sqrt{\frac{(2l+1)}{2}\cdot\frac{(l-|m|)!}{(l+|m|)!}}\,P_l^{|m|}(\cos\vartheta) \tag{4.8}$$

The solution for θ is different from zero only for integer values of the azimuthal quantum number l which must be equal or larger than $|m|$, i.e., $l = |m|, |m| + 1, \dots$ and for discrete values of the constant C in Eq. (4.7),

$$C = -l(l + 1) \tag{4.9}$$

Finally we introduce Eqs. (4.7) and (4.9) into Eq. (4.3) and obtain the differential equation for $R(r)$:

$$\frac{1}{R} \frac{d}{dr}\left(r^2 \frac{dR}{dr}\right) - l(l + 1) + \frac{8\pi^2 m_r}{h^2}(W - V)r^2 = 0 \tag{4.10}$$

where the potential energy V is determined by the Coulomb potential of a central charge eZ,

$$V = -\frac{e^2 Z}{r} \tag{4.11}$$

The solution of Eq. (4.10) for the case that the total energy W is positive warrants brief mention. Here the atom is actually ionized, and the qualitative considerations of Section 3.2.3.3 have shown already that wave functions exist for a continuum of values W. The positive energies are not quantized, therefore, and this conclusion can be verified analytically.

For negative values of W the normalized solution of Eq. (4.9) is obtained in terms of *associated Laguerre polynomials* L of the argument $2Zr/nr_1$:

$$R_{nl}(r) = \left[\frac{4(n - l - 1)!Z^3}{\{(n + l)!\}^3 n^4 r_1{}^3}\right]^{1/2}\left(\frac{2rZ}{nr_1}\right)^l e^{-Zr/nr_1} L_{n+l}^{2l+1}\left(\frac{2rZ}{nr_1}\right) \tag{4.12}$$

where

$$r_1 = \frac{h^2}{4\pi^2 m_r e^2} = 0.53 \times 10^{-8} \text{ cm} \tag{4.13}$$

represents the radius of the smallest orbit in hydrogen [cf. Eqs. (3.37) and (3.39) for $n = 1$ and $Z = 1$]. Equation (4.12) implies that solutions of the radial function $R(r)$ can exist only for $n = l + 1, l + 2, l + 3, \dots$. Hence, negative values of the total energy W are quantized. The values obtained here are identical with those given by the Bohr theory as shown by Eq. (3.39):

$$W_n = -\frac{2\pi^2 m_r Z^2 e^4}{h^2 n^2} \tag{4.14}$$

4.1.2.2 Probability Distribution Functions, Quantum Numbers, and Electron Orbits

Although wave mechanics confirms for hydrogen the major result of the older theories, especially the quantization according to the three position coordinates and the energy values W_n, it does not predict the elliptic electron orbits of the Bohr-Sommerfeld theory. For instance, the wave function for normal hydrogen ($n = 1$, $l = 0$, $m = 0$), reduces to

$$\psi_{100} = \phi_0(\varphi)\theta_{00}(\vartheta)R_{10}(r) = \frac{1}{\sqrt{\pi r_1{}^3}}\, e^{-r/r_1} \tag{4.15}$$

The electron density function (probability per unit volume element) is then

$$\psi\psi^* = \frac{1}{\pi r_1{}^3}\, e^{-2r/r_1} \tag{4.16}$$

hence independent of φ and ϑ, i.e., spherically symmetrical. Thus there results an "electron cloud" of radially decreasing density, a picture which can be visualized only with difficulty in terms of elliptic orbits. Nevertheless, there exists again good agreement for the size of the orbit. The probability of finding the electron of the normal hydrogen atom between r and $r + dr$ (therefore in a spherical shell of volume $4\pi r^2\, dr$) equals

$$4\pi r^2\, dr(\psi\psi^*)_{100} = \frac{4}{r_1{}^3}\, r^2 e^{-2r/r_1}\, dr \tag{4.17}$$

This probability is a maximum for $r = r_1$, i.e., the radius of the Bohr orbit for $n = 1$. Figure 4.3 presents the radial function R_{nl} and the

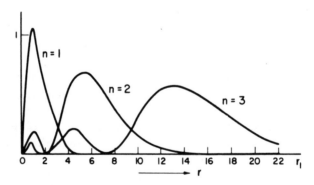

Fig. 4.3.   Radial probability function $4\pi r^2[R_{nl}(r)]^2$ for the hydrogen atom.

corresponding radial probability function $4\pi r^2[R_{n,l}(r)]^2$ for various values of n and l. The Bohr-Sommerfeld theory predicts orbits the dimension of which fit the position of the maxima shown in Fig. 4.3 quite well.

Because of this quantitative agreement in most respects and because of its heuristic plausibility, the model of elliptic orbits is usually invoked, at least qualitatively, for the interpretation of spectra in terms of atomic structure, even though such pictures can have only limited validity (see also Fig. 4.4 for the θ-functions). Thus we can summarize the results of the wave mechanical treatment, which appear as electron wave functions, in terms of the corresponding Bohr-Sommerfeld ellipses. In both descriptions we have the following three quantum numbers which determine the electron configurations and which are rewritten from the conditions set forth in the previous section:

(1) the total quantum number n (proper values of the radial function),

$$n = 1, 2, 3, 4, ... \tag{4.18}$$

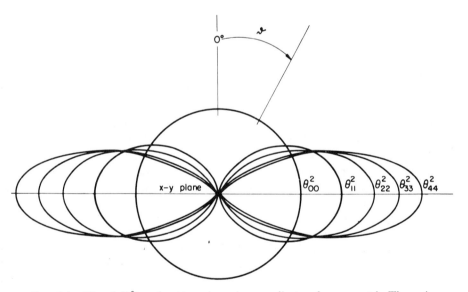

FIG. 4.4. The $\theta_m{}^2(\vartheta)$ — functions in polar coordinates for $m = \pm l$. The polynomials plotted are $\theta_{00} = \frac{1}{2}$, $\theta_{11} = \frac{3}{4} \sin^2 \vartheta$, $\theta_{22} = (15/16) \sin^4 \vartheta$, $\theta_{33} = (35/32) \sin^6 \vartheta$, $\theta_{44} = (315/256) \sin^8 \vartheta$, corresponding to s-, p-, f-, and g-orbitals in planes nearly normal to the z-axis. The x-y plane is the locus of greatest probability for the electron, corresponding to the Bohr-Sommerfeld model for $m = \pm l$. In contrast, the θ^2-functions for $m = 0$ are oriented mainly along the z-axis [cf. H. E. White, "Introduction to Atomic Spectra," p. 63. McGraw-Hill, New York, 1934; L. Pauling and E. B. Wilson, "Introduction to Quantum Mechanics," McGraw-Hill, New York, 1935, p. 149].

(2) the azimuthal quantum number l (from the θ-function),

$$l = 0, 1, 3, 3, ..., n - 1 \qquad (4.19)$$

(3) the magnetic quantum number m (from the ϕ-function),

$$m = 0, \pm 1, \pm 2, \pm 3, ..., \pm l \qquad (4.20)$$

These three numbers determine the electron wave functions of their correlated orbits.

The electron can first of all assume any value n, and this decides the total energy in first order (as shown in Fig. 4.1). For a given n, the electron orbit may have various azimuthal quantum numbers l between 0 and $n - 1$. We may visualize states with different values l as differing in the shape of the orbit (for $l = 0$ an elongated ellipse, for $l = n - 1$ nearly a circle). An angular momentum, designated by the vector* l, has to be assigned to the quantum number l. Quantum mechanics (Weyl[10]) yields the result that the amount l of the angular momentum is determined by

$$|l| = \sqrt{l(l + 1)} \, \frac{h}{2\pi} \qquad (4.21)$$

The effect of the values for l on total energy is quite small in hydrogen, in the order of $1/1000$ of the steps resulting from different n-values. The fine structure indicated in Fig. 4.1 for two degrees of freedom is, therefore, vastly exaggerated, if it should be applied to this case. Finally, the quantum number m for the function ϕ can assume the $(2l + 1)$ values between $-l$ and $+l$. This can be interpreted as a tilting of the electron orbit by quantized amounts. Such orientation is irrelevant for energy considerations, if no fields are present; the levels for $l \neq 0$ are, therefore, degenerate. However, in the presence of a magnetic field, the electron orbit will describe a precession around the field, similar to the motion of a spinning top under gravitational attraction [Fig. 4.5(a)]. It will be discussed in detail later that a magnetic moment μ is associated with the mechanical momentum l of the orbiting electron since, at least from the classical viewpoint, the rotating charge constitutes a current loop. Thus the precession is a result of the interaction between the

---

* By convention, momentum vectors corresponding to the quantum numbers l, s, j, etc., introduced here or in later sections, are indicated by the respective boldfaced letter, l, s, j, etc., and their scalar magnitudes are written as $|l|$, $|s|$, and $|j|$. However, consistent with the notation for scalars elsewhere in this text, the magnitude of the magnetic moment vector μ is shown as μ.

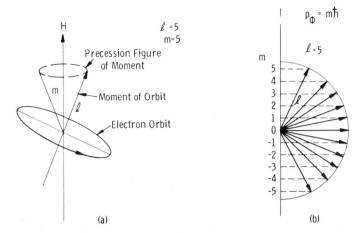

$P_\Phi = m\hbar$

FIG. 4.5. (a) Precession of the electron orbit around the magnetic field direction according to the Bohr-Sommerfeld picture. In this model, the electron describes a current loop which has a mechanical momentum l and a magnetic moment μ. Classically, such a structure precesses around the field axis at any fixed angle. On the atomic scale, the structure can assume only $2l + 1$ quantized positions (b), designated by magnetic quantum numbers $m = 0, \pm 1, \pm 2, ..., \pm l$.

magnetic moment and the magnetic field. The interaction energy is proportional to these two magnitudes, but also depends on the tilt angle between them which can have $2l + 1$ different values. This is the physical mechanism by which the original degeneracy is lifted. The various values m then give rise to quantized "component" mechanical moments p_m which, consistent with the original postulates of the Bohr theory, are given by

$$p_m = m \cdot \frac{h}{2\pi} \tag{4.22}$$

The $(2l + 1)$ possible orientations and resulting magnetic quantum numbers according to the Bohr-Sommerfeld model are shown in Fig. 4.5(b). While in this older presentation the size and shape of the electron orbit as such is completely described by the quantum numbers n and l, it is found in wave mechanics that the electron density distribution is affected by m in shape as well as in angular orientation.

In summary, the number of electronic states for a level with given n and l—its statistical weight—equals $2l + 1$. In hydrogen, the levels of the same total quantum number n lie so close together that the corresponding fine structure usually is not resolved. The intensities of spectral lines (see Section 3.3.3) are then proportional to a total statistical weight g_n of a level n:

$$g_n = 2 \sum_{l=0}^{n-1} (2l + 1) = 2n^2 \qquad (4.23)$$

where the factor of 2 results from the two possible orientations of the electron spin to be discussed in Section 4.1.5.

4.1.3 Term Schemes and Designations

A tradition from the early days of line series allocation has established the following method of designating individual electronic orbits: a number followed by a letter symbolizes the principal quantum number n, and the angular number l, respectively. The letters s, p, d, f ... stand for $l = 0, 1, 2, 3$ ..., respectively (originally they indicated *sharp*, *principal*, *diffuse*, and *fundamental* series of the alkali spectra). We thus have for the lowest terms the designations indicated on the left of Fig. 4.6.

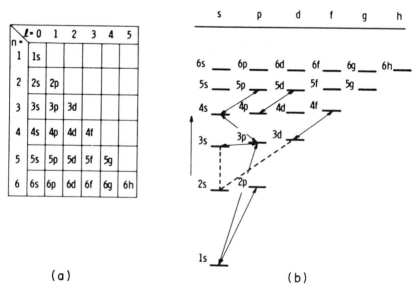

(a)   (b)

FIG. 4.6.   (a) Term designations of electron functions with various quantum numbers n and l. (b) Scheme of energy levels corresponding to the indices as given in the table. Examples of transitions which are permitted by selection rules are indicated by arrowed lines, while some forbidden transitions are shown in dashes.

For example, a 3*d*-electron is in an orbit with $n = 3$ and $l = 2$. The corresponding energy values are presented conventionally in series of increasing n, each series having a constant value of the quantum number l. The lowest quantum numbers determine the lowest energy values; hence the vertical order of the table is inverted in the term scheme as shown on the right of Fig. 4.6.

4.1.4 SELECTION RULES

The assignment of measured spectral lines to transitions between energy levels is vastly simplified by certain well-defined restrictions among the possible combinations. It has been found experimentally— certain exceptions notwithstanding—that only such transitions occur by emission or absorption in which l changes either by $+1$ or by -1. However, no restrictions apply to the change in n. Thus there result the selection rules

$$\Delta n = 0, 1, 2, 3, \ldots \tag{4.23}$$

$$\Delta l = \pm 1 \tag{4.24}$$

Examples for permitted transitions are indicated in Fig. 4.6 by arrowed lines whereas examples of combinations violating the selection rules are presented by dashed lines. Theoretically the selection rules can be derived from the fact that the transition matrix for changes of the electric dipole moment (cf. Section 3.3.2.1) vanishes for $\Delta l \neq \pm 1$. Nevertheless, transitions involving a change of the quadrupole moment (see Section 3.1.1) or magnetic dipole radiation may make a finite, although necessarily very small contribution. Thus the selection rules may be violated so that so-called *forbidden lines* can be observed under exceptional conditions, such as in interstellar gases, at very small intensities. This will be discussed further in Section 4.4.

4.1.5 HYDROGENLIKE ELECTRON STATES INCLUDING SPIN

We have defined a hydrogenlike atom as any configuration in which a single electron is under the influence of a central Coulomb field produced by a shielded, or unshielded, nucleus. This includes the alkali metals which consist of closely bound "shells" of electrons shielding an additional more loosely bound electron from all but the Coulomb field of a single positive charge.

A property of all hydrogenlike spectra—and one particularly pronounced in those of the alkali spectra—is that certain line series exhibit
doublet fine structure. In other words, the lines appear in pairs with a
wavelength interval in the order of angstroms or less. In hydrogen, the
largest doublet spacing only amounts to the order of 0.1 Å so that its
separation requires a spectral resolution of about 10^5 (such as available
from interferometers). On the other hand, the strongest and first excited
transition of the sodium spectrum (by definition, the resonance line,
which gives sodium light its color) is a doublet 6 Å apart, the so-called
D-lines, which are easily resolved in most prism spectrographs.

As a result, it is necessary to assign doublet structure to the term
scheme of hydrogenlike atom configurations, such as shown in Fig. 4.7
for sodium. All terms except those for orbits with $l = 0$ are found to
consist of two levels. Thus lines resulting from combinations between
the s-series and other terms are doublets. However, other transitions
such as between p- and d-orbits should have greater complexity than
doublet structure. Interferometric resolution shows this to be the case;
but because the splitting in one of the doublets is always predominant,
the essential line structure remains that of a doublet.

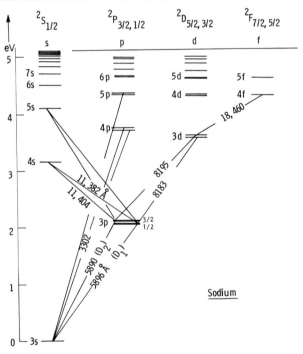

Fig. 4.7.   Term scheme and spectral lines of sodium.

The three-dimensional orbit model and the three-dimensional Schrödinger equation of the radiating electron in hydrogen are not capable of providing an explanation for the fine structure without further refinements. As mentioned in Section 4.1, this modification consisted of the assumption that the electron has a quantized spin around its own axis (Uhlenbeck and Goudsmit[1, 2]). We assign the quantum number* s to the spin. In order to account for doublet structure we must set $(2s + 1) = 2$, just as $(2l + 1)$ possible energy levels result from the interaction of the orbital moment with a magnetic field. The quantum number s of spin must therefore have the constant value of $\frac{1}{2}$. Figure 4.8 shows a model representation of the spin as angular momen-

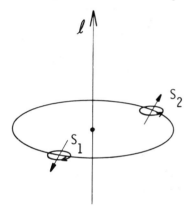

FIG. 4.8. The two possible positions of electron spin in the same orbit. In helium configuration, both positions are occupied.

tum vector **s** which can assume the two positions shown, "parallel" or "antiparallel" to the orbital momentum vector. Thus orbits with quantum number l and spins s can combine to form a resultant momentum of quantum number j which can have two values, if $l > 0$:

$$j = l \pm s. \tag{4.25}$$

The mechanism by which spin and orbit interact to generate different energy values is similar to that which governs the interaction between orbit and magnetic field. Again, the classical viewpoint provides a plausible model. Both the orbit and the spin of the electron produce magnetic fields or, more precisely, magnetic dipole moments. Thus

---

* The symbol s is used in spectroscopy both for the term series with $l = 0$ and for the quantum number of spin since the meaning follows from the connection.

forces are set up between the spin of the electron and its orbital moment and, as a result, each performs a precession around their resultant which we designate by the vector **j**. While the quantum numbers are combined algebraically as in Eq. (4.25), the corresponding moments add up vectorially. This situation is shown in Fig. 4.9 from which the geometric relation follows:

$$\mathbf{j}^2 = \mathbf{l}^2 + \mathbf{s}^2 - 2\mathbf{l}\mathbf{s}\,\cos(\mathbf{l}\mathbf{s}) \tag{4.26}$$

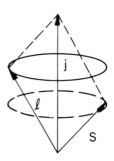

FIG. 4.9.   Vector model for spin-orbit coupling of an electron.

Each of the vectors **l**, **s**, and **j**, respectively, is representative of an angular moment of which the magnitude is given by the solution of the applicable wave equation, viz.:

$$|\,\mathbf{l}\,| = \sqrt{l(l+1)} \cdot \hbar \tag{4.27}$$

$$|\,\mathbf{s}\,| = \sqrt{s(s+1)} \cdot \hbar \tag{4.28}$$

$$|\,\mathbf{j}\,| = \sqrt{j(j+1)} \cdot \hbar \tag{4.29}$$

These equations establish the correlation between angular moment, their vectors, and the corresponding quantum numbers according to the quantum mechanical model. The interaction between spin and orbit vanishes for $l = 0$, as mentioned before. Thus the p-, d-, and f-terms of the hydrogen like spectra are doublets, but the s-terms remain single levels.

4.1.5.1 *Term Notation for Atomic Levels (LS-Coupling)*

To designate an atomic state resulting from spin and orbit contributions one uses by convention capital letters with a system of subscripts and superscripts. Thus S, P, D, F, G, ..., represent resultant orbital quantum numbers of 0, 1, 2, 3, 4, ..., repectively, which in the case of hydrogenlike atoms are given by the orbital states s, p, d, f, g, ..., of the single electron. A superscript on the left of the letter indicates the value $(2S + 1)$, i.e., the multiplicity of the term due to possible orientations of the resultant spin S. Thus a one-electron system $(S = \frac{1}{2})$ has a multiplicity 2 (degenerate for $L = 0$), and the designation ^2P, for example, indicates a term with $L = 1$ and $S = \frac{1}{2}$. L and S can combine to various resultants J, indicated by a subscript on the right of the letter. Thus $^2P_{3/2}$ shows a level with $J = \frac{3}{2}$; $^2P_{1/2}$ a $J = \frac{1}{2}$; and $^3F_2$ a level with $L = 3$, $S = 1$, and $J = 2$. The possible maximum number of levels J within a term is $(2S + 1)$; but if $L < S$, the term has only $(2L + 1)$ levels.

The complete term description has to include the principal quantum number. Thus one writes for $n = 4$, e.g., $4p\,^2P_{3/2}$ or simply $4^2P_{3/2}$. The system of term notations has its greatest usefulness for multielectron spectra which will be discussed later. It is in these cases that resultant values of L and S may be different from the quantum numbers l and s of the individual electron orbits.

4.1.6 Magnetic Moments and Term Splitting

The potential energy of an atomic state may be attributed to two major types of interaction: (1) the electrostatic interaction of electron charges in a Coulomb potential V and (2) the magnetic interaction of spin or orbital moments with each other or an external magnetic field. The first type has been treated quantitatively for hydrogen using the Schrödinger equation (4.1) as a starting point. The second type will be briefly discussed on the basis of classical arguments since these are substantiated by quantum mechanical calculations.

4.1.6.1 *Magnetic Moments of Electron Spin and Orbit*

From a classical viewpoint, an electron orbiting with angular frequency $\omega = 2\pi\nu$ represents a circular current of intensity $i = -ev = -e\omega/2\pi$. It is also known from the theory of electromagnetism that a current in a circular wire loop of radius r produces a magnetic moment

$$\mu = \pi r^2 i/c \tag{4.30}$$

Thus an electron with an orbital momentum vector $l = m_e r^2 \omega$ gives rise to an angular magnetic moment

$$\mu_l = -\frac{e\omega}{2\pi}\frac{\pi r^2}{c} = -\frac{e}{2m_e c}l \tag{4.31}$$

$$\mu_l = \frac{e}{2m_e c}|l| = \frac{eh}{4\pi m_e c}\sqrt{l(l+1)} \tag{4.31a}$$

The value

$$\mu_B = \frac{eh}{4\pi m_e c} = 9.271 \times 10^{-21} \quad \text{erg/gauss} \tag{4.32}$$

represents the basic atomic unit of the magnetic dipole moment and is called the *Bohr magneton*.

The ratio of magnetic moment to mechanical momentum for the electron spin is twice that of the orbit. This follows from spectroscopic and other experimental evidence (Stern-Gerlach experiment) as well as from quantum mechanics. We thus have

$$\mu_s = -\frac{e}{m_e c} \mathbf{s} \tag{4.33}$$

Because $s = \frac{1}{2}$, the magnetic moment of the electron spin is in the order of one magneton,

$$\mu_s = \frac{eh}{2\pi m_e c}\sqrt{\tfrac{1}{2}(1 + \tfrac{1}{2})} = \sqrt{3}\,\mu_B \tag{4.34}$$

4.1.6.2 *The Normal Zeeman Effect*

The values for the magnetic moments, Eqs. (4.31) and (4.34), determine now the various energies of interaction. For instance, the scalar product of the moment vector μ_l and the field $\mathbf{H}$ represents the corresponding energy of precession. Thus the interaction energy of the angular momentum is

$$E_{l,H,m} = (\mu_l \mathbf{H}) = \frac{eh}{4\pi m_e c} mH \tag{4.35}$$

where the magnetic quantum number m has the meaning defined by Eq. (4.22) or Fig. 4.5. A kinetic energy of precession is indeed to be expected on classical grounds. When a magnetic field H is applied to an orbiting electron or, more generally, to an atomic system, the so-called *Larmor* precession frequency ν_L will be superimposed giving rise to such phenomena as diamagnetism or line splitting:

$$\nu_L = \frac{e}{4\pi m_e c} H \tag{4.36}$$

The origin of the Larmor frequency is explained easily for the case of an electron which describes a circular orbit of radius r perpendicular to the magnetic field. As the field is slowly increased from zero to H, an electric field E is produced along the orbit according to Maxwell's equation

$$\text{curl } \mathbf{E} = -\frac{1}{c}\frac{d\mathbf{H}}{dt} \tag{4.37}$$

Since with Stoke's theorem

$$\oint \mathbf{E}dr = 2\pi r E = \iint \text{curl } \mathbf{E}\ df = -\pi r^2 \frac{1}{c}\frac{dH}{dt} \tag{4.38}$$

a field $E = -(r/2c)(dH/dt)$ will change the velocity v along the orbit and with it the angular frequency $\omega = v/r$. We thus have

$$d\omega = \frac{1}{r}dv = -\frac{1}{r}\frac{eE}{m_e}dt = \frac{1}{2}\frac{e}{m_e c}dH \tag{4.39}$$

From this follows the superposition of an angular Larmor frequency

$$\omega_{\text{L}} = \frac{e}{2m_e c}H \tag{4.40}$$

as in Eq. (4.36). The sense of this additional motion of the negative electron is such that it always forms a right-hand screw system with the field H and thus may add or substract from the zero field frequency ν_0.

Similar considerations apply to more complicated atomic systems. Thus observing radiating atoms in the direction of an applied magnetic field one finds the line split into $\nu_0 + \nu_{\text{L}}$ and $\nu_0 - \nu_{\text{L}}$. Oscillations in the direction of the magnetic field are not affected by the field; observations in the *transverse* direction reveal, therefore, the undisturbed frequency in addition to the two satellites (obviously, ν_0 does not appear in longitudinal direction since the latter coincides with the dipole axis). Line pattern changes of this kind—doublets in the longitudinal direction, triplets in transverse observation—are called, after their discoverer,[11] the *normal Zeeman effect*. The line shift ν_{L} as given by Eq. (4.36) can, in fact, be obtained from Eq. (4.35), if one sets

$$\frac{E_{m+1} - E_m}{h} = \nu_L = \frac{e}{4\pi m_e c}H. \tag{4.41}$$

Since, according to the correspondence principle, classical and quantum theoretical results must agree for large quantum numbers, we arrive at the selection rules that in transitions between magnetically oriented states the magnetic component of the momentum can change by $h/2\pi$ or not at all. The selection rules for the magnetic quantum number are then:

$$\Delta m = 1, 0, -1. \tag{4.42}$$

4.1.6.3 *Spin and Orbit Interaction With Weak Magnetic Fields*

Up to now we have considered the magnetic interaction energy of only a single magnetic moment, viz., that associated with the electronic orbital motion. The interaction between the resultant of spin and orbit momentum with the magnetic field is somewhat more complicated. At small fields, there results the picture that orbit *l* and spin **s** precess around their resultant **j**; in turn, then, **j** precesses around the magnetic field. Of importance for the interaction energy are, of course, the magnetic moments and their vectorial addition resulting from the precession around **j** (see Fig. 4.10). Thus we consider in this model a resultant magnetic moment μ_j which in analogy to Eqs. (4.31) and (4.34) is expressed in terms of a Bohr magneton:

$$\mu_j = -g \frac{e}{2m_e c} \mathbf{j} \tag{4.43}$$

$$\mu_j = g \mu_B \sqrt{j(j+1)} = g \frac{eh}{4\pi m_e c} \sqrt{j(j+1)}$$

Here the so-called *Landé factor g* follows from the geometric relationships (cosine law) implicit in Fig. 4.9 and the values for μ_s and μ_l . The detailed calculation yields

$$g = \left(1 + \frac{\mathbf{j}^2 + \mathbf{s}^2 - l^2}{2\mathbf{j}^2}\right) = 1 + \frac{j(j+1) + s(s+1) - l(l+1)}{2j(j+1)} \tag{4.44}$$

We obtain thus for the interaction energy between the spin-orbit resultant and the magnetic field:

$$E_{j,H,m} = (\mu_j \mathbf{H}) = g \frac{eh}{4\pi m_e c} mH \tag{4.45}$$

where the magnetic quantum number *m* represents the product $\sqrt{j(j+1)} \times \cos(\mathbf{j}\mathbf{H})$ and can assume $2j + 1$ integer values according to the possible orientations between **j** and **H**. The selection rules, Eq. (4.42), apply here also; nevertheless, the Zeeman patterns for $g > 1$ have more fine structure lines and are more complex than the normal Zeeman triplets for $g = 1$. The shift from the zero-field frequency for any allowed transition between one of the $2j_2 + 1$ upper levels and one of the $2j_1 + 1$ lower levels for $m = \pm 1$ is, according to Eq. (4.45),

$$\Delta\nu = \pm g \frac{eH}{4\pi m_e c} \tag{4.46}$$

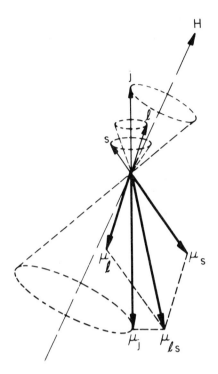

FIG. 4.10. Vector model for spin and orbit interaction with weak magnetic field. Magnetic moments of spin and orbit of negative electron are shown in opposite direction to that of corresponding mechanical momentum vectors. Because of the difference in the individual g-factors, the resultant μ_{ls} of the magnetic moments has a direction different from j, the resultant of the mechanical momenta. The projection of μ_{ls} on the j-direction, i.e., the vector μ_j, precesses around the field **H**.

4.1.6.4 Paschen-Back Effect

When the external magnetic field is so large as to compete with the internal fields of spin and orbital magnetic moments, the coupling between the latter two will weaken and yield to an individual coupling with the external magnetic field.[12] In the limit of high fields, the moments of **s** and *l*, instead of forming a resultant **j**, precess individually around the field direction. The total magnetic energy of interaction then is the algebraic sum of spin and orbit contributions. Applying Eqs. (4.31), (4.33), and (4.35), we obtain

$$E_{l,s,H} = (m_l + 2m_s)\frac{eh}{4\pi m_e c} H \qquad (4.47)$$

In a comparison with the zero-field energy, a small correction has to be taken into account for the interaction between **s** and *l*.

4.1.6.5 Doublet Interval

The interaction between spin and orbit, which causes the fine structure splitting, may be viewed in semiclassical manner as a special case of Zeeman effect. The orbit of the electron is then associated with a magnetic field with which the spin of the electron interacts by an energy of precession. The magnetic field is thus proportional to the angular momentum *l*, and it changes with nuclear distance and principal quantum number as $H \propto 1/r^3 \propto 1/n^3$. If the electrodynamic calculation is carried out, there results for the energy of an electron in the field of an effective nuclear charge Z_{eff} :

$$E = E_{nl} + \frac{\pi^2 e^6}{h^2 c^2} \frac{Z^4_{eff}}{r_1 n^3} \frac{j(j+1) - l(l+1) - s(s+1)}{l(l+\frac{1}{2})(l+1)} \quad (4.48)$$

On the right of Eq. (4.48), E_{nl} represents the first-order term for the energy of an electron with quantum numbers n and l, whereas the second term constitutes the interaction energy between spin and orbit. The total momentum quantum number j can assume two values, $l + \frac{1}{2}$ and $l - \frac{1}{2}$, depending on the orientation of the spin of the electron with respect to its orbit. The two values give rise to the doublet-fine structure of one-electron systems. The separation of the two levels on the wave number scale follows as the difference between the fine structure terms from Eq. (4.48) after numerical substitution of the Bohr radius r_1 and the other constants:

$$\Delta \tilde{\nu} = 5.82 \frac{Z^4_{eff}}{n^3 l(l+1)} \quad (\text{cm}^{-1}) \quad (4.49)$$

For hydrogenlike configurations of the type H, He⁺, Li⁺⁺ etc., Z_{eff} is simply the nuclear charge of 1, 2, 3, respectively. In other cases, in which the electron is shielded from the nuclear charge Ze by shells of electrons (see Section 4.3), Z_{eff} is smaller than Z by an amount corresponding to the effective number of screening electrons which partially compensate the positive nuclear field. In fact, Eq. (4.49) is valid even for those electron orbits which "penetrate" deeply into the core of multielectron configurations provided one sets

$$Z^4_{eff} = Z_0^2(Z_0^2 + Z_i^2) \quad (4.50)$$

where Z_i and Z_0 are the weighed numbers of effective charges that are encountered in the inner and outer part of the orbit, respectively.

For cases in which Z_{eff} is known, Eq. (4.49) proves to be quite accurate. The doublet interval for the lowest values of Z_{eff}, n, and l (i.e., $Z = 1$, $n = 2$, $l = 1$), is seen to amount to 0.36 cm^{-1} or about 10^{-5} of the first order differences in the term E_n which correspond to the lines of the Balmer series. The term splitting decreases with the third power of increasing n and approximately with the second power of l. This is the reason why the lowest p-states exhibit a wider splitting than the other terms so that the transitions from higher to lower states, as mentioned before, have mainly doublet structure. Equation (4.49) also shows that the doublet spacing is proportional to the fourth power of the unscreened center charge. By comparison, the energy gap between terms with various principal numbers n in the sequence H, He$^+$, Li^{++}, etc., increases only as Z^2 [cf. Eq. (3.40)]. The doublet structure is, therefore, even on a relative scale more pronounced for the higher states of ionization than for neutral atoms, so that it was found in the He$^+$ spectrum before it was seen in hydrogen. Similarly, the doublet splitting becomes steadily coarser in the sequence of the neutral alkali atoms with increasing atomic number such that for the resonance line in lithium $\Delta \tilde{\nu} = 0.338 \text{ cm}^{-1}$ or $\Delta \tilde{\nu}/\tilde{\nu} = 2.27 \times 10^{-5}$, whereas in cesium $\Delta \tilde{\nu} = 554 \text{ cm}^{-1}$ or $\Delta \tilde{\nu}/\tilde{\nu} = 4.8 \times 10^{-2}$.

4.2 Nuclear Properties and Hyperfine Structure

In addition to the second-order effect of the electronic spin-orbit coupling, there usually exists a third-order effect owing to the interaction between the electron and the nuclear spin and mass. This reveals itself spectroscopically by a hyperfine structure (hfs) which is superimposed in some lines over their fine structure. The nucleus, as differentiated from a point charge eZ, can perturb electronic spectra in two ways. First, its mass—although it is always large compared with that of the electron—influences slightly the electron orbit around the common center; or from an alternative viewpoint, it enters into the reduced electron mass m_r. Hence, given electron configurations around nuclei of different mass, as they in fact occur in elements with mixed isotopes, will have slightly different energies. A second type of perturbation arises, if the nucleus, composed as it is of protons and neutrons, has a resultant spin and magnetic moment or—because of the spatial distribution of its constituents—also an electric quadrupole moment. Such moments will interact with the magnetic field which the electron produces by virtue of its orbit and spin just as these latter two interact with each other. These effects resulting from the existence of isotopes and nuclear moments are usually much smaller than the electronic spin-orbit splitting for a reason which, as we shall proceed to show, follows basically from the large mass ratio M_n/m_e of nucleus and electron.

4.2.1 ISOTOPE SHIFTS

The hyperfine structure resulting from the ordinary isotope effect is given simply by the Bohr theory which states, according to Eqs. (3.39)–(3.42), that the Rydberg constant and hence the term energy is proportional to the reduced electron mass m_r. Thus the ratio of energies E_1 and E_2 for corresponding levels in isotopes with masses M_1 and M_2 equals

$$\frac{E_2}{E_1} = \frac{m_{r2}}{m_{r1}} = \frac{1 + m_e/M_1}{1 + m_e/M_2} = 1 + \left(\frac{m_e}{M_1} - \frac{m_e}{M_2}\right) \qquad (4.51)$$

The spectra of two isotopes are therefore shifted with respect to each other by the fractional amount

$$\frac{\Delta \tilde{\nu}}{\tilde{\nu}} = \frac{\Delta E}{E} = \frac{E_2 - E_1}{E_1} \cong m_e \frac{M_2 - M_1}{M_2 M_1} \qquad (4.52)$$

It follows that this isotope effect decreases approximately as $1/Z^2$. The isotope mixture of light and heavy hydrogen ($M_2 : M_1 = 2$) shows a fractional line shift of $\frac{1}{2}m_e/M_1 = 1/3657$ (cf. Rydberg constants in Section 3.2.2). This is a relatively large value and, in fact, larger than the fine structure of hydrogen as indicated, by way of example, in Fig. 4.11(b) for the first line of the Lyman series ($2p\ ^2P_{\frac{1}{2},\frac{3}{2}} \rightarrow 1s\ ^2S_{\frac{1}{2}}$). Since spin-orbit coupling increases with a higher power of Z, it quickly surpasses the nuclear effects for the higher elements. Thus the isotope shift $\Delta \tilde{\nu}/\tilde{\nu}$ for $Z = 19$ (K^{39}, K^{41}) is in the order of 10^{-6} whereas the resonance line consists of a doublet at 7665 and 7699 Å with a $\Delta \tilde{\nu}/\tilde{\nu}$ of 4.4×10^{-3} so that the fine structure is here by a factor of about 10^4 larger than the hyperfine structure (see Fig. 4.13).

We mention briefly that there exist other, and more complicated, isotope effects (Kopfermann,[13] p. 161 ff). For example, in multielectron spectra the mass of the nucleus affects all the orbits and therefore their interaction, giving rise to a so-called *coupling* effect. Another type of hfs occurs in heavy elements as the result of nuclear differences, not of mass, but of size: the Coulomb field originates from an extended positive charge distribution, rather than a point source, so that variations of such distribution lead to somewhat different energies of interaction with the electrons.

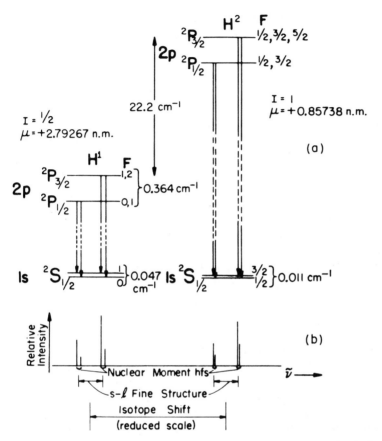

—Fig. 4.11. Isotope shift, fine structure, and hyperfine structure in the hydrogen isotopes H^1 and H^2: (a) Hfs of ground level ($1s\ ^2S_{1/2}$) and fine structure of second level ($2p\ ^2P_{1/2,3/2}$). Isotope shift, shown here reduced, is in hydrogen much larger than the fine structure intervals. Hyperfine splitting is ignored for the ^2P levels, but it is appreciable in the ^2S term where it has been very accurately determined by radio-frequency measurements [P. Kusch, *Phys. Rev.* **100**, 1188 (1955)]. (b) Structure of line (first of Lyman series) calculated from transition between the two levels in (a). Isotope splitting can be readily observed with spectrographs, whereas fine structure separation requires resolution $\nu/\varDelta\nu$ of about 10^5. Hfs ($\nu/\varDelta\nu \approx 10^6$) is obscured by the relatively large Doppler effect in hydrogen. Equal parts of H^1 and H^2 are assumed.

4.2.2 Nuclear Moments and Hyperfine Structure

Magnetic moments result from a nuclear spin $\mathbf{I} = \sqrt{I(I+1)}\hbar$ in analogy to the electron case. There are, however, two important differences: (1) The relationship between magnetic moment and mechanical momentum is modified by a g-factor reflecting the nuclear composition of protons and neutrons; and (2) the magnetic moment μ_n per nucleon must be in the order of

$$\mu_n = \frac{eh}{4\pi M_p c} = \frac{m_e}{M_p} \mu_B \qquad (4.53)$$

Equation (4.53) has been formulated in analogy to Eq. (4.32) and thus defines a *nuclear magneton* μ_n (or n.m) on the basis of a spinning nucleon with positive charge e and proton mass M_p. The value of μ_n is $1/1838$ that of μ_B, 5.0493×10^{-24} erg/gauss. We have then for the total magnetic moment of a nucleus:

$$\mu_I = g_I \frac{eh}{4\pi M_p c} \sqrt{I(I+1)} = g_I \mu_n \sqrt{I(I+1)} \qquad (4.54)$$

Quantitatively, the analogy to the spinning electron can be expected to hold only by order of magnitude. The moment of the proton, with $I = \frac{1}{2}$, was found experimentally to have the value $+2.793$ n.m. rather than simply 1 n.m. Furthermore, the neutron has a negative magnetic moment, $\mu_N = -1.913$ n.m. As protons and neutrons are joined in the structures which form the nuclei of the isotopes of increasing atomic weight A in the periodic table, certain regularities of the nuclear ground states are noticeable: spins of *even-even* nuclei, i.e., with even number Z of protons and even number $(A - Z)$ of neutrons, are zero; *odd-even* and *even-odd*, i.e., odd-A nuclei, have half-integer spins (from $I = \frac{1}{2}$ to $\frac{9}{2}$ have been observed); and odd-odd nuclei yield integer spins (from 0 to 8). All this implies the algebraic addition of positive and negative spin quantum numbers with complete or partial cancellation by pairing of opposite spins. As a result μ_I in Eq. (4.54) does not exceed several nuclear magnetons in order of magnitude.

We now turn to the interaction energy between the nuclear moment μ_I and the magnetic field at the nucleus produced by spin and orbit of the electrons. The latter two are represented by a resultant angular momentum vector **J**, and we may visualize the interaction as a precession of the nuclear spin **I** and the total electronic momentum **J** around a resultant momentum vector **F** (cf. Fig. 4.12). The angle between **I** and **J** is again quantized; there can exist only $2I + 1$ or $2J + 1$ orientations, whichever is the smaller. Thus for $I < J$, the multiplicity of the hfs splitting, which reveals itself by the number of hfs lines, allows the determination of the nuclear spin I. In analogy to Eq. (4.35) the interaction energy is then

$$E = -(\mathbf{\mu}_I \cdot \mathbf{H}) = -|\mathbf{\mu}_I| \cdot |\mathbf{H}| \cos(\mathbf{\mu}_I \mathbf{H}) \qquad (4.55)$$

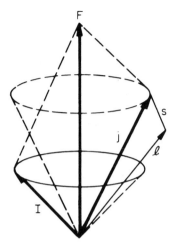

Fig. 4.12. Vector model for nuclear spin interaction with resultant of spin and orbit of electron.

where

$$\cos(\mathbf{\mu}_I \mathbf{H}) = \frac{\mathbf{F}^2 - \mathbf{I}^2 - \mathbf{J}^2}{2\mathbf{I}\mathbf{J}} = \frac{F(F+1) - I(I+1) - J(J+1)}{2IJ} \quad (4.56)$$

For a comparison of nuclear spin interaction with electron spin-orbit energy (which amounts to a comparison of hfs with fine structure), one must form the ratio of the respective spin magnetic moments. By order of magnitude, this ratio is given by that of the reciprocal masses, M_e/M_p. Hence, other things being equal, the nuclear spin effects are typically 10^{-3} as large as the electronic fine structure. In addition, however, the field H at the nucleus may be quite different by magnitude from that with which the electron spin interacts. A knowledge of the nuclear moments I and the field H at the nucleus permits a first-order approximation (neglecting, e.g., quadrupole moment contributions) of the hfs for all the levels with various F and J. From the transition between these levels, follow the position and combination of the hfs line components, subject to the selection rules

$$\Delta F = 0, \pm 1 \quad (\text{except } 0 \to 0) \quad (4.57)$$

The intensities of the various satellite lines are determined by the statistical weights $2F + 1$, just as the intensities of transitions between levels as a whole are given by $2J + 1$, in the absence of an external magnetic field.

An example for these relationships can be seen in Fig. 4.11 which represents the case of the hydrogen isotopes H^1 and H^2. The hfs of the ground state (which with $l = 0$ has no electron spin fine-splitting) amounts to 0.047 and 0.011 cm^{-1}, respectively. Because hydrogen has the lowest nuclear charge and mass, its spectral line pattern is unique in many respects. Thus the hfs of transitions to the ground state, although of typical size, cannot be resolved in the presence of Doppler broadening which is particularly large in hydrogen. The splitting* of the ground state could be measured, however, by beam deflection in the radio-frequency spectrometer (Kusch[14]). For the reasons outlined before, the isotope effect in hydrogen is in the order of 10^3 times larger than the structure due to the nuclear moments, although these two effects are usually of the same magnitude (0.010–0.100 cm^{-1}) in the resonance lines of most other spectra (cf. Fig. 4.13).

4.2.3 EXPERIMENTAL ASPECTS OF HYPERFINE STRUCTURE

The line splitting due to isotope shifts and nuclear moments is important from two points of view. First of all, hfs is of interest simply because it represents a limit to the monochromaticity of a given spectral line. More important, however, is the information which can be gained from the observed pattern regarding nuclear properties, especially spin and magnetic moments.

Hfs in the optical spectrum can be observed at all only if (a) the factors causing line broadening of the light source are sufficiently reduced, and (b) if the spectrometric equipment has the necessary resolution to measure, with the needed accuracy, values of $\Delta \bar{\nu}$ which may be from 10^{-1} to 10^{-2} cm^{-1} or lower.

As to line broadening, the requirements for the light source are often quite severe (cf. Section 3.4 and Table 3.5). In general, pressure and field broadening can be readily eliminated as a disturbing factor, albeit, at some expense to intensity. A much more tenacious limitation is the Doppler width, particularly, if atoms of low mass have to be excited at high temperatures because of a low vapor pressure. However, there exist two methods in which the Doppler broadening of *incoherent* light emission

---

* The transition between the hfs levels for $F = 0$ and $F = 1$ of the $1s\ ^2S_{1/2}$ ground state in H^1 is observed as the 21.1049 cm line in the radio astronomy of galaxies which contain hydrogen. Because this is a magnetic dipole transition at microwave frequencies, the lifetime of the upper state is 11×10^6 years. The line is broadened and shifted by Doppler effect so that conclusions are possible regarding temperature and velocity of the galaxies.

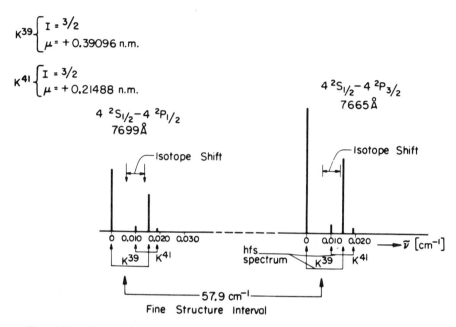

$K^{39}\begin{cases} I = 3/2 \\ \mu = +0.39096 \text{ n.m.} \end{cases}$

$K^{41}\begin{cases} I = 3/2 \\ \mu = +0.21488 \text{ n.m.} \end{cases}$

$4\,^2S_{1/2} - 4\,^2P_{1/2}$
7699Å

$4\,^2S_{1/2} - 4\,^2P_{3/2}$
7665Å

Isotope Shift

Isotope Shift

0 0.010 0.020 0.030

K^{39} K^{41}

0 0.010 0.020 ⟶ $\tilde{\nu}$ [cm⁻¹]

hfs
spectrum K^{39} K^{41}

⟵——— 57.9 cm⁻¹ ———⟶
Fine Structure Interval

FIG. 4.13. Hfs of the resonance line doublet in potassium [from the absorption measurements in atom beams by D. A. Jackson and H. Kuhn, *Proc. Roy. Soc.* **A165**, 303 (1939)]. Intensity ratios follow statistical weights (i.e., go as $2J + 1$ or $2F + 1$) and abundance ratio ($K^{39} : K^{41} = 14 : 1$). Isotope shift and hyperfine structure due to nuclear moments are seen to be of the same order, about 10^{-4} of fine structure interval.

is considerably reduced and which, therefore, have been applied in many hfs investigations:

(1) the *hollow cathode* of Schüler[15] which provides a field-free cavity within walls cooled by liquid nitrogen or hydrogen (see Fig. 4.14). A negative glow is maintained by means of a carrier gas such as helium or neon. Atoms from the specimen deposited on the walls are carried into the discharge by cathode sputtering and experience there a strong excitation by resonance transfer of energy.

(2) light emission from *atomic beams* perpendicular to the direction of observation according to Jackson and Kuhn[16] and Minkowski[17] (see Fig. 4.15). A detailed analysis shows that the distribution of the velocity component in the viewing direction corresponds to a substantially reduced equivalent temperature (a few degrees Kelvin).

It should be remembered, in this connection, that there exist emission processes which do not participate in Doppler broadening. This applies,

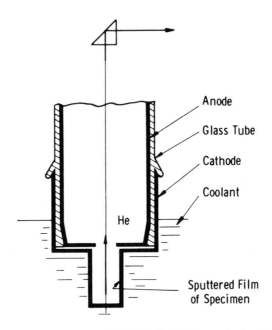

Anode

Glass Tube

Cathode

Coolant

He

Sputtered Film
of Specimen

Fig. 4.14.   Schematic diagram of the Schüler hollow cathode. The observed light originates in the negative glow region from sputtered atoms which are excited by collisions of the second kind with metastable atoms of the carrier gas. Positive column is suppressed by proximity between cathode and anode [see also H. Kopfermann, "Nuclear Moments" (transl. by E. E. Schneider). Academic Press, New York, 1958.] Low temperature and density of radiating atoms in field-free region results in narrow line width.

in particular, to coherence by stimulated emission and recoilless radiation (Mössbauer effect). Where applicable, these phenomena lend themselves particularly well to high resolution spectroscopy.

The value $\tilde{\nu}/\varDelta\tilde{\nu}$ ($= \lambda/\varDelta\lambda = Q'$) of the spectrograph must match that of the source. For prisms of base b and dispersion $dn/d\lambda$, the theoretical resolving power[18] equals $b \times dn/d\lambda$. Practical sizes and materials limit this value to about 10^4. Grating spectrographs and interferometers achieve considerably higher resolutions. An order of magnitude estimate can be arrived at by a simple argument. From the two definitions of the Q-value by Eqs. (1.9) and (1.14) it follows that an oscillation of halfwidth $\varDelta\tilde{\nu}$ may give rise to a wave train of a length equal to $Q/2\pi$ wavelengths. Therefore the resolving power of a spectrograph based on interfering light beams is, by order of magnitude, equal to the maximum number of wavelengths which it can bring to superposition (or more accurately, the phase difference which can be brought to interference). This leads to a theoretical resolution of $m \times N$ for a grating with N slits in the

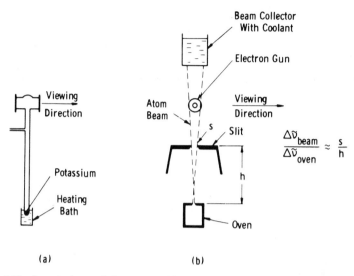

(a)   (b)

Fig. 4.15. Atomic beam light source. (a) Early type for absorption measurements, in potassium vapor by D. A. Jackson and H. Kuhn [*Proc. Roy. Soc.* **A148**, 335 (1935)] defines beam by means of a long tube. (b) Schematic diagram of a later type [W. Z. Paul, *Z. Physik* **117**, 774 (1941)]. Vapor is excited by electron bombardment perpendicular to viewing direction. Doppler width is reduced because only atoms with low velocities in the direction of viewing are passed by the slit.

mth order. For instance, a typical grating may operate with 100,000 slits in third order, resulting in a $\bar{\nu}/\Delta\bar{\nu}$ of 3×10^5. The most effective interferometer is the *Fabry-Pérot etalon* which consists of two glass or quartz plates arranged plane-parallel and partially reflecting at a fixed distance d from each other. When a light beam enters, it experiences multiple reflections. The path difference at normal incidence between a ray that just enters and one which has already undergone two reflections amounts to $2d/\lambda$ in wavelengths. To obtain the resolution, this factor has to be multiplied (in analogy to the number of grating slits) by the *effective number of beams* which represents in essence the number of reflections. The detailed calculation yields then for the resolving power:

$$\frac{\bar{\nu}}{\Delta\bar{\nu}} = \frac{2d}{\lambda} \frac{\pi\sqrt{R}}{(1-R)} \tag{4.58}$$

Here the reflectivity R may be about 0.95 so that with a d/λ, e.g., of 10^5, the resolution equals about 1.2×10^7.

Values of this order represent the limit of definition which has been possible with purely optical spectroscopy. The frequencies of coherent beams, however, can be determined considerably more accurately, at least on a relative scale, by means of *heterodyning* or beating such as by the process of photoelectric mixing. This will be discussed in some detail in Chapter 7.

The energy intervals between hfs levels of interest encompass altogether a frequency region of about 10^6–10^{10} sec^{-1}. Transitions between these levels (i.e., $\Delta F = \pm 1$, $\Delta n = 0$, $\Delta J = 0$) are to be observed, therefore, in the radio- and microwave-frequency region. Such direct measurements of nuclear moment interaction energy are, of course, much more accurate, if less versatile, than the optical methods which observe these effects as perturbations of higher order. Radio-frequency resonance methods on atomic beams for the determination of nuclear moments were first introduced by Rabi[19] and co-workers. This was followed by another important radio-frequency method, that of nuclear magnetic resonance (Bloch,[20] Purcell[21]) and of related resonance effects in solids and liquids. More recently, maser transitions[22] from magnetically or electrically selected spin states in atomic beams permitted a further increase of the accuracy with which nuclear dipole and quadrupole moments could be determined. In general, maser and laser technology is being applied for high definition spectroscopy. Also pumping methods (see Chapter 6) are being used to apply the radio-frequency methods to excited states which were previously inaccessible to them because of the low population ratio to the ground state.

4.3 Atomic Spectra for Multiple-Valence Electrons

4.3.1 ELECTRON CONFIGURATIONS OF THE ELEMENTS

Pauli's principle concerning the nature of electron wave functions was mentioned already in the introduction to this chapter. The part pertinent for our present discussion concerns the statement that in a given configuration space no two electrons can have the same set of quantum numbers. Many features in the buildup of the periodic system can be understood on the basis of this principle. We shall here, however, merely outline the features affecting the spectra of the more complex atoms.

An important result of Schrödinger's equation for hydrogenlike atoms was Eq. (4.19) to the effect that the orbital quantum number l can assume only values between 0 and $n - 1$. Therefore, electrons in the first orbit,

$n = 1$, must be s-electrons. Because of Pauli's principle, there can then be only two electrons in the orbit with $n = 1$,

$$n = 1, \quad l = 0, \quad s = \tfrac{1}{2} \tag{4.59}$$

and

$$n = 1, \quad l = 0, \quad s = -\tfrac{1}{2} \tag{4.60}$$

The element with two electrons, corresponding to an atomic number $Z = 2$, is helium. The configuration of two electrons described by Eqs. (4.59) and (4.60) and also Fig. 4.8 represents the ground state (i.e., the lowest state) of helium. The resultant L vector (the sum of the individual orbit vectors, indicated by capital letter L) is zero: $L = 0$. The same holds for the resultant spin vector, $S = 0$. Thus the ground state of helium is written as $1s^2\,{}^1S_0$, where the upper index of the orbit symbol s indicates that the atom has *two* $1s$ electrons.

Since the quantum numbers in Eqs. (4.59) and (4.60) represent a completed set for $n = 1$, the energy is a minimum, and the configuration of the ground state of helium is particularly strongly coupled and stable. We say that helium by virtue of its position in the periodic system concludes the first shell (the so-called K-shell, $n = 1$, see Table 4.1). According to the aforesaid, the next element, lithium with $Z = 3$ electrons, must begin the next shell (the L-shell, $n = 2$). The third electron, then, must in its lowest state assume a $2s$-orbit. There is, again, room for two electrons in the $2s$ state so that the next element, beryllium, has the configuration $1s^2\,2s^2\,{}^1S_0$ in its ground state. The quantum number $n = 2$ admits now an orbital number $l = 1$ (p-states). The magnetic quantum number m comes into play, because the l-vector can assume $2l + 1$ positions with respect to a magnetic field ($l = -1, 0, +1$). Since each of these may assume one of two spin positions, there can be altogether 6 p-electrons in the L-shell. Neon completes the L-shell with all possible $2s$- and $2p$-electrons in pairs of opposite spin. The next element is again an alkali metal, sodium, which contributes an electron to the third shell ($n = 3$, the M-shell), with an electron configuration $1s^2\,2s^2\,2p^6\,3s\,{}^2S_{1/2}$. Sodium is spectroscopically, as it is chemically, quite similar to lithium. The filled L-shell, with its spin and orbital momenta adding each up to zero, exerts little influence on the properties of the M-electrons. Quite generally the inertness of closed shells provides the reason for the periodicity of physical properties which the elements exhibit when arranged in rows of increasing atom number Z. In other words, it is mainly the valence electrons outside of filled shells which determine the behavior of the element. Table 4.1 lists the ground state configuration of the elements in the periodic system together with the ionization energies. Since the latter represent the work necessary for the removal of the outermost electron they are a measure of the stability of the configuration. A periodicity of the ionization potential with position of the element in the shell is again apparent.

From the element potassium ($Z = 19$) on upward, shells with higher n begin to be occupied before those with lower n are completely filled. Thus the N-shell acquires $4s$-electrons (as in potassium and calcium) while the $3d$-subshell is still empty. Then, with scandium ($Z = 21$), the remaining vacancies of the M-shell begin to be filled up with a series of so-called *transition metals*. These have in common that they contain partially filled d-shells and that they have, for this reason, a relatively large magnetic moment. Furthermore, owing to the fact that the important outermost shell is either a $4s$- or a $4s^2$ configuration, there is considerable physical and chemical likeness between some of these elements. A similar situation exists for the second and third series of transition metals, i.e., that between yttrium and cadmium, in which the $4d$-subshell is being filled up, and that between hafnium and mercury where the corresponding buildup of the $5d$-electrons occurs.

The delay in the filling of available levels is even more pronounced for f-electrons ($l = 3$) which occur first in the N-shell with $n = 4$. Here it is not until the $6s^2$ configuration has been formed (with lanthanum, $Z = 57$) that the $4f$-subshell comes into existence. There are altogether 14 elements, the so-called *rare earths* from cerium ($Z = 58$) to lutetium ($Z = 71$), which differ primarily only in the number of $4f$-electrons while the $6s^2$-subshell remains fixed.* Similarly, the delayed filling of the $5f$-electrons in the O-shell gives rise to a second rare-earth series between thorium ($Z = 90$) and lawrencium ($Z = 103$).

The spectroscopy of atoms and ions with unfilled inner shells has a number of particular features. For instance, if an outer electron is moved into an inner shell, a higher energy level results, although the transition is from higher to lower n. Because transitions are possible within the partially filled f- or d-configurations, the spectra are quite complex. It is a characteristic of these lines that they may be very sharp even in the presence of the strong local fields of crystals owing to the shielding effect of the outer electrons. For this reason, rare earth ions embedded in dielectric host materials offer important possibilities among the solid sources of light.

---

* The limits of the rare-earth series are not sharply defined by known physical properties. For example, one may consider La ($Z = 57$) as the first and Yb ($Z = 70$) as the last of the rare earths so that Lu ($Z = 71$) is the first of the $5d$-transition metals.

TABLE 4.1

ELECTRON CONFIGURATIONS AND GROUND TERMS OF THE ELEMENTS[a]

Period	Number and atom	K $l=0$	L $l=0$	1	M $l=0$	1	2	N $l=0$	1	2	3	O $l=0$	1	Ground state	Ionization energy (ev)
1	1 H	$1s$												$^2S_{1/2}$	13.60
	2 He	$1s^2$												$^1S_0$	24.58
2	3 Li	$1s^2$	$2s$											$^2S_{1/2}$	5.39
	4 Be	$1s^2$	$2s^2$											$^1S_0$	9.32
	5 B	$1s^2$	$2s^2$	$2p$										$^2P_{1/2}$	8.30
	6 C	$1s^2$	$2s^2$	$2p^2$										$^3P_0$	11.26
	7 N	$1s^2$	$2s^2$	$2p^3$										$^4S_{3/2}$	14.54
	8 O	$1s^2$	$2s^2$	$2p^4$										$^3P_2$	13.61
	9 F	$1s^2$	$2s^2$	$2p^5$										$^2P_{3/2}$	17.42
	10 Ne	$1s^2$	$2s^2$	$2p^6$										$^1S_0$	21.56
3	11 Na	$1s^2$	$2s^2$	$2p^6$	$3s$									$^2S_{1/2}$	5.14
	12 Mg	$1s^2$	$2s^2$	$2p^6$	$3s^2$									$^1S_0$	7.64
	13 Al	$1s^2$	$2s^2$	$2p^6$	$3s^2$	$3p$								$^2P_{1/2}$	5.98
	14 Si	$1s^2$	$2s^2$	$2p^6$	$3s^2$	$3p^2$								$^3P_0$	8.15
	15 P	$1s^2$	$2s^2$	$2p^6$	$3s^2$	$3p^3$								$^4S_{3/2}$	10.55
	16 S	$1s^2$	$2s^2$	$2p^6$	$3s^2$	$3p^4$								$^3P_2$	10.36
	17 Cl	$1s^2$	$2s^2$	$2p^6$	$3s^2$	$3p^5$								$^2P_{3/2}$	13.01
	18 A	$1s^2$	$2s^2$	$2p^6$	$3s^2$	$3p^6$								$^1S_0$	15.76
4	19 K	$1s^2$	$2s^2$	$2p^6$	$3s^2$	$3p^6$		$4s$						$^2S_{1/2}$	4.34
	20 Ca	$1s^2$	$2s^2$	$2p^6$	$3s^2$	$3p^6$		$4s^2$						$^1S_0$	6.11
	21 Sc	$1s^2$	$2s^2$	$2p^6$	$3s^2$	$3p^6$	$3d$	$4s^2$						$^2D_{3/2}$	6.56
	22 Ti	$1s^2$	$2s^2$	$2p^6$	$3s^2$	$3p^6$	$3d^2$	$4s^2$						$^3F_2$	6.83
	23 V	$1s^2$	$2s^2$	$2p^6$	$3s^2$	$3p^6$	$3d^3$	$4s^2$						$^4F_{3/2}$	6.74
	24 Cr	$1s^2$	$2s^2$	$2p^6$	$3s^2$	$3p^6$	$3d^5$	$4s$						$^7S_3$	6.76
	25 Mn	$1s^2$	$2s^2$	$2p^6$	$3s^2$	$3p^6$	$3d^5$	$4s^2$						$^6S_{5/2}$	7.43
	26 Fe	$1s^2$	$2s^2$	$2p^6$	$3s^2$	$3p^6$	$3d^6$	$4s^2$						$^5D_4$	7.90
	27 Co	$1s^2$	$2s^2$	$2p^6$	$3s^2$	$3p^6$	$3d^7$	$4s^2$						$^4F_{1/2}$	7.86
	28 Ni	$1s^2$	$2s^2$	$2p^6$	$3s^2$	$3p^6$	$3d^8$	$4s^2$						$^3F_4$	7.63
	29 Cu	$1s^2$	$2s^2$	$2p^6$	$3s^2$	$3p^6$	$3d^{10}$	$4s$						$^2S_{1/2}$	7.72
	30 Zn	$1s^2$	$2s^2$	$2p^6$	$3s^2$	$3p^6$	$3d^{10}$	$4s^2$						$^1S_0$	9.39
	31 Ga	$1s^2$	$2s^2$	$2p^6$	$3s^2$	$3p^6$	$3d^{10}$	$4s^2$	$4p$					$^2P_{1/2}$	6.00
	32 Ge	$1s^2$	$2s^2$	$2p^6$	$3s^2$	$3p^6$	$3d^{10}$	$4s^2$	$4p^2$					$^3P_0$	7.88
	33 As	$1s^2$	$2s^2$	$2p^6$	$3s^2$	$3p^6$	$3d^{10}$	$4s^2$	$4p^3$					$^4S_{3/2}$	9.81
	34 Se	$1s^2$	$2s^2$	$2p^6$	$3s^2$	$3p^6$	$3d^{10}$	$4s^2$	$4p^4$					$^3P_2$	9.75
	35 Br	$1s^2$	$2s^2$	$2p^6$	$3s^2$	$3p^6$	$3d^{10}$	$4s^2$	$4p^5$					$^2P_{3/2}$	11.84
	36 Kr	$1s^2$	$2s^2$	$2p^6$	$3s^2$	$3p^6$	$3d^{10}$	$4s^2$	$4p^6$					$^1S_0$	14.00
5	37 Rb	$1s^2$	$2s^2$	$2p^6$	$3s^2$	$3p^6$	$3d^{10}$	$4s^2$	$4p^6$			$5s$		$^2S_{1/2}$	4.18
	38 Sr	$1s^2$	$2s^2$	$2p^6$	$3s^2$	$3p^6$	$3d^{10}$	$4s^2$	$4p^6$			$5s^2$		$^1S_0$	5.69

TABLE 4.1 (*continued*)

Period	Number and atom	K $l=0$	L $l=0$	L 1	M $l=0$	M 1	M 2	N $l=0$	N 1	N 2	O $l=0$	O 1	Ground State	Ionization energy (ev)
5	39 Y	$1s^2$	$2s^2$	$2p^6$	$3s^2$	$3p^6$	$3d^{10}$	$4s^2$	$4p^6$	$4d$	$5s^2$		$^2D_{3/2}$	6.38
	40 Zr	$1s^2$	$2s^2$	$2p^6$	$3s^2$	$3p^6$	$3d^{10}$	$4s^2$	$4p^6$	$4d^2$	$5s^2$		$^3F_2$	6.83
	41 Nb	$1s^2$	$2s^2$	$2p^6$	$3s^2$	$3p^6$	$3d^{10}$	$4s^2$	$4p^6$	$4d^4$	$5s$		$^6D_{1/2}$	6.88
	42 Mo	$1s^2$	$2s^2$	$2p^6$	$3s^2$	$3p^6$	$3d^{10}$	$4s^2$	$4p^6$	$4d^5$	$5s$		$^7S_3$	7.13
	43 Tc	$1s^2$	$2s^2$	$2p^6$	$3s^2$	$3p^6$	$3d^{10}$	$4s^2$	$4p^6$	$4d^5$	$5s^2$		$^6S_{5/2}$	7.23
	44 Ru	$1s^2$	$2s^2$	$2p^6$	$3s^2$	$3p^6$	$3d^{10}$	$4s^2$	$4p^6$	$4d^7$	$5s$		$^5F_5$	7.36
	45 Rh	$1s^2$	$2s^2$	$2p^6$	$3s^2$	$3p^6$	$3d^{10}$	$4s^2$	$4p^6$	$4d^8$	$5s$		$^4F_{3/2}$	7.46
	46 Pd	$1s^2$	$2s^2$	$2p^6$	$3s^2$	$3p^6$	$3d^{10}$	$4s^2$	$4p^6$	$4d^{10}$			$^1S_0$	8.33
	47 Ag	$1s^2$	$2s^2$	$2p^6$	$3s^2$	$3p^6$	$3d^{10}$	$4s^2$	$4p^6$	$4d^{10}$	$5s$		$^2S_{1/2}$	7.57
	48 Cd	$1s^2$	$2s^2$	$2p^6$	$3s^2$	$3p^6$	$3d^{10}$	$4s^2$	$4p^6$	$4d^{10}$	$5s^2$		$^1S_0$	8.99
	49 In	$1s^2$	$2s^2$	$2p^6$	$3s^2$	$3p^6$	$3d^{10}$	$4s^2$	$4p^6$	$4d^{10}$	$5s^2$	$5p$	$^2P_{1/2}$	5.78
	50 Sn	$1s^2$	$2s^2$	$2p^6$	$3s^2$	$3p^{6.}$	$3d^{10}$	$4s^2$	$4p^6$	$4d^{10}$	$5s^2$	$5p^2$	$^3P_0$	7.33
	51 Sb	$1s^2$	$2s^2$	$2p^6$	$3s^2$	$3p^6$	$3d^{10}$	$4s^2$	$4p^6$	$4d^{10}$	$5s^2$	$5p^3$	$^4S_{3/2}$	8.64
	52 Te	$1s^2$	$2s^2$	$2p^6$	$3s^2$	$3p^6$	$3d^{10}$	$4s^2$	$4p^6$	$4d^{10}$	$5s^2$	$5p^4$	$^3P_2$	9.01
	53 I	$1s^2$	$2s^2$	$2p^6$	$3s^2$	$3p^6$	$3d^{10}$	$4s^2$	$4p^6$	$4d^{10}$	$5s^2$	$5p^5$	$^2P_{3/2}$	10.44
	54 Xe	$1s^2$	$2s^2$	$2p^6$	$3s^2$	$3p^6$	$3d^{10}$	$4s^2$	$4p^6$	$4d^{10}$	$5s^2$	$5p^6$	$^1S_0$	12.13

Period	Number and atom	K L M N	N $l=3$	O $l=0$	O 1	O 2	P $l=0$	P 1	Ground State	Ionization energy (ev)
6	55 Cs	Filled up to		$5s^2$	$5p^6$		$6s$		$^2S_{1/2}$	3.87
	56 Ba	4d-subshell		$5s^2$	$5p^6$		$6s^2$		$^1S_0$	5.19
	57 La			$5s^2$	$5p^6$	$5d$	$6s^2$		$^2D_{3/2}$	5.59
	58 Ce		$4f^2$	$5s^2$	$5p^6$		$6s^2$		$^3H_4$	(6.54)
	59 Pr		$4f^3$	$5s^2$	$5p^6$		$6s^2$		$^4I_{11/2}$	(5.76)
	60 Nd		$4f^4$	$5s^2$	$5p^6$		$6s^2$		$^5I_4$	(6.31)
	61 Pm		$4f^5$	$5s^2$	$5p^6$		$6s^2$		$^6H_{5/2}$	(6.3)
	62 Sm		$4f^6$	$5s^2$	$5p^6$		$6s^2$		$^7F_0$	(5.6)
	63 Eu		$4f^7$	$5s^2$	$5p^6$		$6s^2$		$^8S_{7/2}$	(5.64)
	64 Gd		$4f^7$	$5s^2$	$5p^6$	$5d$	$6s^2$		$^9D_2$	6.7
	65 Tb		$4f^9$	$5s^2$	$5p^6$		$6s^2$		^6H	(6.74)
	66 Dy		$4f^{10}$	$5s^2$	$5p^6$		$6s^2$		^5I	(6.82)
	67 Ho		$4f^{11}$	$5s^2$	$5p^6$		$6s^2$		^4I	
	68 Er		$4f^{12}$	$5s^2$	$5p^6$		$6s^2$		^3H	
	69 Tm		$4f^{13}$	$5s^2$	$5p^6$		$6s^2$		$^2F_{7/2}$	
	70 Yb		$4f^{14}$	$5s^2$	$5p^6$		$6s^2$		$^1S_0$	6.2
	71 Lu		$4f^{14}$	$5s^2$	$5p^6$	$5d$	$6s^2$		$^2D_{5/2}$	5.0
	72 Hf		$4f^{14}$	$5s^2$	$5p^6$	$5d^2$	$6s^2$		$^3F_2$	5.5
	73 Ta		$4f^{14}$	$5s^2$	$5p^6$	$5d^3$	$6s^2$		$^4F_{3/2}$	6.0
	74 W		$4f^{14}$	$5s^2$	$5p^6$	$5d^4$	$6s^2$		$^5D_0$	7.94
	75 Re		$4f^{14}$	$5s^2$	$5p^6$	$5d^5$	$6s^2$		$^6S_{5/2}$	7.87

TABLE 4.1 (*continued*)

Period	Number and atom	K L M N	N l=3	O l=0	O 1	O 2	P l=0	P 1	Ground State	Ionization energy (ev)
6	76 Os		$4f^{14}$	$5s^2$	$5p^6$	$5d^6$	$6s^2$		$^5D_4$	8.7
	77 Ir		$4f^{14}$	$5s^2$	$5p^6$	$5d^7$	$6s^2$		$^4F_{3/2}$	9.2
	78 Pt		$4f^{14}$	$5s^2$	$5p^6$	$5d^9$	$6s$		$^3D_3$	8.96
	79 Au		$4f^{14}$	$5s^2$	$5p^6$	$5d^{10}$	$6s$		$^2S_{1/2}$	9.22
	80 Hg		$4f^{14}$	$5s^2$	$5p^6$	$5d^{10}$	$6s^2$		$^1S_0$	10.43

Period	Number and atom	K L M N	O l=0	O 1	O 2	O 3	P l=0	P 1	P 2	Q l=0	Ground State	Ionization energy (ev)
6	81 Tl	Filled shells	$5s^2$	$5p^6$	$5d^{10}$		$6s^2$				$^2P_{1/2}$	6.12
	82 Pb	up to 4f	$5s^2$	$5p^6$	$5d^{10}$		$6s^2$	$6p^2$			$^3P_0$	7.42
	83 Bi	subshell	$5s^2$	$5p^6$	$5d^{10}$		$6s^2$	$6p^3$			$^4S_{3/2}$	(8.8)
	84 Po	containing	$5s^2$	$5p^6$	$5d^{10}$		$6s^2$	$6p^4$			$^3P_2$	(8.2)
	85 At	60 electrons	$5s^2$	$5p^6$	$5d^{10}$		$6s^2$	$6p^5$			$^2P_{3/2}$	(9.6)
	86 Rn		$5s^2$	$5p^6$	$5d^{10}$		$6s^2$	$6p^6$			$^1S_0$	10.75
7	87 Fr		$5s^2$	$5p^6$	$5d^{10}$		$6s^2$	$6p^6$		$7s$	$^2S_{1/2}$	4.0
	88 Ra		$5s^2$	$5p^6$	$5d^{10}$		$6s^2$	$6p^6$		$7s^2$	$^1S_0$	5.25
	89 Ac		$5s^2$	$5p^6$	$5d^{10}$		$6s^2$	$6p^6$	$6d$	$7s^2$	$^2D_{3/2}$	
	90 Th		$5s^2$	$5p^6$	$5d^{10}$		$6s^2$	$6p^6$	$6d^2$	$7s^2$	$^3F_2$	
	91 Pa		$5s^2$	$5p^6$	$5d^{10}$	$5f^2$	$6s^2$	$6p^6$	$6d$	$7s^2$	(^4K)	
	92 U		$5s^2$	$5p^6$	$5d^{10}$	$5f^3$	$6s^2$	$6p^6$	$6d$	$7s^2$	$^5L_6$	
	93 Np		$5s^2$	$5p^6$	$5d^{10}$	$5f^4$	$6s^2$	$6p^6$	$6d$	$7s^2$	(^6M)	
	94 Pu		$5s^2$	$5p^6$	$5d^{10}$	$5f^6$	$6s^2$	$6p^6$		$7s^2$	(^7K)	
	95 Am		$5s^2$	$5p^6$	$5d^{10}$	$5f^7$	$6s^2$	$6p^6$		$7s^2$	^8H	
	96 Cm		$5s^2$	$5p^6$	$5d^{10}$	$5f^7$	$6s^2$	$6p^6$	$6d$	$7s^2$	^9D	
	97 Bk		$5s^2$	$5p^6$	$5d^{10}$	$5f^8$	$6s^2$	$6p^6$	$6d$	$7s^2$		
	98 Cf		$5s^2$	$5p^6$	$5d^{10}$	$5f^{10}$	$6s^2$	$6p^6$		$7s^2$		
	99 Es		$5s^2$	$5p^6$	$5d^{10}$	$5f^{11}$	$6s^2$	$6p^6$		$7s^2$		
	100 Fm		$5s^2$	$5p^6$	$5d^{10}$	$5f^{12}$	$6s^2$	$6p^6$		$7s^2$		
	101 Md		$5s^2$	$5p^6$	$5d^{10}$	$5f^{13}$	$6s^2$	$6p^6$		$7s^2$		
	102 No		$5s^2$	$5p^6$	$5d^{10}$	$5f^{14}$	$6s^2$	$6p^6$		$7s^2$	?	
	103 Lw		$5s^2$	$5p^6$	$5d^{10}$	$5f^{14}$	$6s^2$	$6p^6$	$6d$	$7s^2$	?	

[a] See A. Unsöld, "Physik der Sternatmosphären," 2nd ed., p. 90, Springer, Berlin, 1955; M. Born, "Atomic Physics," 7th ed. p. 183, Hafner, New York, 1962; W. Finkelnburg "Einführung in die Atomphysik," 4th ed., p. 151. Springer, Berlin, 1956. In the case of the rare earths ($Z = 58$ to 71 and $Z = 90$ to 103), the configuration and ground term assignments have mostly not been determined spectroscopically; hence the data are speculative. This applies, in particular, to the number of f-electrons if the next higher d-subshell is assumed empty.

4.3.2 The Coupling of Multiple-Valence Electrons

Completely filled shells affect the spectra of the valence electrons only to the extent of their penetration. For example, in the case of magnesium ($Z = 12$) we have to concern ourselves only with the two 3s-electrons in the M-shell (see Table 4.1). Nevertheless, the spectra of multiple valence electrons are generally quite complex, and we shall limit ourselves to the description of some salient features.

4.3.2.1 *LS-(Russell-Saunders) Coupling*

Although the spins and orbits of a plurality of electrons can interact in a great variety of possible modes, two idealized schemes account for the majority of observed spectra. In the first scheme the orbits of individual electrons interact with each other much stronger than they interact individually with the spins. Thus the orbit vectors *l* precess around their resultant **L**, while the spin vectors **s** precess around their resultant **S**. This type of system is shown schematically in Fig. 4.16 for two-electron

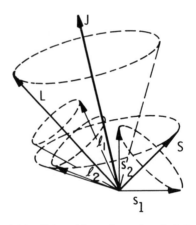

Fig. 4.16.   Spin-orbit configurations in LS-coupling.

configurations such as helium. The resultant orbital moment **L** and spin moment **S** precess around each other forming a total moment **J**. The interaction scheme described here is called LS- or Russell-Saunders-coupling. This represents the *normal* vector arrangement, valid for the lighter elements and, generally, the simple spectra.

4.3.2.2 *jj-Coupling*

Obviously, there exists another extreme. This is the case in which spin and orbit of each individual electron are much tighter coupled than with either moment of any of the other electrons. Each electron then contributes a combined moment **j** of orbit and spin, and the various **j**-vectors form a resultant **J** around which they execute their precession. An example of the *jj*-coupling mechanism is shown in Fig. 4.17 for the case of three valence electrons. In complete *jj*-coupling, the quantum numbers L and S have no meaning. Intermediate coupling conditions are represented by the *Racah* scheme of notation (see Section 6.3.2.3).

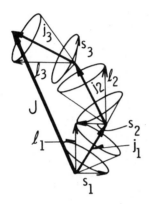

Fig. 4.17.   Vector diagram for three electrons in *jj*-coupling.

4.3.2.3 *The Spectrum of Helium*

We have seen that, as a consequence of Pauli's principle, two *s*-electrons in the same orbit have to assume opposite spins. Thus the normal state of helium is characterized by the symbol $1s^2\,{}^1S_0$; it has no resultant spin. However, the spins can have the same direction if the two electrons are in orbits of different principal quantum number; they can, for instance, form the configuration $1s2s\,{}^3S_1$. Here, $S = 1$, hence for $L > 0$ all levels can have 3 (i.e., $2S + 1$) fine-structure terms, and thus a triplet system results.

The term scheme of helium, shown in Fig. 4.18, gives evidence of pronounced LS-coupling. The spins of the two electrons strongly interact to form a resultant spin of either 0 or 1. Thus there result two term schemes consisting of singlets and triplets, respectively, with transitions occurring either within the singlet or within the triplet system. It is noteworthy that no combinations between the singlet and triplet system are observed; there are no lines of the type ${}^3P \rightarrow {}^1S$. Apparently the strong LS-coupling does not permit a flipping of the spins. One speaks, therefore, of *parhelium* if the two electron spins are antiparallel ($S = 0$), and of *orthohelium*, if the spins are parallel ($S = 1$). The lowest state of orthohelium is that in which the two electrons are in *s*-states ($l = 0$), but necessarily in different orbits, namely, $1s2s\,{}^3S_1$. There is no transition possible to the parhelium state $1s^2\,{}^1S_0$ by a process of spontaneous emission. Such a level is called metastable. It has usually a much longer lifetime than normal dipoles and depends for its return to the ground state on collision processes with walls or with other atoms to which it may transfer its excitation energy in a process called collision of the second kind (or resonance transfer). The lifetime is therefore a function of gas mixture and pressure.

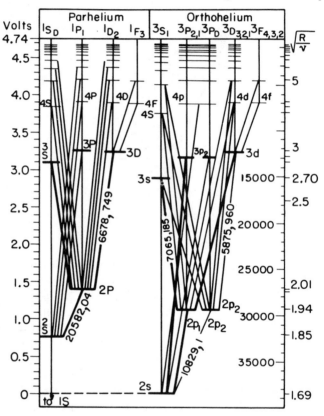

Fig. 4.18.   Energy levels and spectral lines of helium.

4.4 Metastable States, Forbidden Lines, and Multipole Radiation

Metastable states such as the lowest level of orthohelium result whenever the selection rules which have been discussed so far for electric dipole radiation (see Section 4.1.4) exclude a transition to a lower level. If electric dipole radiation were the only process by which such an excited level could lose its energy, its lifetime would be virtually infinite. However, the mechanism by which the metastable atom was generated, e.g., by way of an ionized state, is reversible even though the occasion may be exceedingly rare. In the *laboratory*, with upper limits of travel distance and lower limits of pressure for an observable discharge, the metastable atom returns to the ground state mainly by collisions with other atoms in the gas or at the walls or by intermediate excitation to a higher level. Lifetimes, then, are typically in the order of milliseconds as was first established by measurement for the case of the rare gases and mercury. Indeed, a number of interesting and practically important effects are based on just this property of metastable atoms: to wit, of storing excitation energy to be sooner or later transferred to the energy level of a foreign atom nearly in resonance with, or somewhat lower than, that of the first.

However, a different situation exists for metastable atoms in high vacua under conditions of large mean free paths. Particularly, in interstellar space the density may vary from 1000 to 10 atoms/cm³ so that each particle will undergo collisions from about once a day to three or four times a year. If, also, the radiation density which could be absorbed is negligible, then an indirect return of the metastables to the ground state cannot occur at a rate faster than about 10^{-5}–10^{-7} sec⁻¹, respectively. It is under such conditions that radiative transitions of extremely small probabilities (A-values) can be detected. What is observed is that the spectra of interstellar atmospheres contain *forbidden lines* so defined because they violate the selection rules for electric dipoles. These lines were first attributed to foreign elements until they were also observed under special conditions in the laboratory.

There are, of course, exceptional circumstances in which electric dipole radiation may be involved even though the pertinent transition matrix of the *unperturbed* atom is zero. An example for this are two-quantum processes. It is in principle possible for a radiative transition to occur in two or more simultaneous steps rather than in a single jump (see e.g. Shapiro and Breit [23]). Thus two or more photons may be emitted as the atom goes through intermediate *virtual* levels, and since the latter are not quantized, a continuum rather than sharp lines will result. The transition from the metastable hydrogen level $2s\ ^2S_{1/2}$ to the ground term $1s\ ^2S_{1/2}$ (forbidden since $\Delta l = 0$) is ascribed primarily to this process with an A-value of 8.2 sec⁻¹. Aside from two-quantum transitions, there exists also electric dipole radiation which is induced by external fields, such as from surrounding ions, or by internal fields, such as from the nucleus.

In many instances, however, forbidden lines occur as the result of *multipole radiation.* Such lines are in higher order less intense than electric dipole transitions, but they become important in the absence of the latter. We are concerned here mainly with magnetic dipole and electric quadrupole transitions since the next order, corresponding to magnetic quadrupoles and electric octopoles, is again very much less intense. The factor of line strength reduction, from order to order of multipoles, is given by $(2\pi a/\lambda)^2$ where a is the radius of the atom and λ the wavelength.[24] This has been discussed in Section 3.3.2.3.

4.4.1 Transition Probabilities for Multipole Radiation

The formalism used to describe the intensities and line strengths for electric dipole radiation (see Sections 3.3.2 and 3.3.3) can be extended to "second order" multipole radiation by replacing the matrix containing the dipole moment $\mathbf{P}$(or $\mathbf{X}$) with either the magnetic dipole moment $\boldsymbol{\mu}$ or the electric quadrupole moment $\mathbf{Q}$, respectively. Adopting the notation of Eqs. (3.106) and (3.107), we have then for the spontaneous transition probabilities:

$$A_m = \frac{64\pi^4\nu^3}{3hc^3}\frac{1}{g_A}\sum_{a,b}|(a\,|\,\boldsymbol{\mu}\,|\,b)\,|^2 \quad \text{(magnetic dipole)} \tag{4.61}$$

$$A_q = \frac{32\pi^6\nu^5}{5hc^5}\frac{1}{g_A}\sum_{a,b}|(a\,|\,\mathbf{Q}\,|\,b)\,|^2 \text{ (electric quadrupole)} \tag{4.62}$$

where again the sum of the squared matrix elements is extended over the components a of an upper level A and the components b of a lower level B. For the case of LS-coupling, these values have been computed explicitly by Pasternack[25] and Shortley[26] (see also the review articles by Borisoglebskii[27] and Garstang[28]). We indicate briefly the results. We assume in both types of radiation a transition from an upper level αJ to a lower level $\alpha' J'$, where the α's represent specific values for L and S. The statistical weight g_A in Eqs. (4.61) and (4.62) is then $2J + 1$. The magnetic moment $\boldsymbol{\mu}$ of the atom due to orbital momenta l_i and spins $\mathbf{s}_i$ of a number i of valence electrons is given according to Eqs. (4.31) and (4.33) by

$$\boldsymbol{\mu} = -\frac{e}{2m_ec}\sum_i (l_i + 2\mathbf{s}_i) \tag{4.63}$$

A change of $\boldsymbol{\mu}$ does not imply a change in L or S or the n and l-value of any individual electron; indeed, for pure LS-coupling, only transitions between levels of the same term are allowed. This means that either

L and S change their orientation with respect to each other with $\Delta J = \pm 1$ and (for the total magnetic quantum number) $\Delta M = 0, \pm 1$, or that L, S, and J remain constant and only a jump between magnetically split levels occurs, $\Delta M = \pm 1$. Shortley evaluated the line strengths in Eq. (4.62) with Eq. (4.63) for these transitions, in which the states a and b are defined as follows:

$$\sum_{a,b} |(a \mid \mathbf{\mu} \mid b)|^2 = S_m(\alpha J, \alpha' J') = S_m(\alpha' J', \alpha J) = \sum_{M,M'} |(\alpha J M \mid \mathbf{\mu} \mid \alpha' J' M')|^2$$

$$(4.64)$$

The solutions for transitions with $\Delta J = \pm 1$ and $\Delta J = 0$ are then, respectively,

$$S_m(SL\ J,\ SL\ J + 1) =$$
$$\frac{(J - S + L + 1)(J + S - L + 1)(J + S + L + 2)(S + L - J)}{4(J + 1)} \left(\frac{eh}{4\pi m_e c}\right)^2$$

$$(4.65)$$

and

$$S_m(SLJ,\ SLJ) = [g(SLJ)]^2\ J(J + 1)(2J + 1)\left(\frac{eh}{4\pi m_e c}\right)^2 \qquad (4.66)$$

where g is the Landé factor for Zeeman splitting and the line strengths have been summed over the various combinations between the magnetically separated levels M and M'.

The computation of electric quadrupole transition probabilities by means of Eq. (4.62) is somewhat more complex than the magnetic dipole case because $\mathbf{Q}$ is a tensor. We refer, for the solution of specific cases, to Shortley,[26] Racah,[29] and Garstang.[30]

The line intensities of multipole radiation follow, of course, relationships similar to those given in Section 3.3.3 for electric dipole radiation. Thus the line intensities [cf. Eq. (3.87)] are given by

$$I = N \cdot h\nu \cdot A \qquad (4.67)$$

where N is the number of atoms in the upper (metastable) level and A represents the probability of the spontaneous multipole transition.

4.4.2 Selection Rules for Multipole Radiation

Just as the vanishing of the matrix elements in Eq. (3.106) provides the theoretical basis for the selection rules known from experiment, so are equations of the type (4.62) to (4.67) also able to predict selection rules for multipole radiation. In this latter case, however, it is more difficult to establish experimental verification because of the small line strengths and intensities involved. Not only are such lines often suppressed by nonradiative transitions, but if they do appear, their presence may actually be caused by first order electric dipole transitions for which the restrictions have been partially relaxed by perturbations and other second order processes. The selection rules as first derived by Rubinovicz,[24] Brinkman,[31] and Blaton[24,32] for multipole radiation are listed together with those for electric dipoles in Table 4.2 (see also Borisoglebskii[27]).

TABLE 4.2

Selection Rules for Atomic Transitions of Various Multipolarity k

Multipolarity k	$\Delta \mathcal{J}$	ΔM	Parity Change	ΔS	ΔL
$E1$ Electric dipole	$0, \pm 1$ $\mathcal{J}_1 + \mathcal{J}_2 \geqslant 1$	$0, \pm 1$	Yes	0	$0, \pm 1$
$M1$ Magnetic dipole	$0, \pm 1$ $\mathcal{J}_1 + \mathcal{J}_2 \geqslant 1$	$0, \pm 1$	No	0	$0, \pm 1$
$E2$ Electric quadrupole	$0, \pm 1, \pm 2$ $\mathcal{J}_1 + \mathcal{J}_2 \geqslant 2$	$0, \pm 1, \pm 2$	No	0	$0, \pm 1, \pm 2$
Ek Electric 2^k-pole	$0, \pm 1, \pm 2, ..., \pm k$ $\mathcal{J}_1 + \mathcal{J}_2 \geqslant k$	$0, \pm 1, \pm 2, ..., \pm k$	For k odd: Yes For k even: No	0	$0, \pm 1, \pm 2, ..., \pm k$
Mk Magnetic 2^k-pole	$0, \pm 1, \pm 2, ..., \pm k$ $\mathcal{J}_1 + \mathcal{J}_2 \geqslant k$	$0, \pm 1, \pm 2, ..., \pm k$	For k odd: No For k even: Yes	0	$0, \pm 1, \pm 2, ..., \pm k$

If radiation is due to 2^k electric or magnetic poles, one speaks of the *multipolarity k*. Thus for magnetic and electric quadrupoles (although these are usually counted as different orders), k equals 2. The selection rules cited in Table 4.2 refer to the quantum numbers of the entire atomic system. In addition, there exist restrictions for the transitions of individual electrons i, as described in previous sections for the electric dipole case. These rules, which include $\Delta n = 0$ (no electron jump between various n) for magnetic dipoles and $\Delta l = 0, +2$ for electric quadrupoles, are, however, not rigorous since exceptions occur whenever perturbations and interactions affect the individual electron orbit and spin. Similarly, the restrictions for ΔS and ΔL are relaxed to the extent to which the term scheme deviates from LS coupling. More rigid are the rules regarding changes of J, of the magnetic quantum number M, and of the sum Σl_i which is composed of the individual orbital quantum numbers l_i. This sum may be odd or even. It is an important criterion for the type of radiation whether transitions occur from *odd* to *even terms* or if, instead, they go from odd to odd and even to even. In other words, one differentiates between transitions occurring with or without a change of the so-called *parity*, defined as $(-1)^{\Sigma l_i}$. In Table 4.2 the presence or absence of parity changes is indicated by "yes" or "no," respectively.

By way of summary, we conclude that the validity of selection rules is in no case absolute. Forbidden transitions from metastable states may appear—albeit with much smaller transition probabilities than the lines arising from neighboring unstable states*—even in electric dipole radiation because of two-quantum processes or perturbations arising from internal and external fields. Multipole radiation may be allowed where electric dipole radiation is forbidden and may be quite strong under exceptional circumstances. Such rules as $\Delta S = 0$ have only relative validity, depending on the strength of the LS coupling: intercombination lines do not exist in helium; but the resonance line of mercury at 2537 Å is the intercombination transition $6s^2\ ^1S_0 - 6s6p\ ^3P_1$. Thus, selection rules reveal atomic structure by the extent of their violation.

4.5 Molecular Spectroscopy

4.5.1 STRUCTURE AND SPECTRA

When two atoms are brought together from large to relatively small distances, their electronic energy levels and the transitions between these levels experience in general a major transformation in two respects. First, the levels themselves undergo a change. They assume a minimum value for certain internuclear distances, if, as is usually the case, there exists an attractive potential, particularly between the ground states of free atoms (see Fig. 3.17). Secondly, additional terms and lines appear, increasing greatly the complexity of the spectrum.

If the total energy of the constituent atoms passes through a minimum, they form at that location a more or less stable molecule. The original

---

* The criterion of small intensity compared with transitions from neighboring unstable levels is sometimes used as the definition of "forbidden lines."

transitions ($h\nu_0$ in Fig. 3.17) between two electronic levels in any of the atoms lose their identity, and an entirely new electronic spectrum results. All this can be best understood for the case of the diatomic molecule (Herzberg[33]). Here an axis of orientation exists which is provided by the electric bonding field between the two atoms. This field is usually very strong, and as a result a decoupling of spins and orbits occurs not unlike that of the Paschen-Back effect in strong magnetic fields (cf. Section 4.1.6.4) or a similar decoupling of atomic configurations in strong electric fields. The individual orbital momenta now add up to a component Λ in the direction of the interatomic field (see Fig. 4.19) whereby

$$\Lambda = 0, 1, 2, ..., L \tag{4.68}$$

Instead of the terms S, P, D, etc., corresponding to $L = 0, 1, 2$, etc., we now have Σ, π, Δ, etc., corresponding to $\Lambda = 0, 1, 2$, etc. Since the energy of interaction between the angular momentum component vector Λ and the electric field is the same whether Λ is parallel or antiparallel to the interatomic field, the electronic levels for $\Lambda \neq 0$ are doubly

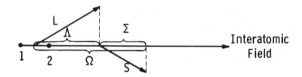

Fig. 4.19.   Orientation of resultant orbit and spin vectors with respect to axial field in diatomic molecules.

degenerate. In atomic spectra a resultant spin vector **S** gives rise to a multiplicity $2S + 1$ of levels in each term. Similarly, there exist $2S + 1$ different orientations of the spin component* Σ in molecular spectroscopy:

$$\Sigma = 0, \pm 1, \pm 2, ..., \pm S \tag{4.69}$$

However, these quantum numbers arise as a result of precessions of **S**, not around the electric field, but around the axially directed magnetic field caused by the orbital moment. Hence no orientation or precession occurs if $\Lambda = 0$, whereas $2S + 1$ values of $\Lambda + \Sigma$ exist for any given value of $\Lambda \neq 0$. Furthermore, because Λ and Σ are both parallel to the

---

* Just as the letter S in atomic spectroscopy, so Σ serves, by convention, two meanings, viz. the molecular term designation for $\Lambda = 0$, and the notation for the total spin component.

interatomic field, these two values can be added algebraically (rather than vectorially, as is the case for **L** and **S** in atomic spectra) to form the resultant Ω (see Fig. 4.19).

The features of molecular levels outlined so far originate basically from the force or, more specifically, the electric field between the constituent atoms of the molecule. The second differential property of molecular spectra, the increase in the number of levels and possible transitions, follows directly from the increase in the number of degrees of freedom owing to the more complex structure. From this point of view, it is possible to deduce the various combinations which can give rise to spectral terms and lines other than those typical of free atoms.

(1) First, bound atoms are capable of rotating around one or more common axes giving rise to quantized *rotational states*. If transitions between such levels are associated with a change of an electric dipole moment, then there results a *rotational spectrum*. Classical arguments already show that the corresponding frequencies are relatively small, and that in fact the rotational spectrum of the molecules lies in the far infrared at wavelengths above about 25μ.

(2) The atoms furthermore can perform vibrations with respect to each other, and these occur at considerably higher frequencies than the rotations. If a change of dipole moment is involved, a *vibrational* spectrum is observable for wavelengths larger than about 1μ. Each vibrational transition may be "modulated" by a finer structure due to superimposed rotational levels. In other words, instead of one or more lines corresponding to jumps between vibrational terms, there will appear bands of lines which may, or may not, be resolved, their intervals corresponding to the rotational frequencies. The frequency of a fine structure line is therefore

$$\nu = \nu_v + \nu_r \tag{4.70}$$

where ν_v represents the frequency of the purely vibrational transition in the absence of all rotational excitation and ν_r constitutes a small additional frequency corresponding to the change in rotational levels.

(3) Finally, electrons can undergo transitions within the molecular bond structure, and these carry the largest energy among the three cases. As in electron jumps between atomic levels, the short wavelength limits lie near 0.1μ in the ultraviolet. Thus electronic, vibrational, and rotational transitions represent magnitudes of descending order. This is evident in the patterns of molecular spectra (see Fig. 4.20) which contain, for each electron jump, bands of superimposed vibrational transitions, of which everyone has a fine structure from the rotational frequencies.

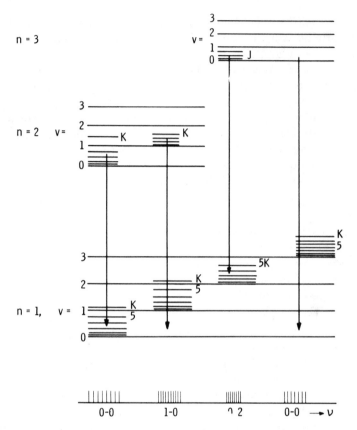

FIG. 4.20. Superposition of rotational levels K and vibrational levels v on electronic levels n. Part of the resulting band spectrum is indicated schematically at the bottom.

Thus, the frequency of each rotational line in the electronic spectrum consists of three parts:

$$\nu = \nu_e + \nu_v + \nu_r \tag{4.71}$$

Equation (4.71) is valid for emission and absorption, and since in such jumps vibrational or rotational energy may increase while the electronic energy decreases, etc., all combinations of positive and negative signs for the three frequencies on the right are possible.

4.5.2 THE QUANTIZED ROTATOR AND FAR INFRARED RADIATION

The dumbbell model of the diatomic molecule (see Fig. 4.21) serves to illustrate the most important features of quantized rotation. We

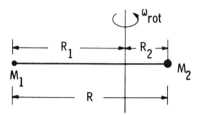

Fɪɢ. 4.21. Dumbbell model of diatomic molecule.

visualize two atoms of masses M_1 and M_2 as rotating around an axis that is perpendicular to the internuclear field direction, with radii R_1 and R_2 which, according to the laws of elementary mechanics, are inversely proportional to the respective masses. Similar to the electron case treated in Section 3.2.2, we define for such a rotator a reduced mass M_r :

$$M_r = \frac{M_1 M_2}{M_1 + M_2} \tag{4.72}$$

In terms of M_r one then obtains simple expressions for the moment of inertia θ and the energy of rotation E_{rot} ,

$$\theta = M_1 R_1^2 + M_2 R_2^2 = M_r R^2 \tag{4.73}$$

and

$$E_{\text{rot}} = \tfrac{1}{2}\theta \omega_{\text{rot}}^2 \tag{4.74}$$

where R is the internuclear distance and ω_{rot} the angular frequency of rotation. Unlike oscillation, the process of rotation does not depend on a periodic interchange between kinetic and potential energy. The angular frequency therefore does not correspond to a state of resonance; in classical mechanics, it can assume a continuous range of values.

The quantum mechanical treatment reveals a somewhat different behavior of the rotator. If the Schrödinger equation (4.1) is solved for the rigid structure in which the potential energy V is zero, R constant, and the electron mass m_r is replaced by the reduced atomic mass M_r , there result solutions only for discrete energy values E_{rot} :

$$E_{\text{rot}} = \frac{h^2}{8\pi^2\theta} K(K + 1) \tag{4.75}$$

with the *rotational quantum number* K which can assume the integer values

$$K = 0, 1, 2, \ldots \tag{4.76}$$

To discrete values K belong also discrete angular momenta $\mathbf{K}$ of the amount

$$|\mathbf{K}| = \sqrt{K(K+1)}\,\frac{h}{2\pi} = \theta\omega_{rot} \tag{4.77}$$

Selection rules limit transitions between various rotational levels to those in which K changes by $+1$ or -1:

$$\Delta K = \pm 1 \tag{4.78}$$

The rotational spectrum for all possible values K is therefore obtained by forming the difference of the energy values, Eq. (4.75), for K and $K + 1$. There results after division by hc

$$\tilde{\nu}_r = \frac{h}{4\pi^2 c\theta}\cdot(K+1) \tag{4.79}$$

Thus the rotational spectrum consists of a series of equidistant lines separated by the constant interval

$$\Delta\tilde{\nu}_r = \frac{h}{4\pi^2 c\theta} \tag{4.80}$$

A comparison between Eqs. (4.77) and (4.79) yields the result that the frequency of rotation ν_{rot} of a level is approximately equal to the frequency ν_r of the light emitted by a transition from that level:

$$\nu_{rot} = \frac{\omega_{rot}}{2\pi} = \sqrt{K(K+1)}\,\frac{h}{4\pi^2\theta} \cong \tilde{\nu}_r c = \nu_r \tag{4.81}$$

The position of the rotational lines in the spectrum is found by substituting the necessary atomic constants into Eqs. (4.72), (4.73), and (4.79). We obtain then

$$\tilde{\nu}_r = 3.34 \times 10^{-5}\cdot\frac{1}{R^2}\frac{A_1 + A_2}{A_1 A_2}(K+1)\quad \text{cm}^{-1} \tag{4.82}$$

where A_1 and A_2 represent the atomic weights of the two atoms forming the molecule. The internuclear distance R is in the order of 10^{-8} cm. Hence, the rotational spectrum of diatomic molecules containing hydrogen (HF, HCl, etc.) lies between 10 and 100 cm^{-1}, whereas that of CO occupies the region of a few cm^{-1} (i.e. the millimeter wavelength region). Molecules such as H_2, N_2, etc., have no dipole moment (see Section 3.1.1) and therefore no infrared spectrum except for multipole radiation at extremely small intensities.

4.5.3 THE QUANTIZED OSCILLATOR AND ITS INFRARED SPECTRUM

The model of the classical oscillator constructed, e.g., of an electron bound to a central charge point by a restoring force $-Gx$, is capable of explaining many features of light emission (see Sections 1.2.2., 3.1.3, and 3.1.4). One of the characteristics of the oscillator is simply that it has a resonant frequency $(G/M)^{1/2}/2\pi$. The assumption based on Hooke's law of a force $-Gx$, *linear* in the displacement x, *leads* to the *harmonic oscillator*; but, of course, such a law is only a first order approximation for electronic and atomic bonding. This is at once obvious since the particle removed from its equilibrium position, at $x = 0$, will experience for large displacements x away from the central charge, not a large $-Gx$, but a vanishing force (ionization or dissociation); whereas approaching the central charge, it will be ultimately repelled by a force much larger than Gx (see Fig. 4.22 which shows an example of two potential curves for diatomic molecules). The actual relationship between force and displacement is therefore nonlinear. It is better described by a power series in x so that the differential equation of this *anharmonic oscillator* has the form

$$M\ddot{x} + G_1 x + G_2 x^2 + G_3 x^3 + \cdots = 0 \tag{4.83}$$

where we have omitted the damping term $R\dot{x}$.

The solution of Eq. (4.83) is an anharmonic oscillation which can be expressed in terms of a Fourier series

$$x = x_1 e^{2\pi\nu_0 t} + x_2 e^{4\pi\nu_0 t} + x_3 e^{6\pi\nu_0 t} + \cdots \tag{4.84}$$

Here the "overtone" amplitudes x_2, x_3, etc., vanish for vanishing higher order G-values in Eq. (4.83), and ν_0 will depend on the amplitude.

Electronic spectra, as was seen in Section 3.2, cannot be explained classically as harmonics of a fundamental vibration. The classical approach is successful, however, for diatomic spectra in the infrared. Indeed, the classical result is in first order agreement with the solution of the Schrödinger equation

$$\frac{\partial\psi^2}{\partial x^2} + \frac{8\pi^2 M_r}{h^2}\,[W - (\tfrac{1}{2}G_1 x^2 + \tfrac{1}{3}G_2 x^3 + \tfrac{1}{4}G_3 x^4 + \cdots)]\psi = 0 \tag{4.85}$$

where the potential energy term corresponds to the nonlinear restoring force implicit in Eq. (4.83). For small values of G_2, G_3, etc., Eq. (4.85) can be solved by the first-order perturbation theory of quantum mechanics (see e.g. Pauling and Wilson,[9] p. 160). The result (Herzberg,[33] p. 100) can be written as

$$W_v = W_{0v} + W_{pv} = (v + \tfrac{1}{2})h\nu_0 + x_e(v + \tfrac{1}{2})^2 h\nu_0 + y_e(v + \tfrac{1}{2})^3 h\nu_0 + \cdots \tag{4.86}$$

where W_{0r} represents the energy of the unperturbed harmonic oscillator (for which all G-values except G_1 are zero so that also the constants x_e and y_e vanish), $v = 0, 1, 2$, etc., the vibrational quantum number, and W_{pv} the perturbation energy.

It is seen that quantum mechanics yields, for $v = 0$, a so-called *zero-point energy*; a vibration is present even when the oscillator is at its lowest state. In preceding discussions we postulated integer multiples

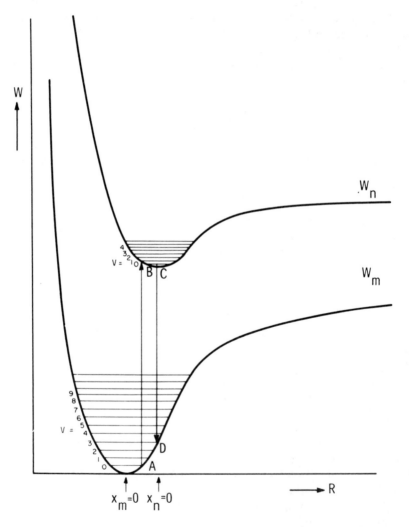

FIG. 4.22. Potential energy W and vibrational levels v of quantized oscillator. The energy scheme shown here is characteristic for diatomic molecules.

of $h\nu_0$ and ignored zero-point energy [cf. Eq. (1.23) and Section 2.3.1]. Since the frequency of radiation is given by the *difference* of energy levels rather than by their absolute amount, the zero-point energy is not directly observable in any spectrum. Derivations of intensity distributions which ignore this minimum oscillator energy remain, therefore, valid and afford greater mathematical ease.

It is of interest to analyze Eq. (4.86) and its implications for the vibrational spectrum in further detail. For the linear case, Eq. (4.86) reduces to the harmonic term

$$W_{0v} = (v + \tfrac{1}{2})h\nu_0 \tag{4.87}$$

Here ν_0 equals $(G_1/M_r)^{\frac{1}{2}}/2\pi$ as can be verified by substitution into Eq. (4.83). The zero energy is $\tfrac{1}{2}h\nu_0$. Selection rules for this simple case are

$$\Delta v = \pm 1 \tag{4.88}$$

Hence transitions occur only from v to $v + 1$ or vice versa with a frequency

$$\nu_{0v} = \frac{W_{0v+1} - W_{0v}}{h} = \nu_0 \tag{4.89}$$

In agreement with the classical result for the harmonic oscillator, Eq. (4.89) states that there exists only one resonant frequency. The next approximation which includes the nonlinear term $G_2 x^2$ yields a series of (weak) equidistant lines. The complete set of transitions between the levels determined by Eq. (4.86) is further augmented by relaxing the selection rules, inasmuch as transitions with $\Delta v = \pm 2, \pm 3$, etc., are allowed with descending probability. The vibrational spectrum then can be written for various $\Delta v \equiv v'$ as

$$\nu_{v+v', \, v} = \frac{W_{v+v'} - W_v}{h} = \nu_0 v' + a\nu_0 v' + b\nu_0 vv' + c\nu_0 v'^2 + \cdots \tag{4.90}$$

In this equation v' represents an integer number, and the constants a, b, etc., decrease in rapid order. A strong successive reduction applies also to the intensities so that the only really intense transition is that of the first terms on the right of Eq. (4.90) for $v' = 1$. The relationship between the quantum mechanical result and the classical solution in Eq. (4.84) has, therefore, the character of a slight correction. Eq. (4.90) just as Eq. (4.84) represents a fundamental frequency and its harmonics; but the latter are here somewhat shifted by the higher order terms.

The *observed* absorption spectra confirm the quantum theoretical treatment in all its refinements. The vibrational transitions are accompanied, of course, by the rotational structure. Thus diatomic molecules show a single strong band corresponding to the transition from v to $v + 1$ of the linear oscillator. Furthermore, the location of the band for such molecules as HF, HCl, HBr, HI, and CO is in the intermediate or near infrared where it should be expected, namely, at 2.52μ, 3.46μ, 3.90μ, 4.33μ, and 4.66μ, respectively (Herzberg,[33] p. 56). If enough molecules are in the optical path, additional harmonics may be observed, or more accurately, bands which, with rapidly vanishing intensities, are located at frequencies given approximately by the relation

$$\nu = \nu_0 v' - \nu_1 v'^2 \tag{4.91}$$

It will be seen that this agrees with Eq. (4.90) for $v = 0$ (absorption from ground state), if the constant a and the orders beyond those shown are negligible.

The radiative lifetimes of purely vibrational transitions are found to be much larger than those of electronic transitions. Typical values for the lowest levels are 10^{-1} sec. An allowed "ionic dipole" transition probability 10^7 times smaller than that of the excited electron can be explained by the larger mass and lower frequency of the molecular oscillator [see, e.g., Eq. (3.19)].

4.5.4 Electronic Transitions and Raman Effect

Light emission of highest energy in molecules is due to transitions of the electrons. The other types of configuration changes, involving vibrational and rotational quantum numbers, are superimposed and result in bands and fine structure lines. The main features of the purely electronic transition, those not involving a superimposed vibrational and rotational spectrum, have been discussed in Section 4.5.1. The main difference with respect to atomic spectra lies in the existence of the interatomic binding forces and the orientation of spin and orbital moments along the internuclear axis. The subject next to be discussed is the effect of the rotational and vibrational structure on the electronic configurations and their transitions.

We first consider the modification experienced by a fixed electronic state due to the existence of discrete vibrational levels. The energy of each electronic state is a function of internuclear distance such as shown for two levels n and m by the potential curves in Fig. 4.22. The discreteness of the vibrational levels is indicated by the horizontal lines which indicate the allowed energy values. Only certain internuclear distances R are possible or, to be more precise, most probable.

We may interpret the intersections of the vibrational levels with the potential curves as positions of rest with relatively large dwelling times of the oscillator, whereas elsewhere along the horizontal line the molecule has a kinetic energy given by the height above the corresponding curve.

Selection rules for the transitions in diatomic molecules depend on the type of coupling and include $\Delta J = 0, \pm 1, \Delta S = 0, \Delta v = 0, \pm 1, \pm 2, ...$, and $\Delta K = 0, \pm 1$ (with the exceptions mentioned later).

An important principle stated by Franck[34] and Condon [35a,b] applies to the electronic transitions between the levels W_m and W_n : the electron motion occurs so fast compared to vibrational changes that the relative position and velocity of the nuclei remains virtually unchanged during the photon emission or absorption. Therefore, transitions have to be shown as vertical lines in the so-called Franck-Condon diagrams such as Fig. 4.22. This principle has an immediate bearing on the relative intensity of the vibrational bands. For instance, under the conditions presented in Fig. 4.22, absorption by the ground state at point A ($v = 0$) is most intense for the transition to point B ($v = 1$) so that the absorption band $0 \rightarrow 1$ is considerably stronger than $0 \rightarrow 0$ or $0 \rightarrow 2$. In emission, however, when the upper state has relaxed to the lowest vibrational state at point C, the band $0 \rightarrow 3$ (from C to D) is the most intense, while neigboring transitions become less likely. The maxima of intensity, of course, change position for different potential curves W_m and W_n . In fact, two maxima may appear, particularly in emission for a broad minimum of W_n . Finally, it must be remembered that each transition represents a band of fine structure lines due to the rotational frequencies.

In the following we shall consider the vibrational conditions as part of the electronic state, i.e., we shall deal with the electronic energy at a fixed internuclear distance, and proceed to discuss the effect of the superimposed rotational momentum.

There are, of course, various sequences of coupling energies with which the spin, orbital, and rotational energies can interact. One important case is that in which the electronic orbit and with it the spin are strongly coupled to the internuclear field forming a resultant Ω (see Section 4.5.1). The rotational momentum and Ω precess then around a total resultant $\mathbf{J}$. Another important possibility is strong coupling between Λ and the rotational momentum, their resultant interacting weakly with spin $\mathbf{S}$ to form $\mathbf{J}$.

Of interest is the band pattern to which various combinations of the rotational K-values give rise in electronic transitions. We consider the electronic and vibrational quantum numbers of the upper and lower state as fixed and vary the rotational quantum numbers K in the initial and final state. Thus the individual line frequencies in the band are given by

$$\nu = (\nu_e + \nu_v) + \nu_r = (\nu_e + \nu_v) + F_n(K_n) - F_m(K_m) \qquad (4.92)$$

where $F(K)$ is the rotational energy (with $F(0)$ being zero) and F_n and F_m refer, respectively, to the upper and lower electronic state. Thus $F_n(K)$ and $F_m(K)$ are different functions; since they correspond to different electronic and vibrational levels, the rotational energies will be different for a given quantum number K. This remains true, even if we consider this energy only in first order as given by Eq. (4.75)

$$F(K) = \frac{h^2}{8\pi^2\theta_v} K(K + 1) \qquad (4.93)$$

where θ_v represents the moment of inertia and depends on the internuclear distance R determined by the vibrational state v [cf. Eq. (4.73)].

Selection rules for transitions between the rotational levels K of upper and lower electronic states are

$$\Delta K = 0, \pm 1 \qquad (4.94)$$

An exception applies for the transition $\Lambda = 0 \rightarrow \Lambda = 0$ for which $\Delta K = 0$ is forbidden. In general, the three alternatives allowed by Eq. (4.94) apply to a multiplicity of values K and thus give rise to three series of lines or *band branches*. In other words, the transition from a higher electronic state W_n to a lower level W_m can occur with a change in rotational energy that is either upward (*P-branch*), or zero (*Q-branch*), or downward (*R-branch*). The sequence of line frequencies of the three branches can be computed in first order with the aid of Eqs. (4.92) and (4.93):

P-Branch: $\nu = (\nu_e + \nu_v) - (B_n + B_m)K + (B_n - B_m)K^2$ (4.95a)

Q-Branch: $\nu = (\nu_e + \nu_v) + (B_n - B_m)K + (B_n - B_m)K^2$ (4.95b)

R-Branch: $\nu = (\nu_e + \nu_v) + 2B_n + (3B_n - B_m)K + (B_n - B_m)K^2$ (4.95c)

where the B's stand for the constants in Eq. (4.93), i.e.,

$$B_n = \frac{h}{8\pi^2\theta_n} \qquad (4.96)$$

Higher order approximations change these values somewhat. The main features of Eqs. (4.95a) to (4.96) are in excellent agreement with the observed molecular spectra. In fact, one of the earliest empirical facts in the analysis of band spectra was just that the graphical representation (*Fortrat diagram*) of a line sequence number K as ordinate, versus line frequency as abscissa, resulted usually in three branches, called P, Q, and R, all of which outlined segments of parabolas. This is clearly understood on the basis of Eqs. (4.95a,b,c).

We conclude our discussion of molecular spectra with a brief reference to the *Raman effect*. A light ray generally experiences *scattering* (see Chapter 5) when it encounters any kind of inhomogeneities of the medium in its path. For most disturbances, such as dust particles or fluctuations in the density of atoms, no change of light frequency is involved, although the scattering process favors higher frequencies so that blue light is scattered in the average over larger angles than red light. In the case of the Raman effect, however, a different type of scattering takes place. Monochromatic light encountering molecules may excite vibrational levels subject to the selection rule $\Delta v = \pm 1$. As a rule energy will be rendered from the photon to the molecule (which is usually in the ground state) rather than vice versa. Thus the phenomenological aspect of the *vibrational* Raman effect consists of one satellite line on the long wavelength side of the incident scattered spectral line at an interval typical for the molecule but invariant with the primary frequency. There exists also a *rotational* Raman effect subject to the selection rule $\Delta K = 0, \pm 2$. Because of the low energies involved (in the order of kT or smaller), a series of equidistant rotational lines appear on *both* sides of the incident spectral line. Molecular electronic and Raman spectra have the feature in common that quantized rotational or vibrational energy is added to, or subtracted from, a quantum of higher order energy representing electric dipole radiation. Therefore, rotational and vibrational transitions may be seen which have no dipole moment and therefore no infrared spectra.

4.6 The Spectra of Solids

To a certain extent it is possible to treat the optics of the solid state as if it had evolved from atomic and molecular spectroscopy. Such a treatment proceeds as before in three steps: (1) it postulates a model of the radiating structure; (2) it derives the energy states consistent with such structure; (3) it obtains the spectrum as transitions of various probability between the energy levels. This procedure is easily followed in atomic and molecular spectroscopy, and it represents also the only method of deriving the spectra of solids.

However, there are important differences in emphasis and complexity. Atomic spectroscopy represents the most important, if not the only, facet through which the structure of the atom can reveal itself, through which, in other words, the reality of the postulated model can be tested. This is true, to a lesser extent, also for molecular spectroscopy. On the other hand, the role which spectroscopy plays for the understanding of the solid state is far less significant, and this for two reasons.

First, the electrical, thermal, and mechanical properties of the solids already offer numerous clues for the exploration of binding forces and structure. Secondly, solid state spectra, with notable exceptions, consist of more or less extended continua because of the close spacing of a large number of the emitting or absorbing centers, if for no other reason. It may often neither be possible nor, in fact, necessary to interpret the continua in terms of detailed energy structure. For example, black body radiation can be explained by relatively simple assumptions regarding the participating oscillators of which the individual frequencies are actually irrelevant. In a manner of speaking, such spectra hide rather than reveal complexities of structure and energy states far larger than those existing in free atoms and molecules.

These are the reasons why the characteristics of the solid state are treated as a field by itself. We refer to the standard texts on the subject of solid state physics[36, 37] and discuss in the following only the details pertinent for solid state spectra.

4.6.1 Structure and Energy in Solids

The bonding of the atoms within a solid may in itself be considered as an extension of those types of forces which are responsible for the structure of various molecules. In particular, we shall cite examples of spectra for two kinds of classes: (1) the *ionic crystals* (best represented by the alkali halides); and (2) the *covalent crystals* (such as germanium, silicon, or silicon carbide). In the first of these, an element with nearly filled shell (F, Cl, Br) combines with an element at the start of a shell (Li, Na, etc.) depriving it of its outer electron. Thus, there result cubic crystal structures in which negative ions (F^-, Cl^-, Br^-) alternate with positive ions (Li^+, Na^+, etc.) and so are held together mainly by electrostatic forces. In the covalent bond, two atoms contribute electrons of opposite spin to the common configuration space so that each electron shares in the binding of the two atoms. This may be repeated for each of the valence electrons. For instance, a carbon atom in diamond shares electrons with each of its four nearest neighbors.

There exist other types of bonding which deserve to be discussed in terms of their contribution to solid state spectra. We mention here the molecular crystal (e.g., carbon hydrides) which are weakly bonded by mutually induced electric dipole moments. A detailed account of molecular spectra has been given by McClure[38]. [39] *Metal crystals* can be considered as positive ion lattices held together by a medium of free electrons, and precisely for this reason, they have little to contribute to the spectra of bound electrons. Of course, the conductivity of the free electrons is important to metal optics in other aspects, namely in the phenomena of propagation, and these will be described in the next chapter.

4.6.1.1 *Ionic Bonding Forces*

The binding energy of an atom with respect to its neighbors in the solid is of the same order as for molecules, namely, 8–12 eV in ionic and covalent crystals. Also, just as in molecules, vibrations around the equilibrium position are possible with a frequency given by the stiffness G and the reduced ionic mass M_r. Of course, with N ions in a given solid volume, there exist $3N$ degrees of freedom and therefore $3N$ possible modes of elastic standing waves. However, only part of these are associated with a macroscopic dipole moment; in fact, the interaction with light is possible only within a relatively narrow region around the classical frequency, giving rise to the *optical branch* of the crystal vibrations. All other modes belong to the *acoustical branch*; they do not emit or absorb the electromagnetic spectrum which corresponds to their frequencies. The optical band is of particular importance to the propagation of light through crystals. Not only absorption, but dispersion and reflection undergo strong changes in this region (see Chapter 5).

4.6.1.2 *Electron Energies and Band Structure in Solids*

As might be expected, the energy schemes and transitions possible for electrons in solids exhibit a large diversity. Just as in the case of ionic vibrations of solids, we deal here too with an oscillator continuum or rather with continua. In addition, many new features are added owing to the Pauli exclusion principle which governs the statistical distribution of electron energies.

The formation of continua or bands can be understood on the basis of the so-called *tight binding approximation*, a method which superimposes individual electron wave functions as atoms are brought from infinity to the close spacing of the lattice. The effect can be explained qualitatively on the basis of Fig. 4.23, which, of course, is analogous to Fig. 1.4, except that we are now discussing electronic states rather than vibrational frequencies in general as in Section 1.2.3. The example presented in Fig. 4.23 shows how the electronic energy levels of lithium atoms ($1s^2\,2s$) are altered as the atoms are brought close together. If the process is first applied to two atoms only (shown in black shading), each electronic state splits into two levels, in analogy to the coupling of two oscillators. Quantum-mechanically this corresponds to the superposition of the two wave functions ψ^A and ψ^B which may form a resultant $\psi^A + \psi^B$ or $\psi^A - \psi^B$. The two states are characterized by different energy values which diverge more and more as the two atoms are brought closer together. The effect is strongest for the outer electronic states; the inner shielded core, such as that of the $1s^2$ electrons in lithium, requires a yet tighter approach for a corresponding perturbation.

The individual levels have finite width, and thus they overlap and

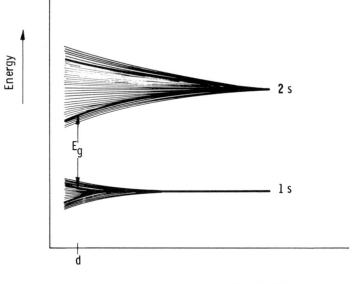

Interatomic Distance

FIG. 4.23.   Variation of electronic energy levels with interatomic distance. Black shading indicates mutual approach of only two atoms; gray shading represents the bands which form as many atoms are brought together to a lattice distance d. Original energy difference of $1s$ and $2s$ electrons becomes the forbidden gap width, E_g. Perturbation of $1s$-electrons is smaller than that in the $2s$-states because of shielding by the latter. An example for the energy conditions shown in this diagram is the $1s^2\,2s$-configuration of lithium.

spread into a band as more atoms are added and the configuration of a solid is attained. Nevertheless, the discrete structure of the originally sharp levels may stay intact. As shown for the 1s- and 2s-states, they form energy bands between which there remains a *forbidden gap*. In other words, the electrons of the solid can occupy only certain zones of energy.

In this situation the Pauli exclusion principle plays a decisive role. N monovalent atoms exhibit $2N$ electron states of opposite spin but otherwise equal quantum numbers, just as in the case of N free atoms. This has the consequence that for crystal atoms with only one s-electron in the outer shell (as in the monovalent metals) the corresponding band is only half filled. In the presence of an external field, therefore, electrons can assume slightly higher energies within the half-filled band, i.e., they can be accelerated and carry a current; the solid is a conductor. An important feature is the possibility of overlap between bands as shown for the 2s- and 2p-bands of beryllium in Fig. 4.24. This endows even divalent crystal atoms with metallic properties since the overlapping p-band in effect augments the number of available states so that conduction is again possible. However, in cases in which the electrons fill all states in the occupied bands, a nonmetallic behavior results, in that no current can be carried by these filled bands. Of course, there exist

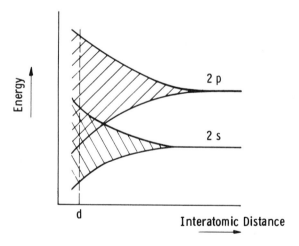

FIG. 4.24. Overlap of 2s- and 2p-bands in solids. At interionic distance d, bands may overlap instead of maintaining a gap as in Fig. 4.23. This is of importance for $2s^2, 3s^2, ..$, configurations which occupy filled bands. The overflow into the 2p-band makes metallic conduction by holes or electrons possible in group II elements such as beryllium and magnesium.

higher, empty bands, corresponding to higher quantum numbers. If the gap separating the filled so-called *valence band* from the next higher *conduction band* is relatively small (e.g., in the order of 1 eV), some electrons can be elevated into these unfilled states by thermal energy at normal or higher temperatures (see Fig. 4.25a). There results a so-called *intrinsic semiconductor* characterized by moderate conductivity (10^{-9} to $10^2 \, \Omega^{-1} \, cm^{-1}$) which increases with temperature. However, if the forbidden gap below the conduction band is large, only a negligible amount of electrons will be raised to it by heat energy. If unperturbed, the solid will then be an insulator.

The presence of impurities may alter considerably the electric and optical behavior of the solid (*extrinsic case*). Atoms with easily removable electrons form so-called *donors* when they are introduced into the lattice of an insulator or semiconductor, i.e., they require only a relatively small energy to ionize and surrender their electron to the conduction band (see Fig. 4.25b). Conversely, if an impurity atom is situated close to the

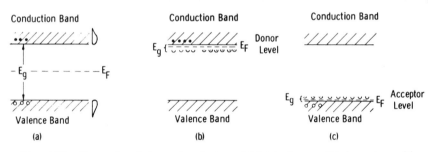

Fig. 4.25. Energy bands in semiconductors: (a) intrinsic case; (b) donor impurities; (c) acceptor impurities. Fermi level is indicated in each case by dashed line, and carrier energy distribution is shown in (a).

valence band in the energy diagram, i.e., if it is capable of taking, with little energy, an electron from the valence band, it forms an *acceptor* (see Fig. 4.25c). In this case, vacancies are created in the filled band so that so-called *hole conductivity* results. Thus in extrinsic semiconductors the current is carried either by a majority of holes or of electrons, corresponding to so-called *p-type or n-type material*, respectively. A relatively. small number of minority carriers—electrons in *p*-type and holes in *n*-type semiconductors—are, of course, also present since they are constantly generated in a thermal or radiation equilibrium. We shall resume this subject at the occasion of an important example (Section 4.6.5.1).

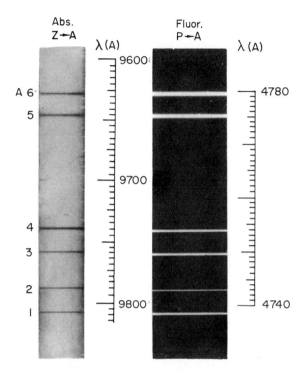

FIG. 4.26. Crystal field line structure of 1 % Er^{3+} in Y_2O_3 in absorption and fluorescent emission. In both cases the so-called A group is involved, corresponding to the $^4I_{11/2}$ level of the free Er^{3+} ion at about 10,000 cm^{-1} (see Fig. 4.27). On the left, absorption from the ground state $^4I_{15/2}$ (Z group) to $^4I_{11/2}$ yields the same pattern as the fluorescent spectrum from $^2P_{3/2}$ (P group) to $^4I_{11/2}$. The six lines seen correspond to the $J + \frac{1}{2} = 6$ states resolved by the crystal field in the $^4I_{11/2}$ level. This simplicity results from the fact that at low temperatures only the lowest state of the ground term is occupied and that the primary light in fluorescence populates only the lowest state of the $^2P_{3/2}$ level [courtesy of G. H. Dieke and H. M. Crosswhite, *Appl. Optics* 2, 675 (1963)].

4.6.1.3 *Density of States and Fermi-Dirac Distribution Law*

A subject which deserves a brief description here concerns the *number* or *density of states* within a given energy or momentum interval and the probability for such states of being occupied by an electron. The product of the density and probability yields the electron energy distribution function by a process similar to the manner in which the photon distribution function has been determined for black body radiation in Section 2.3. Indeed, the analogy is far-reaching: the electron in the solid exhibits periodicity in time and space with a de Broglie wavelength λ given by the momentum p, $\lambda = h/p$ (see Section 1.3.3). Therefore we can postulate a pattern of standing waves (as shown in Fig. 1.5 for the one-dimensional case) such that they intersect the sides of an enclosure with an integral number of halfwaves (cf. Eq. 1.16). This condition is expressed conventionally in terms of a *wave vector* $\mathbf{k}$, of which the amount equals $k = p/\hbar = 2\pi/\lambda$, and of a triplet of integers $\mathbf{n} = (n_x, n_y, n_z)$,

$$\mathbf{k} = \frac{2\pi}{L}\,\mathbf{n} \tag{4.97}$$

In all, the same considerations apply that lead to the number of "distinguishable cells" G_j in Eq. (2.45) except for the polarization factor of 2. Therefore, if there are no restrictions on the allowable energy E, as for instance in the case of a simplified model of a metal, the density of states $g(E)dE = G_j$ follows from Eq. (2.45) or (1.32) by substitution of $p = (2m\,E)^{\frac{1}{2}}$ and $\Delta p = (m/2E)^{\frac{1}{2}}\,\Delta E$:

$$g(E)\,\Delta E = \frac{4\pi(2Em^3)^{\frac{1}{2}}V}{h^3} \cdot \Delta E \tag{4.98}$$

The *probability* of being occupied for any one of the G_j (degenerate) states of energy E is computed by a procedure similar to that used for the Bose-Einstein distribution in Section 2.3.2, except for one fundamental difference. That is the Pauli principle, to wit, that any cell (or state given by four quantum numbers including spin) can be occupied by no more than one electron (whereas, of course, "Bosons" such as photons face no such occupancy restriction). This, as already mentioned in the preceding section, leads to the filling-up of bands to an extent depending on the number of available electrons. The detailed calculation of the most probable energy distribution by the method of Langrangian multipliers is presented in several texts on solid state physics (see also Born,[40] p. 261). The probability for a state to be occupied is given by the Fermi-Dirac function

$$f(E) = \frac{1}{\exp[(E - E_F)/kT] + 1} \tag{4.99a}$$

Thus we have, for instance, for metals after applying Eq. (4.98) per unit volume

$$dN = g(E)f(E)\,dE = \frac{4\pi(2Em^3)^{\frac{1}{2}}}{h^3} \frac{dE}{\exp[(E - E_F)/kT] + 1} \tag{4.99b}$$

Here E_f, the *Fermi level*, is an energy parameter determined by the boundary condition that the total number of electrons equals that available:

$$\int dN = g(E)f(E) \, dE = N_0$$

The significance of the Fermi-Dirac function (4.99a) is most plausible for small temperatures T. For energies larger than the Fermi level, $E > E_f$, the probability for states of that energy to be occupied is small and tends to zero as $T \to 0$. For $E < E_f$ the probability approaches unity as $T \to 0$. A clear definition of E_f emerges from Eq. (4.99a), if we set $E = E_f$: the Fermi-level is that energy at which the probability of occupation is $\frac{1}{2}$. Of course, the density of states at E_f may be zero, the level therefore virtual, as shown for the semiconductor models of Fig. 4.25.

4.6.2 CLASSIFICATION AND GENERAL PROPERTIES OF SOLID STATE SPECTRA

Because the ions of a solid provide fields of multifold symmetry, and because the electrons may belong to localized or mobile states, the variety and complexity of solid state spectra is far more imposing than that of free atoms and molecules. The possible radiative transitions may be resonant or they may correspond to continua, and they may be classified according to several, more or less arbitrary, points of view. Table 4.3 lists three major divisions of solid state spectra, each further subdivided* so that there result altogether 14 classes of radiative interactions, and these again are subject to further subdivision. Obviously some of the interactions are significant, not as processes of emission or absorption, but for their secondary effects—photoconduction and photoemission, for instance—and their discussion is therefore deferred to the later chapters. At any rate, the field is so large that only a cursory treatment is possible here for the detailed classes of solid state spectra.

Despite the great variety of the underlying mechanism and appearance of solid state spectra, a few more or less general rules apply to their behavior. For example, an important relationship exists between spectral bandwidth and temperature. Within certain limits, such bandwidth will be approximately proportional to $\sqrt{T}$. Although the same dependence exists also for Doppler broadening [see Eq. (3.132)], the immediate physical cause is different, viz., the perturbation of the oscillator by its

---

* These subdivisions may be assigned differently according to the point of view. Thus types 6 and 7 in Table 4.3 could be counted under group III, and type 14 under II.

Table 4.3

Survey of Solid State Spectra

Type	Occurrence	Characteristic phenomena	Spectral width and region**
I. Plasma oscillations			
1. Optical crystal vibrations	All ionic crystals	Reststrahlen, absorption	$Q_{abs} \sim 2\text{--}30$ $$\nu \approx \frac{1}{2\pi}\sqrt{\frac{G}{M}},$$ 15–300 μ;
2. Crystal vibrations superimposed on electronic transitions	Anion lattice containing rare earth or transition metal ions	Increase or repetition of line pattern	$Q \sim 10^4\text{--}10^5$, visible and UV region
3. Free electron oscillations	Metals	High reflectivity	Continuum. Transparent in the UV at $$\nu > \frac{1}{2\pi}\sqrt{\frac{4\pi n e^2}{m}}$$ for alkali metals
4. Electron plasma oscillations	Metals under electron bombardment-speculative	Light spot	Bands
II. Transitions between localized states			
5. Transitions between unfilled shell configurations with $\Delta n = 0$	$4f$ terms (lanthanides) $5f$ terms (actinides) $3d$, $4d$, $5d$ terms (transition metals)	Sharp lines in fluorescence	$Q \sim 10^4\text{--}10^5$ $\Delta_{\tilde{\nu}_h} \sim 0.1$ cm^{-1},
6. Exciton spectra	Alkali halides, silicon, SiC, Cu_2O, TlCl, etc.	Absorption, enhancement of photoemission	$Q \sim 10^4$, $h\nu < E_g$
7. Electronic excitation of ionic crystals	Alkali halides	Fundamental absorption bands	Bands of 20 to 50 Å widths in the UV
8. Electronic spectra of organic molecules	Benzene, naphtalene, etc.	Absorption	UV bands of 10 to 100 cm^{-1} width ($Q \sim 3 \times 10^2$– 3×10^3)
9. Absorption by color centers	Defects in ionic crystals	Absorption bands (coloring of crystals)	F-centers in the visible region, V-centers in the UV, $Q \sim 5$

TABLE 4.3 (*continued*)

III. Transitions between nonlocalized states

Type	Occurrence	Characteristic phenomena	Spectral width and region
10. Band-to-band transitions	Intrinsic semiconductors	Photoconductivity, absorption ($k = 10^4$–10^6 cm^{-1}), recombination radiation	Continuum for absorption at $h\nu > E_g$; narrow bands (kT), $Q \sim 10$–100, in emission
11. Impurity $\rightleftharpoons$ band absorption	Extrinsic semiconductors	Photoconductivity at weak absorption ($k = 10^{-2}$–10^2 cm^{-1})	Continuum for $h\nu > E_g$
12. Impurity $\rightleftharpoons$ band emission	Activators in phosphors	Luminescence and fluorescence	Continuum or bands
13. Free carrier transitions	Semiconductors	Weak absorption	Continuum
14. Impurity-to-impurity transitions	Semiconductors and insulators	Emission and absorption of lines	$Q \sim 10^4$

**Q stands for $\nu/\varDelta\nu_h$. Spectral region is cited only if characteristic for the process.

nearest neighbors. The disturbance fluctuates in time because of the thermal agitation and will, therefore, increase with some power of the temperature. This type of statistical broadening (see also Section 3.4.3.4) exists even at $T = 0$, for instance because of zero point vibrations, so that the $\sqrt{T}$ law is not valid for very low temperatures. Besides broadening, a shift of the lines or bands occurs with increasing temperature. Increase of pressure will also produce a shift and—as to be expected—in general, in the same direction as decreasing temperature (Paetzold[41, 42]). For lines of the Cr^{3+} ion in the crystals of aluminum oxide or magnesium oxide, width and shift were found to follow the temperature behavior of the lattice thermal energy; in other words, they were proportional to the square of the lattice vibration amplitude (Schawlow[106]). Here a square-law relationship between static deformation and line shift must be assumed.

As pointed out before, solid state spectra appear as gaslike lines or even as narrow bands only for a few types of materials and even then only at low temperatures. The lanthanides—rare earths in which

$4f$-electrons can combine to produce various excited terms—and the actinides*—in which the corresponding spectra are produced by $5f$-electrons—have line widths below about 0.1 Å in the visible region at 4.2°K. Various other spectra to be discussed in the next sections such as connected with exciton or impurity-to-impurity transitions typically have 1 Å ($Q \approx 10^4$) bandwidth. The absorption bands of the alkali halides exhibit widths of about 20 Å at liquid helium temperatures. The demands made on the experimental equipment are therefore less rigorous with regard to the resolution of the spectrograph than with respect to the operating conditions of the sample. This is in contrast to the conditions most common for gaseous spectroscopy in which lines have typical halfwidths of 0.1 Å or less at room temperature, if Doppler broadening represents the most important limitation. Crystals, however, must be kept usually at liquid helium or at least nitrogen temperatures to obtain sufficient pattern resolution. To this end, the crystal is either directly immersed in the coolant, or, if that procedure disturbs the light passage, the sample is tightly clamped in a metal block or frame in good thermal contact with the coolant. Furthermore, surface preparation and orientation with respect to the light beam are of importance in crystals, although some solids can be obtained and observed only in powder form—albeit, at a corresponding loss of structural information.

Observation by absorption is far more applicable to solids as a whole than by emission. Compared to gases, it is more difficult to excite various distinct energy levels in solids and to determine their width and position by re-emitted spectral lines, partly because so-called *radiationless* transitions may replace quantum jumps accompanied by light emission, partly because of the overlap of the broader energy bands. Nevertheless, in specific instances information can be obtained only by emission spectra, excited, for instance, with electron bombardment in *cathodoluminescence* and with primary visible or ultraviolet light in *fluorescence*.

Absorption spectroscopy of solids has to deal with the fact that there are in the order of 10^{22} atoms/cc in the radiation path. For electric dipole transitions from the ground state, the average absorption coefficient in a band may be as high as 10^6 cm^{-1}, according to Eq. (3.101c), if all atoms can participate. It is necessary in such cases to work with samples about 10^{-6} cm thick which are, for instance, evaporated on a metallic substrate and viewed in reflection. In a large number of pertinent cases, however, the absorbing centers are diluted in a lattice which itself is

---

* The rare earths are named after lanthanum ($Z = 57$) and actinium ($Z = 89$), respectively, which precede them in the periodic table, although these two elements are usually not considered as belonging to their group.

transparent in the observed wavelength region; or the transition of interest is forbidden in the free ion or atom, but can occur at weak intensities owing to electric dipoles induced by the field; or, in a majority of cases, both of these circumstances may obtain. Here, then, the problem may be the inverse: it may be necessary to obtain single crystals of several centimeter length. Finally, a type of "cascaded" absorption process is of interest in which primary light *pumps* the optical centers from the ground state to an excited level and in which the absorption of this level in transitions to a third state is studied. Such experiments usually combine widely different wavelength regions. For instance, if the transition of interest occurs between Zeeman levels of an excited state, the pumping radiation may consist of visible or ultraviolet light while the transition itself is studied in the microwave region, typically by observing the effect on the Q of a cavity which contains the sample.

4.6.3 PLASMA OSCILLATIONS

The term "plasma oscillations" is used here in a very general sense, i.e. in reference to the periodic motion of multiple charges—electrons or ions—in solids. These processes have been classified as the first group in Table 4.3.

4.6.3.1 *Optical Lattice Vibrations*

The most significant aspect of ion oscillations which give rise to a change in dipole moment concerns reflection in combination with absorption. The subject is discussed under *reststrahlen* in Chapter 5.

However, we mention here the existence of spectra which are ascribed to the superposition of crystal vibrations on electronic transitions (Fick and Joos,[43] p. 260), in a process not unlike the Raman effect. The spectral pattern consists of a repetition of line groups at distances corresponding to the lattice vibration frequencies. An example is the spectrum of $Nd(NO_3)_3$ in which the electronic transitions show a type of fine structure accountable by "internal" vibrations of the NO_3 ion (Joos and Ewald[44]).

4.6.3.2 *Electron Plasma Oscillations*

It is conceivable that electrons bound quasi-elastically to positive ion sites are capable of coherent oscillations in a manner similar to the vibrations of negative ions. In a metal, for instance, electrons can be considered free only in first order. Aside from collisions, they find themselves in a potential periodic with the lattice constant, and therefore tend to exhibit a periodic space charge.

A phenomenon which is explained plausibly in terms of electron plasma oscillations is that of *Lilienfeld radiation*,[45,46] which is a light spot caused by electron bombardment of metals such as in X-ray tubes, typically at 5 kV and 20 μamp, and which is visible to the dark-adapted eye. An interpretation can be given first on the basis of electron plasmas (Ferrell[47]); but one must hold against that the possibility of so-called *transition radiation* (Ginsburg and Frank[48]). The existence of the latter is postulated for the case that an electron is transmitted through the boundary between media of different dielectric constant. In this case the charge and its image generate a time-variant dipole field which collapses after the electron passes through the boundary. Some experimental results[49] favor this explanation, although an interpretation is possible which employs features of both the plasma and the transition radiation theory.[50,51]

4.6.3.3 *Free Electron Oscillations*

If one considers a model as in Section 1.2.1 in which charge carriers are free, a response to incident radiation is obtained which is nonresonant and which, as shown in Chapter 5, results in reflection up to a characteristic frequency limit and to transmission beyond such limit. This, in first order, holds true for metals and, in particular, for the alkali metals. Free carrier absorption in semiconductors, too, may be listed under this heading to the extent to which a large number of electrons or holes react coherently to the incident electromagnetic wave. However, if individual transitions of free carriers occur, the process should be classified under the third group (cf. Table 4.3).

4.6.4 Transitions Between Localized States

The spectra discussed in the first group have in common that they originate in multiple charges moving in cooperative or coherent fashion. The exact opposite applies to the second group listed in Table 4.3, the transitions between localized states. Here, centers undergo changes in energy individually, restricted more or less to their own configuration space. This, at any rate, brings about a commensurate limitation of spectral bandwidth because the broadening effect of exposure to the whole lattice is missing or at least reduced. For these reasons, the spectra of the second group resemble to a varying degree those of free atoms such as in gases.

4.6.4.1 *Transitions between Unfilled Shell Configurations With* $\Delta n = 0$.

A group of spectra referred to, in a narrower sense, as *crystal field spectra* arises from transitions between the unfilled f- and d-states of the rare earths and transition metals, respectively. A number of particular conditions apply to the ions of these elements and their electron configurations when they are embedded in a crystal lattice. The fact, first of all, that the f- and d-electrons are in *unfilled* shells makes possible various combinations of the individual orbits (additions of the l-vectors) with respect to each other, i.e., the various $4f$- or $5f$-, $3d$-, etc., electrons can give rise to a finite number of terms. Transitions between these terms are ordinarily forbidden by the selection rule $\Delta l = 0$. However, in the lattice these rules are partially relaxed because electric dipole moments are induced by the crystal field. In fact, this field is large enough so that the states of outer electrons would be broadened into bands. Of course, as Table 4.1 shows, the f-electrons of the lanthanides and actinides are shielded by the protecting O- and P-shell, respectively, even for trivalent ions. Thus there results a delicate balance of conflicting effects such that solid state spectra of gaslike appearance become possible: the crystal field affects the f-electrons sufficiently so as to make transitions possible, even if weakly, for which $\Delta n = 0$ and $\Delta l = 0$ and in which, furthermore, the line structure is resolved; yet the field is sufficiently weak and the transitions because of $\Delta n = 0$ are sufficiently localized so that the resulting lines are very narrow with Q-values in the order of 10^5.

Crystal field spectra may be observed in emission and absorption at low oscillator strengths, viz., of the order 10^{-6}–10^{-5}. Therefore, the lines are much less intense than the allowed transitions such as from an f- to a d-orbit which also are much broader because they involve unscreened electron configurations. As a consequence, the crystal field spectra cannot be observed in competition with the allowed lines or bands wherever these occur. In the lanthanides, this is the energy region of the vacuum ultraviolet so that the $4f \rightarrow 4f$ lines have to be observed in the visible and near ultraviolet window. The choice is still further limited for fluorescent transitions because of the competition by radiationless recombinations which take place when the coupling between the f-states and the lattice is relatively large.

Another feature which is particular for crystal field spectra concerns level splitting. The various configurations of the individual electron orbits and spins give rise to corresponding terms with given values L and S. Each term consists of a multiplet with up to $(2S + 1)$ different levels $J = L + S$ (assuming the validity of Russell-Saunders coupling) with spacings in the order of 10^3 cm^{-1} (0.1 eV). This is the case for the free rare earths ions also; but in the lattice each level J is split by the crystal field into $2J + 1$ states for even numbers of electrons in the unfilled shell and $(J + \frac{1}{2})$ states, if the occupation number is odd.

It will be clear that the occurrence of forbidden lines and their splittings, aside from their implications for crystal field theory, can be of great value to the identification of the spectroscopic energy level scheme which is most complex for the rare-earth configurations. Such identifications based on comparisons between the free ion and crystal term schemes have been undertaken by Dieke[52, 53] and his co-workers. That such comparisons can be made at all is, of course, a unique property of the crystal field spectra, although even here the term schemes of free and bound atoms are not identical but shifted with respect to each other by several hundred cm^{-1}. A number of basic relationships and theories are of aid in the assignment of energy terms as discussed in the following.

4.6.4.2 Analysis of Crystal Field Spectra

On the theoretical side, the foundation for crystal spectra was laid by Bethe.[54] Of similar basic significance are the papers by Racah[55a,b, 56a,b] which deal with that most complex case of atomic spectra, the rare earths. Since then a large amount of material has appeared on the combination of the two subjects, and we must refer here to the bibliography cited in the specialized treatments of crystal field spectra.[38, 39, 44, 53]

The starting point of the quantum mechanical treatment is again the Schrödinger equation in which in this case the Hamiltonian [cf. Eq. (3.54)] has the form[57]

$$H(p_i, r_i) = \sum_i^n \left[\frac{\mathbf{p}_i^2}{2m_i} - Z\frac{e^2}{r_i} + \zeta(r_i)\mathbf{s}_i\mathbf{l}_i \right] + \sum_{i>j=1}^n \frac{e^2}{r_{ij}}$$

$$+ \sum_{ij}' q_j \frac{e^2}{r_{ij}} \tag{4.100}$$

In this equation the symbols have a significance analogous to those of Eq. (3.54), except that they refer to multiple electron systems with index numbers i and j; to wit, momenta p_i, nuclear distances r_i, interelectron distances r_{ij} (in the repulsion term). Furthermore, Eq. (4.100) contains an energy term for the normal spin-orbit interaction which is proportional to the radial potential gradient of the atom through

$$\zeta(r_i) = \frac{\hbar^2}{2m_i{}^2c^2r_i{}^2} \left[\frac{\partial V(r_i)}{\partial r_i} \right] \qquad (4.101)$$

Finally, the last term of Eq. (4.100) refers to the effect of the crystal field. The problem then consists of the evaluation of the terms contained in Eq. (4.100) and the solution of the Schrödinger equation to arrive at the theoretical energy terms of the rare earth ion spectra.

On the experimental side, certain simple rules may be used for the tentative assignment of spectral terms so that, in combination with the theory, the complete energy scheme can be derived. Some of these rules are the following:

Hund's Rule.[58] (1) Among terms with the same L, that with highest multiplicity (S-value) will lie deepest. (2) Among these terms, that with the highest L-value will lie deepest (see also White,[59] p. 259). These guiding principles, empirically discovered, have been found highly reliable in the classification of complex free-atom spectra.

Sequence of Ionization in Spectra. In a comparison between spectra of crystals with those of free atoms, it is necessary to ascertain the degree of ionization in the latter. A plausible, empirically well-confirmed rule is that in electrical excitation higher states of ionization are favored by higher potential gradients. Arc discharges will produce predominantly neutral atom or "second" (singly ionized) spectra; sparks tend to produce higher orders of ionization commensurate with the maximum current developed.[53] By comparing the spectra of free atoms under systematically changed electrical discharge conditions, it is, in general, possible to assign the various line systems to the proper ions. In addition, it may be informative to compare the spectra of successive elements of increasing stages of ionization, provided their electron configurations are alike. This condition does not hold if inner rather than outer electrons are ionized.

Separation of Terms for Higher Degrees of Ionization. The existence of unfilled shells implies that electrons are bound to the inner shells with smaller bonding energy than to certain outer states. This fact leads to an overlap in the energy of different electron configurations. For instance, $4f^n$ configurations may overlap states in which one or two of the $4f$-electrons are in $5d$ or $6s$ orbits. However, as the degree of ionization increases, a differentiation in energy sets in: terms with $5d$- and $6s$-electrons assume consecutively higher values.

4.6.4.3 *Energy Levels of the Rare Earth Ions*

As pointed out before, the sharpness of transitions between $4f$-electron configurations endows the rare earths with a unique position in solid state spectroscopy. This is the case not just because these transitions provide an important link between atomic and solid state spectra and, among other information, yield clues to the crystalline charge and field distributions. The sharp lines emitted in fluorescence are of practical importance as well, notably for laser operation as we shall see in Chapter 6.

Emission and absorption spectra are used in combination for the identification of the levels and the transitions between them. An example is shown for Er^{3+} in Fig. 4.26 (p. 105) with lines of emission and absorption transitions to the same group of states resulting from the splitting of a level by the crystal field. Figure 4.27 shows the term scheme of trivalent lanthanide ions as derived by Dieke and his co-workers.[53, 60] Divalent ions exhibit significant differences in their spectra with respect to the triply ionized lanthides. They show strong broad absorption bands in combination with sharp fluorescent lines. The former appear to belong to $4f \rightarrow 5d$ transitions[61] which should be strong because they are allowed and broad because the $5d$-state of the divalent ion is exposed to perturbations by the crystal field; the latter are again due to transitions between the $4f$-configurations.

A situation very much related to that of the lanthanides with their incomplete $4f$-configurations exists for the actinides, rare earths in which the $5f$-shell is being filled up. The actinides, which extend from thorium ($Z = 90$) to lawrencium ($Z = 103$) are all radioactive, although thorium and uranium have a relatively long lifetime. Among the lanthanides only promethium ($Z = 61$), a fission product of uranium, is radioactive.

Radioactivity in rare earth crystals produces a number of particular effects. Aside from the difficulty of handling, the materials are often too short-lived for accurate observations. Some radioactive crystals are self-luminiscent, such as in the case of promethium,[62] as a result of excitation by beta ray bombardment. The emission of radioactive particles, especially electrons, produces the further effect of blackening by F-centers and the like. By this process, opaqueness might develop within a matter of hours, but it can be bleached out again by heating of the crystal.

The spectra of the actinides show narrow lines, as do the lanthanides, owing to transitions between f-configurations. The sharpness of these transitions may apparently exceed even that of the lanthanides, although

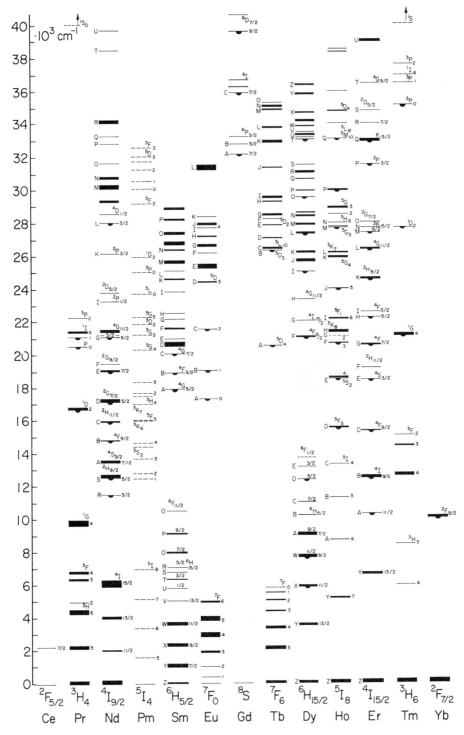

FIG. 4.27. Observed energy levels of the free trivalent lanthanide ions [from G. H. Dieke and H. M. Crosswhite, *Appl. Optics* **2**, 675 (1963)].

the 5*f*-electron can be expected to have a larger orbit than the 4*f*-configuration and should be less shielded by the next shell. We mention in passing the spectra of the transition metals which also exhibit relatively sharp, gaslike lines, the result of transitions between unfilled *d*-configurations. Here again transitions occur with $\Delta n = 0$ so that the radiative electron remains in a tight orbit around the nucleus. However, the feature of shielding by outer electrons is missing for the crystal ions of the transition metals because the electrons outside the *d*-shell (2*s*-electrons) have been removed by ionization. Thus, oscillator strengths are relatively high while the spectral lines are still quite narrow.

4.6.4.4 *Exciton Spectra*

In 1931, Frenkel[63, 64] predicted a type of crystal excitation in semiconductors or insulators in which an electron bound to an ion is brought into a higher energy state without reaching the conduction band. This process, of course, is very similar to excitation of electronic levels in the free atoms of gases before ionization is reached; but it is significant for solid state phenomena that the energy can be coupled to the lattice neighbors and thus is propagated through the crystal. An equivalent description of the process treats the electron as bound to a hole, and this configuration, which is called an exciton, may or may not move through the crystal. A number of optical phenomena of semiconductors and insulators has been attributed to the creation and destruction of excitons. Most persuasive are patterns of absorption lines seen, e.g., in Cu_2O, which resemble the hydrogen series in their convergence into a continuum.[65, 66] This suggests an energy term scheme for excitons such as shown in Fig. 4.28. In analogy to the hydrogen system, Rydberg-

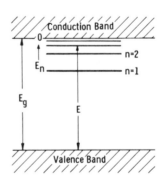

FIG. 4.28.   Exciton level structure. In analogy to the hydrogen term scheme, zero energy is defined for $n \to \infty$ (onset of ionization).

type energy levels can be expected:

$$E_n = -\frac{2\pi^2 m_r e^4}{\epsilon^2 h^2} \cdot \frac{1}{n^2} \tag{4.102}$$

This expression results from Eq. (3.39) for $Z = 1$, for a reduced hole-electron mass m_r, and a dielectric constant ϵ of the lattice medium. Good agreement of Eq. (4.102) with observation has been obtained in a number of cases.

Exciton structure has been found in many semiconductors and insulators, in particular the alkali halides, cuprous compounds, silicon, germanium, and silicon carbide. The lines are often remarkably sharp, particularly for bound excitons.

The existence of excitons reveals itself also by phenomena other than absorption. For example, photoemission (see Chapter 7) can be enhanced by excitons as intermediary factors.[67] In general, the destruction of excitons, i.e., the recombination of hole and electron, sets free the energy

$$E = E_g + E_n \tag{4.103}$$

where E_g is gap energy and E_n the excited level given by Eq. (4.102). Aside from enhanced photoemission, other conversion mechanisms are possible, the production of phonons, F-centers, and photons being the most important examples.

A feature distinguishing excitons from electron-hole pair production across the gap is the fact that they transport excitation energy, but no net charge, through the lattice. Hence excitons do not contribute to photoconductivity. However, they may aid photoconductivity indirectly, just as in the case of photoemission, by liberating an F-center electron with the energy set free during recombination. It is interesting to reflect that these processes represent occurrences which are removed in time and space from the origin of the excitons. Unless immobilized, excitons move through the crystal with a velocity corresponding to an energy in the order of kT, 10^7 cm/sec at room temperature. Assuming an unperturbed lifetime of 10^{-8} sec (e.g., from classical considerations as in Section 3.1.3.1), we arrive at a path length of 1 mm, although elastic scattering with a mean free path of an estimated 10 Å will limit the diffusion length to 10^{-4} cm as a result of random step addition (cf. Seitz[68, 69]). However, in general the exciton lifetime will be shortened by collisions with imperfections in radiationless processes such as the generation or destruction of F-centers.

The existence of excitons is apparently not limited to intrinsic materials. Excitons bound to impurity states can give rise to sharp lines in absorption and fluorescence at wavelengths longer than the edge of the fundamental region. The occurrence of both types of excitons, depending on the presence of impurities was demonstrated by Haynes[70] for silicon (see Fig. 4.29).

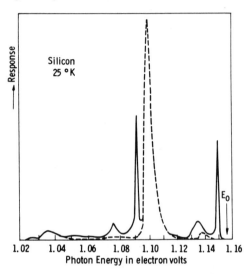

FIG. 4.29. Fluorescent emission spectrum in silicon with negligible amount of impurity (dashed curve) and with 8×10^{16} cm^{-3} arsenic atoms (solid curve). The sharp peaks of the solid trace are ascribed to impurity exciton fluorescence. The broader spectral lines of the pure sample are apparently due to intrinsic exciton recombination from J. R. Haynes, *Phys. Rev. Letters* 4, 361 (1960)].

The halfwidth of exciton lines is smallest for immobilized states; in propagating excitons it cannot be smaller than that given by their velocity distribution. In the alkali halides the exciton spectra have a width of about the same order as the transitions from the valence to the conduction band with which they form a complex called the *fundamental absorption band*.

Figure 4.30 shows examples for the fundamental absorption band of various alkali halides. The first and perhaps also the second peak correspond to excitons. The spectrum at shorter wavelengths, however, represents electron-hole pair production across the gap and as a result is associated with photoconductivity. Since the width of these bands is still quite narrow, at least at low temperatures, the excitation may be considered as a localized process and has been assigned, for this reason, to group II in Table 4.3.

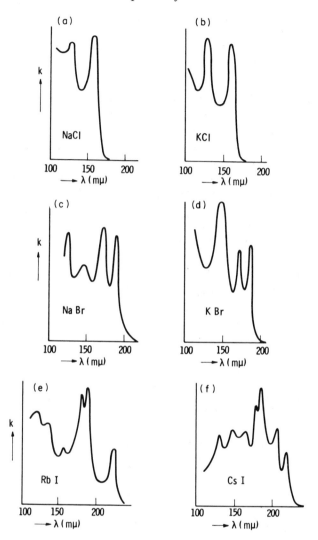

FIG. 4.30. Fundamental absorption bands of alkali halides. [E. Fick and G. Joos, Kristallspektren, *in* "Handbuch der Physik" (S. Flügge, ed.), Vol. 28, p. 286, Springer, Berlin, 1957].

This atomistic interpretation implies that the oscillator strengths of the exciton and band-to-band transitions should add up to the order of unity. According to Eq. (3.101c) the absorption coefficient should be then in the order of 10^6. This is found to be the case for exciton and band-to-band spectra. In other words, the mean free path of photons is in the order of 100 Å, and therefore this represents the correct magnitude of sample thicknesses through which absorption measurements should be taken.

4.6.4.5 Color Centers

When alkali halides or generally ionic crystals are exposed to short-wavelength photons or particle bombardment of sufficient energy, they develop absorption bands in the visible and other spectral regions and transmit the complimentary color. Thus sodium chloride turns yellow under ultraviolet, X-ray, or electron bombardment, whereas a silver chloride crystal will become violet merely after prolonged illumination with blue light. The phenomenon was first described by Goldstein[71] in 1896 in connection with cathode rays and explored systematically several decades later by Pohl and his co-workers.[72] Detailed discussions of the subject are presented in the reviews of Seitz[69, 73] and the books of Mott and Gurney,[66] and Schulman and Compton.[74]

A typical absorption spectrum resulting from irradiation of alkali halides with X-rays is shown Fig. 4.31. In the main, one distinguishes three bands: (1) a strong peak in the visible (F-band); (2) one or more peaks in the ultraviolet (V-band); and (3) a weak absorption region in the red or near infrared (M-band). A number of phenomena are associated with the coloring of the crystals. For example, irradiation of the crystal with light in the F-band region reduces the color density but

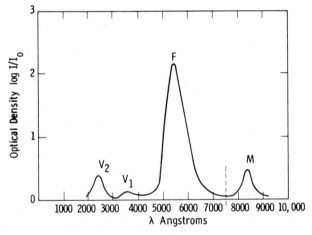

Fig. 4.31. Typical shape and position of V, F, and M absorption bands of KCl after X-ray or electron bombardments [H. Dorendorf, Z. Physik **129**, 317 (1951); see also R. Casler, P. Pringsheim, and P. Yuster, Phys. Rev. **18**, 1564 (1950) for V- and F-bands; C. Z. Van Doorn and Y. Haven, ibid. **100**, 753 (1955) for M-Bands].

creates new absorption bands in the red (*R*-bands). Bleaching can occur in some crystals spontaneously; in other samples the colorless state is restored by heating. More durable are *F*-bands produced by chemical means—heating of the alkali halide in the vapor of the alkali metal. Here heating does not bleach the *F*-band, but creates additional absorption regions. The phenomenon that is perhaps most suggestive of the underlying processes is that of electrolytic coloration.

This experiment is best performed with heated alkali halide crystals which are exposed to an electric field between a point electrode and a flat plate. Under these conditions a color cloud develops at the region of highest gradient near the point and gradually grows and extends towards the positive plate.

The sum total of these and related phenomena can be understood with a model based on *color centers* which are assumed to result from certain lattice defects.[75] Lattice imperfections exist naturally in any crystal or they may be introduced by some treatment such as the exposure to alkali vapor. The pertinent defect in this case is the absence of an ion from its normal lattice site—the absence of an anion from the negative or of a cation from the positive site. For instance, an *F*-center

creates new absorption bands in the red (*R*-bands). Bleaching can

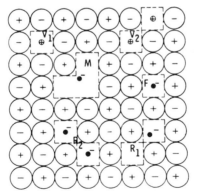

F𝐼𝐺. 4.32. Model of color centers in ionic crystals. Various observed absorption bands are ascribed to lattice imperfections some of which are shown here: the V_1-center (a hole bound to a cation vacancy; the V_2-center (two associated V_1-centers), the *M*-center (an electron associated with two negative and one positive vacancies); the *F*-center (an electron replacing an anion), the R_2-center (two coagulated *F*-centers); and the R_1-center (an *F*-center associated with an anion vacancy).

results when an electron is trapped at the site of a missing anion (see Fig. 4.32). The interpretation of the other absorption bands is somewhat less certain. A V-center, according to Seitz,[69, 76] is the complimentary configuration of the F-center, namely, a hole substituting a vacant cation. An M-center is believed to be an L-shaped junction of an F-center surrounded by a vacancy of a cation-anion pair (see Fig. 4.32).

The location of absorption bands in the spectrum for various crystals appears to be directly related to the lattice constant. This is shown in Fig. 4.33 which presents the F-bands of the alkali halides with face-

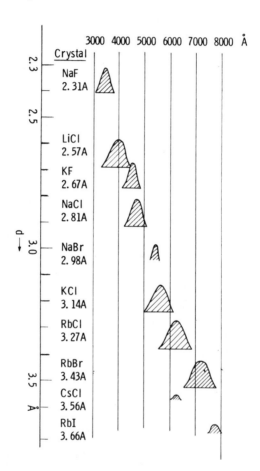

FIG. 4.33. F-bands of some alkali halides. The maximum of each band is plotted so as to show the dependence of λ_{max} with interionic distance d, the latter being indicated below the crystal. Halfwidth data are those of H. Pick [*Nuovo Cimento* [10] **7**, Suppl., 498 (1958)]; where they are not known, the bands are only partially drawn.

centered cubic structure. Mollwo,[77] on empirical grounds, expressed this relationship between the maxima of the F-bands and the interatomic distance d (in angstroms) by the equation

$$\lambda_{max} = 667d^2 \quad (\text{Å}) \tag{4.104}$$

More accurate empirical relationships were derived by Ivey[78] for various types of bands (see Table 4.4).

TABLE 4.4

EMPIRICAL RELATIONSHIPS BETWEEN BAND MAXIMA λ_{max} (IN ANGSTROMS) AND INTERIONIC DISTANCES d (IN ANGSTROMS) FOR VARIOUS COLOR CENTERS ACCORDING TO IVEY[a]

Band	F	U	R_1	R_2	M
Maximum λ_{max}	$703d^{1.84}$	$615d^{1.10}$	$816d^{1.84}$	$884d^{1.84}$	$1400d^{1.56}$

[a] H. Ivey, *Phys. Rev.* **72**, 341 (1947).

Although quantum-theoretical derivations exist for the $\lambda_{max}(d)$-relationships of F-centers that agree with observation within about 10–20% (see e.g., Gourary and Adrian[79] or Martino[80]), it may be useful to show here that an interpretation is also possible by means of simple classical models. For instance, if we assume that an anion vacancy in a face-centered cubic lattice (see Fig. 4.34) can be represented by a positive charge e uniformly distributed over a cube with edge $d' = \sqrt{2}d$, an elastic restoring force $-Gr$ will be experienced by the trapped electron. This condition applied to F-centers is described by the linear oscillator equation

$$m\ddot{r} + Gr = m\ddot{r} + \frac{4\pi}{3\sqrt{8}} \frac{e^2}{d^3} r = 0 \tag{4.105}$$

where m and e are electronic mass and charge, respectively, and d is the interionic distance. The resonance wavelengths follow from Eq. (4.105) as

$$\lambda_{max} = 5.16 \frac{c}{e} \sqrt{md^3} = 955d^{3/2} \tag{4.106}$$

Table 4.5 presents a comparison between observed values of λ_{max} and those following from Eq. (4.106).

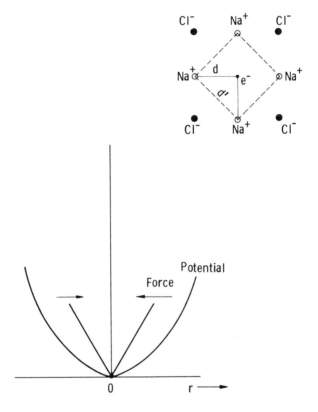

FIG. 4.34. A classical model for F-centers in fcc alkali halide crystals.

According to Section 3.3.3.2, the absorption coefficient k_{max} in the center of a band should be proportional to the oscillator strength f, the density N of color centers and the reciprocal bandwidth $(\Delta\nu_h)^{-1}$. Essentially such a relationship was derived by Smakula[81] for crystals with a refractive index n_0 and which, somewhat simplified for narrow bands of Gaussian shape, has the form

$$k_{max} = \frac{2}{9} \frac{e^2}{mc} Nf \frac{(n_0^2 + 2)^2}{n_0} \frac{1}{\Delta\nu_h} \tag{4.107}$$

Since the oscillator strengths for the F-centers in most crystals are not far from unity, the maximum absorption coefficients are by order of magnitude

$$k_{max} \approx 10^{-16} \frac{N}{W} \tag{4.108}$$

where W is the halfwidth of the band in electron volts and k_{max} in cm^{-1}. An approximate value of the color center density can be obtained from Eq. (4.107) or (4.108), if the absorption is known from measurement.

TABLE 4.5

SPECTRAL MAXIMUM OF F-BANDS FOR FACE-CENTERED CUBIC ALKALI HALIDES

Crystal	Interionic Distance (Å)	From Eq (4.106) (Å)	Max observed (Å)
LiF	2.01	2730	2550
LiCl	2.57	3930	3850
LiBr	2.75	4350	—
LiI	3.00	4970	—
NaF	2.31	3350	3380
NaCl	2.81	4500	4640
NaBr	2.98	4930	5400
NaI	3.23	5560	5880
KF	2.67	4160	4550
KCl	3.14	5320	5600
KBr	3.29	5690	6280
KI	3.53	6040	6910
RbF	2.82	4510	—
RbCl	3.27	5640	6165
RbI	3.66	6700	7650
CsF	3.00	4960	—

4.6.5 TRANSITIONS BETWEEN NONLOCALIZED STATES

The third major group of solid state spectra consists of transitions in which the electron or hole changes position from one ion or atom to the next along with the change in energy. In fact, in some cases we may consider the "location" as being distributed over the entire lattice. That such differentiation from the previously discussed spectra of solids is somewhat arbitrary has already been mentioned in the introduction to this subject. Thus the optics of metals certainly deals with non-localized states although we have found it convenient to list it among the plasma oscillations. Again, the spectra of mobile excitons were classified in the second group only because it could be argued that the transitions themselves are localized and, at any rate, behave more like the processes listed in group II than like any in group III. Nevertheless, certain properties which appear together, give the transitions we have to discuss in the following a common basis. One such characteristic is the limitation of this subject more or less to semiconductor optics. In other words, we shall be dealing with transitions of electrons and holes involving bands. This leads to another feature of this group, the appearance of continua in absorption; and although some emission spectra are not continua, their widths will still be in the order of kT or more, corresponding to Q-values of 10^2 or less. A related property is that the spectra are featureless in the sense that they do not reveal their origin in a particular ion or atom species. This "anonymity" is in contrast, for instance, even to the fundamental absorption band of the alkali halides which show a strong resemblance among crystals with the same anion. It will be seen that one or the other of the properties cited here are shared by some of the previously described spectra; but the occurrence of all of them is limited to group III.

4.6.5.1 Band-to-Band Transitions and Injection Electroluminescence

The generation of a hole and electron pair by a photon of energy larger than the forbidden gap and the inverse event of recombination are of interest from several points of view. The intrinsic absorption process is perhaps most important because of the secondary effect it produces with virtually 100% quantum efficiency, that of photoconductivity, and therefore will be discussed in greater detail along with the latter in Chapter 7. Both emission and absorption across the gap reveal, of course, detailed interactions and structure of semiconductors in general. We shall comment in the following only on aspects of recombination which are most important for theory and application.

Direct and Indirect Recombination. It is shown in solid state theory that in a crystal, the electron energy, which in the foregoing (cf. Figs. 4.23–4.25) has been indicated summarily as being spread into bands, in detail depends within such bands on the momentum vector $\mathbf{p}$ of the electron, i.e., on amount and direction of a motion which corresponds to a discrete energy value. This situation is analogous to the continuum of oscillators described in Section 1.2.3 which has been represented as composed of a finite number of discrete modes. Thus, in general, the dependence of energy is three-dimensional, within a fixed coordinate system of the lattice:

$$E = E_0 + E_k(k_x, k_y, k_z) = E_0 + E_k(p_x/\hbar, p_y/\hbar, p_z/\hbar) \qquad (4.109)$$

Here E_0 is a fixed reference value of the band, e.g., its edge, and $\mathbf{p}$ has been expressed in terms of a wave vector $\mathbf{k}$ already introduced in Section 4.6.1.3,

$$p_x = k_x \cdot \hbar, \qquad p_y = k_y \cdot \hbar, \qquad p_z = k_z \cdot \hbar \qquad (4.110)$$

The kinetic energy of a free electron is given by $E_k = mv^2/2 = p^2/2m$. Hence the simplest type of energy surface should be given by

$$E_k = \frac{h^2}{8\pi^2 m^*} (k_x{}^2 + k_y{}^2 + k_z{}^2) \qquad (4.111)$$

where m^* represents the (scalar) *effective* electron mass which compensates for the fact that unspecified internal fields affect the equation of motion. The energy for electrons in the conduction band and holes in the valence band according to Eq. (4.111) is shown for one dimension in Fig. 4.35a in which two different masses for the two bands have been assumed. This shape of the energy bands of spherical symmetry, if all directions are considered, with the extrema both occurring at $k = 0$, constitutes the simplest, if rarely realistic case. This simplicity extends also to considerations of recombination radiation. If pairs of electrons and holes have been generated by such means as absorption, they will assume values of lowest energy, hence positions near the extrema, except for the modifying effect of the Boltzmann distribution. Recombinations then can be *direct*, without a change of momentum, since the photon impulse h/λ can be essentially ignored for wavelengths in the order of 1μ or more (cf. Section 3.4.2.1). Such transitions are indicated by vertical lines in the $E(k)$ diagrams.

Actually, the effective mass m^* will in general depend on the direction of the **k** vector, and this fact destroys at once the symmetry of the energy surfaces. Furthermore an additional, although small, energy term is contributed by spin-orbit interaction of the electrons. For most cases, therefore, Eq. (4.111) is not valid, and as a result the $E(k)$ surfaces are more complicated, a fact first shown by Herman.[82, 83, 84] In particular, for most semiconductors the lowest values of energy in the conduction band will be located at k-values different from zero as shown for one dimension in Fig. 4.35b. At low temperatures electrons tend to dwell in this region, although they may have been produced by absorption in a direct transition at $k = 0$. Recombination with holes at $k = 0$ then requires a change in the electron momentum by $\Delta k \neq 0$. Because the total momentum must be conserved, such *indirect transitions* are accompanied by the generation or absorption of a phonon. The probability of such an event—involving the simultaneous interaction with a photon and a phonon—is considerably smaller than that of a direct transition. The alternatives for the recombination spectrum are (1) that it is in fact very weak; (2) or, at not too low temperatures, direct transitions are preferred despite of the unfavorable Boltzmann factor; or (3) that the transitions occur by way of impurities, producing light of usually considerably longer wavelength than that corresponding to the gap.

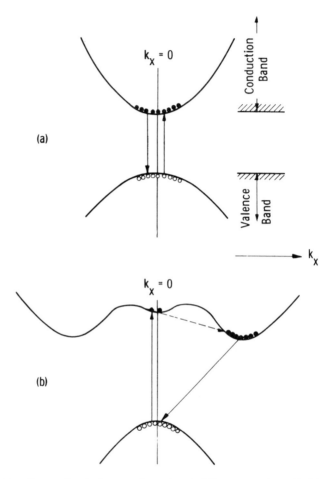

FIG. 4.35. Energy bands in one-dimensional $E(k)$ presentation. (a) Simplest case in which effective mass does not depend on direction and in which both extrema occur at $k = 0$; (b) the usually occurring configuration in which the lowest energy value of the conduction band is at $k \neq 0$ making indirect recombination transitions likely.

Nonradiative transitions are of even larger importance for pair recombination than for the solid state spectra we have discussed so far. The usual situation is in fact that the pair energy is ultimately dissipated in the crystal as heat through phonons, long-wavelength photons and excitation of intermediate centers. Thus recombination radiation for hole-electron pairs is the exception rather than the rule unless specially designed conditions prevail such as the production of large carrier densities.

Electroluminescence. The problem of recombination radiation is, of course, of vast practical importance since it offers the potential for light sources with optimal properties (high brightness and efficiency, small size, operation at low lattice temperatures). Various excitation mechanisms exist for the production of holes and electrons, and among these the first to be used were (1) photon conversion, mainly in the ultraviolet, in fluorescence and phosphorescence and (2) electron bombardment or cathodoluminescence. These processes operate with recombination through impurities and their efficiency is actually quite low. More recently, however, another mode of excitation, viz., injection electroluminescence, has proven highly successful in terms of efficiency and recombination across the gap, and we shall now discuss the pertinent aspects in greater detail.

The emission of light from a solid in response to an applied voltage is generally called *electroluminescence*. The field as a whole has been reviewed by Ivey,[85] Henisch,[86] and Hahn,[87] among others. Aside from a number of miscellaneous mechanisms, deemed at present less important, two major processes stand out: (1) the high-field effect (Destriau[88]) typified by ZnS : Cu (zinc sulphide activated by copper) for which experimentation is relatively simple, but the theory complex; and (2) injection luminescence (Lossew[89]) which is easily understood, but more difficult to demonstrate. The first of these usually involves impurity transitions and, partly for that reason, represents a more or less inefficient conversion of electrical energy into light. Injection luminescence, however, relies essentially on band-to-band recombination, and the efficiency of this process can be in the order of 100%.

The operation of the injection electroluminescence process with which we are concerned here depends basically on the introduction of minority carriers into a semiconductor, i.e., of electrons into p-type and of holes into n-type material. Radiation in energy about equal to that of the band gap results then by recombination of the injected and the majority carriers.

This process is possible with so-called p-n junctions which result when p-type and n-type semiconductors (usually of the same material, cf. Fig. 4.25b and c) are brought into contact with each other. In solid state theory it is shown that junctions between different materials develop a contact potential such that the Fermi level of the components is at the same potential. For this to be the case, electrons from the n-type material move across the junction into the p-type zone, from which, in turn, holes move across the boundary into the n-type region (see Fig. 4.36a and b). As a result the n-type region is charged positively with respect to the p-type material, and an electric field is formed across the junction.

It is not difficult to see that as a further consequence the junction is now a rectifier. First of all, the junction field presents a barrier to the majority carriers, i.e., the junction retards the electrons in the n-type side and the holes in the p-type region, and since the majority carriers are, of course, responsible for most of the conductivity, the region has become highly resistive. If a *reverse bias* is applied, i.e., if a negative voltage is applied to the p-region (see Fig. 4.36c), the junction field and hence the barrier height is further increased; hence conductivity is quite small. Conversely, if a *forward bias* is impressed on the junction such that a

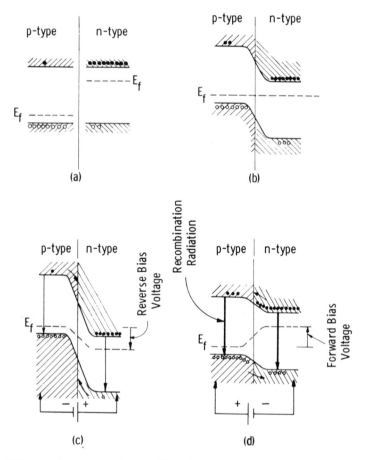

Fig. 4.36. P-n junctions with and without bias voltage. (a) Conditions before contact is made; (b) energy levels in p- and n-materials after contact is made (Fermi levels of both components come into equilibrium); (c) equilibrium disturbed by reverse bias; (d) forward bias and recombination radiation.

negative voltage drives the n-region, the Fermi levels are so shifted with respect to each other that field and barrier are reduced (see Fig. 4.35d). Hence, relatively good conduction is now established: electrons are drifting in large numbers across the junction into the p-region, just as holes are diffusing into the n-region, and both types of carriers are recovered from the electrodes to the external voltage. Thus the conditions are realized which have been mentioned as favorable for injection luminescence, i.e., the introduction of a large number of minority carriers which can recombine with the majority carriers under emission of light.

Nevertheless, although the theoretically possible conversion efficiency exceeds 100% with the aid of heat energy, in practice a number of additional conditions have to be fulfilled before the potential of this type of solid state light source can be fully utilized. This includes requirements which are relatively easy to satisfy, viz., ohmic (nonrectifying) contacts to the exterior circuit, good conductivity through the junction, and prevention of overflow of the minority carriers to the electrodes before recombination—in all, conditions which provide for an ample supply of holes and electrons. However, two more conditions have to be fulfilled for good light output, namely, (1) a high efficiency of radiative recombination, and (2) high photon escape probability. This last requirement, which amounts to a demand for low self-absorption, is not more stringent than usual for the emission of resonance radiation. We are left then with the photon yield as the last hurdle, and indeed it has been mentioned before that the incidence of radiative recombinations is usually negligible compared to nonradiative transitions. It appears plausible from the foregoing that high carrier densities, operation at low temperatures, the possibility of direct transitions or at least radiative transitions by way of impurities will all be favorable to a high conversion efficiency of electrical energy into recombination radiation.

These conditions were first obtained (see Fig. 4.37) with p-n junctions of gallium arsenide, usually prepared[90] by diffusing zinc (from a dilute solution in gallium) into n-type single crystal wafers of GaAs (doped e.g., with 10^{18} to 10^{19} tellurium atoms which enter as donors). The current in the forward direction of a junction diode increases exponentially with voltage until saturation begins, and this occurs typically at 10^3 to 10^4 amps/cm^2 over an area of about 10^{-3} cm^2 so that pulsed operation may become necessary. In gallium arsenide, recombination takes place with almost theoretical efficiency—upward from 0.8 photon per carrier. A beam of very high brightness—in the order of several kw/cm^2 · sterad—is therefore generated, its spectrum mainly concentrated near 8500 Å with a high energy tail in the red region, as reported by several observers.[90, 91, 92] The width of the recombination spectrum is quite narrow,

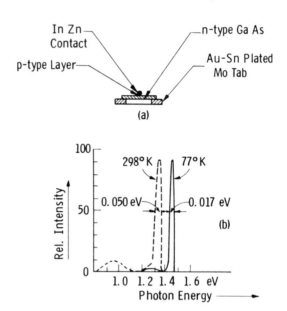

FIG. 4.37.   (a) Cross section of GaAs diode; (b) injection electroluminescence spectrum of GaAs at 77°K and 298°K, both on independent relative scale [from R. J. Keyes and T. M. Quist, *Proc. IRE* **50**, 1822 (1962)].

e.g., 0.017 eV at 77°K because the carriers are concentrated near the band edge (see Fig. 4.25a). The $E(k)$ surfaces in GaAs are such as to allow direct transitions. Actually the peak emission occurs at an energy somewhat less than the gap; presumably these processes are not distinctly defined because of the density of impurity levels close to the band edge. Two other features stand out in this remarkable development of solid state light sources. The recombination rate is high enough so that the radiation intensity can be modulated at frequencies of the order 10^8 to 10^9 sec^{-1}, a capability which had been out of reach for other bright light sources by orders of magnitude. The second feature is the adaptability to laser operation, and this is described in Chapter 6.

After these results had been obtained with GaAs, injection electroluminescence was applied successfully also to other semiconductors under similar conditions, again by the use of p-n junctions in the forward direction. An example is indium phosphide[93] with a peak emission near 9100 Å—about 150 Å longer than the value expected from direct band-to-band recombination across a gap of 1.395 eV. This diode, too, could subsequently be adapted to laser action.[94]

We mention briefly electroluminescence produced by injection in the reverse direction (see Fig. 4.36c). Conductivity in this case is quite low, but the minority carriers may acquire a high kinetic energy as they pass across the junction. In fact, even after many collisions this energy may still be larger than the binding energy when the electrons reach the surface, and thus a small current may be emitted from this kind of cold cathode.[95, 96] Therefore, a so-called "hot electron" distribution results which, in combination with the low carrier density, is responsible for a low efficiency of recombination radiation. The material first and most extensively studied[89, 97, 98, 99] under reverse bias conditions is silicon carbide, which forms highly refractory and, in the pure state, transparent crystals with rectifying properties. According to Patrick,[99, 100] the junctions which occur normally in industrially grown SiC crystals are more complex; the p- and n-regions are separated by an insulating zone of about 10^6 Ω-cm resistivity which may be the seat of recombination (p-i-n junction).

4.6.5.2 Optical Transitions Involving Impurity Levels

The role of impurities in recombination radiation discussed in the foregoing paragraphs has been limited essentially to rendering semiconducting media p- or n-type, i.e., to providing holes or electrons, respectively. In addition, of course, the impurity atoms themselves provide energy terms which lie in the forbidden gap of the intrinsic material. This follows from the fact that it requires an energy smaller than the gap energy to remove an electron from a donor state and set it free in the conduction band. In turn, the ionized donor state can capture a free electron, and the energy available from this trapping process may be emitted as a photon. Similarly, holes may be excited into the valence band from acceptor sites; or in the inverse process, a free hole may be captured with accompanying emission of light. Thus it will be seen that impurity levels may represent themselves the initial or final states of radiative transitions in emission or absorption. In fact, some of the injection luminescence phenomena discussed before may not be real band-to-band recombinations but transitions through impurity levels close to the band edge.

Extrinsic absorption is mostly small compared to the intrinsic case. This follows from Eq. (3.101c) which shows, e.g., that for an impurity density of 10^{15} atoms/cm^3 and a halfwidth of 0.1 eV, the absorption coefficient in the center of the band is about 1 cm^{-1}. Absorption by impurity levels is indirectly of importance, however, through its secondary effect of photoconductivity which results here from a change in conductivity by a variation in the number of free holes or in the number of free electrons. This subject is discussed in detail in Chapter 7.

The emission process involving impurity levels can be initiated by a variety of exciting mechanisms, such as electron bombardment in cathodoluminescence, electric fields in the Destriau-type electroluminescence, and photon excitation in fluorescence (and in *phosphorescence* which is defined somewhat arbitrarily, for the case that the secondary light emission process lasts beyond the 10^{-8} secs typical for electric dipole radiation in the visible region).

Fluorescence of the kind discussed here for semiconductors, unlike the emission processes of the rare earth ions or *F*-centers, is associated with carrier motion in the valence and conduction bands. This follows from the fact that photoconductivity can be observed to accompany and actually outlast the luminescent emission. The basic process is pair production across the gap and subsequent recombination processes with the impurity levels forming intermediate steps. A simple explanation has been proposed by Lambe and Klick,[101] and typically consists of the following sequence (see Fig. 4.38): (1) absorption of photon (which may be in the visible or ultraviolet region) and pair production; (2) diffusion of hole and electron through the lattice more or less independently of each other; (3) capture of hole by impurity center with fluorescent emission of photon (e.g., in the visible region); (4) capture of electron by the neutralized impurity center some time later, again with emission of photon (e.g., in the infrared), completing the cycle. As in previous examples, one or both of the transitions to the center may be radiationless resulting in phonons rather than in photons (see also Kittel,[36] p. 525). The model of Lambe and Klick is an alternative to a mechanism previously proposed by Schön[102] and Klasens.[103, 104] These two models and their further developments are, however, quite similar, and a choice between them on the basis of experiment appears to be difficult.

There exist, of course, more complicated mechanisms of ultimate recombination, involving simultaneously several types of impurity

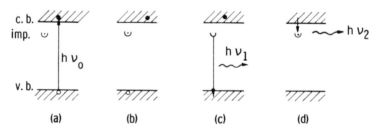

Fig. 4.38.   Four steps in luminescence process [according to J. Lambe and C. C. Klick, *Phys. Rev.* **98**, 909 (1955)].

centers, for example. Almost all conceivable combinations of the various transitions and their sequences have been suggested—probably correctly —for diverse luminescent phenomena. Further details are treated in the texts and symposia reports[105,106] on luminescence.

REFERENCES

1. G. E. UHLENBECK and S. A. GOUDSMIT, *Naturwissenschaften* **13**, 953 (1925).
2. G. E. UHLENBECK and S. A. GOUDSMIT, *Nature* **117**, 264 (1926).
3. W. PAULI, *Naturwissenschaften* **12**, 741 (1924).
4. H. N. RUSSEL, see in W. F. Meggers and K. Burns, *J. Opt. Soc. Am.* **14**, 449 (1927).
5. D. M. DENNISON, *Proc. Roy. Soc.* **A115**, 483 (1927).
6. S. A. GOUDSMIT, *Physics Today* **14**, 18 (1961).
7. W. PAULI, *Z. Physik* **31**, 765 (1925).
8. A. SOMMERFELD, *Ann. Physik* [4] **51**, 1 (1916).
8a. H. E. WHITE, "Introduction to Atomic Spectra." McGraw-Hill, New York, 1934.
9. L. PAULING and E. B. WILSON, "Introduction to Quantum Mechanics," McGraw-Hill, New York, 1935.
10. H. WEYL, "Gruppentheorie und Quantenmechanik," 2nd ed. Hirzel, Leipzig, 1933.
11. P. ZEEMAN, *Phil. Mag.* [5] **43**, 226 (1897).
12. F. PASCHEN and E. BACK, *Physica* **1**, 261 (1921).
13. H. KOPFERMANN, "Nuclear Moments" (transl. by E. E. Schneider). Academic Press, New York, 1958.
14. P. KUSCH, *Phys. Rev.* **100**, 1188 (1955).
15. H. SCHÜLER, *Z. Physik* **59**, 150 (1930).
16. D. A. JACKSON and H. KUHN, *Proc. Roy. Soc.* **A148**, 335 (1935).
17. R. MINKOWSKI, *Z. Physik* **95**, 284 (1935).
18. See e.g. J. STRONG, "Concepts of Classical Optics," p. 212f. Freeman, San Francisco, California, 1958.
19. I. RABI, S. MILLMAN, P. KUSCH, and J. R. ZACHARIAS, *Phys. Rev.* **55**, 526 (1939).
20. F. BLOCH, W. W. HANSEN, and M. PACKARD, *Phys. Rev.* **69**, 680 (1946).
21. E. M. PURCELL, H. C. TORREY, and R. V. POUND, *Phys. Rev.* **69**, 37 (1946).
22. See e.g. F. RAMSEY, *Microwave J.* **6**, 89 (1963).
23. J. SHAPIRO and G. BREIT, *Phys. Rev.* **113**, 179 (1959).
24. A. RUBINOWICZ and J. BLATON, *Ergeb. Exakt. Naturwiss.* **11**, 176 (1932).
25. S. PASTERNACK, *Astrophys. J.* **92**, 129 (1940).
26. G. H. SHORTLEY, *Phys. Rev.* **57**, 225 (1940).
27. L. A. BORISOGLEBSKIĬ, *Soviet Phys.—Usp.* **66**, No. 1, 211 (1958).
28. R. H. GARSTANG, *in* "Atomic and Molecular Processes" (D. R. Bates, ed.), p. 1. Academic Press, New York, 1962.
29. G. RACAH, *Phys. Rev.* **62**, 438 (1942).
30. R. H. GARSTANG, *Proc. Cambridge Phil. Soc.* **53**, 214 (1957).
31. H. C. BRINKMAN, Dissertation, Utrecht, 1932.
32. J. BLATON, *Z. Physik* **89**, 155 (1934).
33. G. HERZBERG, "Molecular Spectra and Molecular Structure," Vol. I: Diatomic Molecules. Prentice-Hall, Englewood Cliffs, New Jersey, 1939.
34. J. FRANCK, *Trans. Faraday Soc.* **21**, 536 (1926).
35a. E. U. CONDON, *Phys. Rev.* **28**, 1182 (1926)

35b. E. U. Condon, *Phys. Rev.* **32**, 858 (1928)
36. C. Kittel, "Introduction to Solid State Physics," 2nd ed. Wiley, New York, 1953.
37. R. A. Smith, "Semiconductors." Cambridge Univ. Press, London and New York, 1959.
38. D. S. McClure, *Solid State Phys.* **8** (1959).
39. D. S. McClure, *Solid State Phys.* **9** (1960).
40. M. Born, "Atomic Physics." 5th ed. Hafner, New York, 1951.
41. H. K. Paetzold, *Ann. Physik* [5] **37**, 470 (1940).
42. H. K. Paetzold, *Z. Physik* **129**, 123 (1951).
43. E. Fick and G. Joos, *In* "Handbuch der Physik" (S. Flügge, ed.), Vol. 28, p. 205. Springer, Berlin, 1957.
44. G. Joos and H. Ewald, *Göttinger Nachr. Math.-Physik. Fachgruppe II* **3**, 71 (1938).
45. J. E. Lilienfeld, *Physik Z.* **20**, 280 (1919).
46. J. E. Lilienfeld, *Physik. Z.* **21**, 249 (1920).
47. R. A. Ferrell, *Phys. Rev.* **111**, 1214 (1958).
48. W. Ginsburg and I. Frank, *Zh. Eksperim. i Teor. Fiz.* **16**, 15 (1946).
49. H. Boersch, C. Radeloff, and G. Sauerbrey, *Z. Physik* **165**, 464 (1961).
50. R. H. Ritchie and H. B. Eldridge, *Phys. Rev.* **126**, 1935 (1962).
51. A. L. Frank, E. T. Arakawa, and R. D. Birkhoff, *Phys. Rev.* **126**, 1947 (1962).
52. G. H. Dieke and L. Heroux, *Phys. Rev.* **103**, 1227 (1956).
53. G. H. Dieke and H. M. Crosswhite, *Appl. Optics* **2**, 675 (1963).
54. H. A. Bethe, *Ann. Physik* [5] **3**, 133 (1929).
55a. G. Racah, *Phys. Rev.* **61**, 186 (1942).
55b. G. Racah, *Phys. Rev.* **62**, 438 (1942).
56a. G. Racah, *Phys. Rev.* **76**, 1352 (1949)
56b. G. Racah, *Phys. Rev.* **63**, 367 (1943).
57. J. B. Gruber, Thesis UCRL-9203, University of California, 1961.
58. F. Hund, "Linienspektren und Periodisches System der Elemente." Springer, Berlin, 1927.
59. H. E. White, "Introduction to Atomic Spectra." McGraw-Hill, New York, 1934.
60. G. H. Dieke, H. M. Crosswhite, and B. Dunn, *J. Opt. Soc. Am.* **51**, 820 (1961).
61. S. Sugano, *Appl. Optics* **1**, Suppl., 92 (1962).
62. J. G. Conway and J. B. Gruber, *J. Chem. Phys.* **32**, 1586 (1960).
63. J. Frenkel, *Phys. Rev.* **37**, 17 (1931).
64. J. Frenkel, *Phys. Rev.* **37**, 1276 (1931).
65. S. Nikitine, *Helv. Phys. Acta* **28**, 307 (1955).
66. N. F. Mott and R. W. Gurney, "Electronic Processes in Ionic Crystals." Oxford Univ. Press, London and New York, 1940.
67. L. Apker and E. A. Taft, *Phys. Rev.* **79**, 964 (1950).
68. F. Seitz, *in* "Imperfections in Nearly Perfect Crystals" p. 3. Symposium Paper. Wiley, New York, 1952.
69. F. Seitz, *Rev. Mod. Phys.* **26**, 7 (1954).
70. J. R. Haynes, *Phys. Rev. Letters* **4**, 361 (1960).
71. E. Goldstein, *Z. Instrumentenk.* **16**, 211 (1896).
72. R. W. Pohl, *Physik. Z.* **39**, 36 (1938).
73. F. Seitz, *Rev. Mod. Phys.* **18**, 384 (1946).
74. J. H. Schulman and D. Compton, "Color Centers in Solids." Pergamon Press, New York, 1962.
75. J. H. de Boer, *Rec. Trav. Chim.* **56**, 301 (1937).
76. F. Seitz, *Phys. Rev.* **79**, 529 (1950).

77. E. MOLLWO, *Nachr. Ges. Wiss. Göttingen* p. 97 (1931).
78. H. F. IVEY, *Phys. Rev.* **72**, 341 (1947).
79. B. S. GOURARY and F. J. ADRIAN, *Phys. Rev.* **105**, 1180 (1957).
80. F. MARTINO, *Phys. Rev.* **131**, 605 (1963).
81. A. SMAKULA, *Z. Physik* **59**, 603 (1930).
82. F. HERMAN, *Phys. Rev.* **88**, 1210 (1952).
83. F. HERMAN, *Proc. IRE* **43**, 1703 (1955).
84. F. HERMAN and J. CALLAWAY, *Phys. Rev.* **89**, 518 (1953).
85. H. F. IVEY, "Electroluminescence and Related Effects." Academic Press, New York, 1963.
86. H. K. HENISCH, "Elektroluminescence." Pergamon Press, New York, 1962.
87. D. HAHN, *Ergeb. Exact. Naturwiss.* **31**, 1 (1959).
88. G. DESTRIAU, *J. Chim. Phys.* **33**, 587 (1936).
89. O. W. LOSSEW, *Telegrafiya i Telefoniya* No. 18, 61 (1923).
90. R. J. KEYES and T. M. QUIST, *Proc. IRE* **50**, 1822 (1962).
91. S. MAYBURG, post-dead line paper, Am. Phys. Soc. Meeting, Baltimore, March 1962.
92. J. I. PANKOVE and M. MASSOULIÉ, *Abstr. Electrochem. Soc., Electronics Div.* **11**, 71-75 (1962).
93. R. BRAUNSTEIN, *Phys. Rev.* **99**, 1892 (1955).
94. K. WEISER and R. S. LEVITT, *Appl. Phys. Letters* **2**, 178 (1963).
95. J. A. BURTON, *Phys. Rev.* **108**, 1342 (1957).
96. W. J. CHOYKE and L. PATRICK, *Phys. Rev. Letters* **2**, 48 (1959).
97. K. LEHOVEC, C. A. ACCARDO, and E. JAMGOCHIAN, *Phys. Rev.* **83**, 603 (1951).
98. K. LEHOVEC, C. A. ACCARDO, and E. JAMGOCHIAN, *Phys. Rev.* **89**, 20 (1953).
99. L. PATRICK, *J. Appl. Phys.* **28**, 765 (1957).
100. L. PATRICK and W. J. CHOYKE, *J. Appl. Phys.* **30**, 236 (1959).
101. J. LAMBE and C. C. KLICK, *Phys. Rev.* **98**, 909 (1955).
102. M. SCHÖN, *Z. Physik* **119**, 463 (1942).
103. H. A. KLASENS, *Nature* **158**, 306 (1946).
104. H. A. KLASENS, *J. Electrochem. Soc.* **100**, 72 (1953).
105. See e.g. H. P. KALLMANN and G. MARMOR SPRUCH, editors, "Luminescence of Organic and Inorganic Materials." Wiley, New York, 1962.
106. A. L. SCHAWLOW *in* "Proceedings of Quantum Electronics," 3rd International Congress (P. Grivet and N. Bloembergen, eds.), Columbia University Press, New York, 1964.

5

PHENOMENA OF PROPAGATION

In this chapter, the interaction of radiation with matter is treated insofar as it affects the properties of propagation. When electromagnetic energy, in general, is transmitted through gases, liquids, and solids, radiation intensity, wavelength, direction of propagation, and plane of vibration may be affected by one or more of the following propagation phenomena (see Fig. 5.1):

(1) *Reflection* of radiation at the interface between two media which differ optically. In this process, part or all of the incident beam of radiation is returned back into the medium from which it came.

(2) *Refraction* at the interface between the two media such that the direction of the beam undergoes a discontinuous change. The amount of the deflection depends on the ratio of the propagation velocities in the two media. Again, the propagation velocities, therefore also the deflections, are, in general, functions of wavelength (dispersion).

(3) *Absorption* by which the intensity of the beam is more and more reduced as it propagates through the absorbing medium. In this process, the energy of the radiation is converted into other forms, usually into

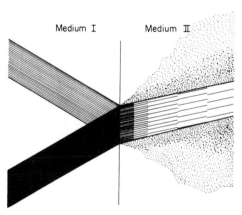

Fig. 5.1. Propagation of light through two media and their boundary. The diagram indicates schematically the phenomena associated with reflection, refraction, absorption, and scattering.

heat, but also—and most important to optical physics—into radiation of different wavelengths (fluorescence) or into electronic types of energy. The inverse process, negative absorption or stimulated emission (see Section 3.3.1), represents a special case. It is of such importance in its own right that it will be treated separately in the next chapter.

(4) *Scattering* of the radiation by particles of random distribution. This interaction diffuses the incident radiation, but does not transform its energy or change its frequency.

(5) *Polarization* which changes, or discriminates against, the direction of the oscillating electric and magnetic fields.

These phenomena of propagation are, of course, the basis of geometric and physical optics; they can be discussed to a large extent without reference to atomistic properties. Our concern in this enormous field is thus quite limited. It is, in fact, restricted to just those interactions which bring about the fundamental behavior of light propagation in the various media. First, we shall deal with the phenomenological aspects of propagation in terms of certain optical parameters or indices. Next, these indices are discussed in their own right, and in particular, they are correlated to the atomistic structure of matter. Later, we shall take up the case in which the assumption of linear relationships between field and induced moments no longer holds, in other words, the case of nonlinear optics which is of practical importance for high intensities. Some of the interaction processes which are understandable on the basis of individual atomic mechanisms have been described already in the preceding chapters. Here, however, the interaction with large numbers of centers will be of primary interest, and partly for this reason, the electromagnetic wave concept will govern most considerations.

For the aspects treated in this chapter, the question of coherence will be of little or no importance; the properties discussed, with few exceptions, are equally valid for single waves as for the superposition of random fields. However, there exist propagation phenomena which are singularly characteristic of, and dependent on, the degree of coherence. This subject has become an important field by itself and will be dealt with in the next chapter.

5.1. Electromagnetic Waves in Various Media

It follows from the preceding remarks that the phenomena of propagation can be explained largely by the theory of electromagnetism. A pertinent aspect of the theory was discussed in Section 3.1.2 where a vibrating dipole was shown to emit transverse waves of electric and magnetic fields. The subject matter of this chapter is, of course, not the origin of electromagnetic waves, but their behavior in vacuum, dielectric materials, and metals. Thus we are concerned with material "constants": the dielectric constant ϵ, the magnetic permeability μ, and the electric conductivity σ. In general, we cannot expect these values to be constant, even for a given frequency and material; they may depend on a particular direction in the material and also on time in a complex manner. However, we shall restrict ourselves in Section 5.1 to homogeneous and isotropic materials, i.e., materials in which the relevant properties are the same everywhere and independent of direction. Furthermore, our subject matter is limited to media which are at rest or move with velocities negligible compared to that of light. With these assumptions, then, it is possible to relate the electric field $\mathbf{E}$ and the magnetic field $\mathbf{H}$ to the corresponding quantities in the presence of matter, namely, the electric displacement $\mathbf{D}$ and the magnetic induction $\mathbf{B}$, viz.,

$$\mathbf{D} = \epsilon\mathbf{E} \tag{5.1}$$

and

$$\mathbf{B} = \mu\mathbf{H} \tag{5.2}$$

Here belongs also Ohm's law in terms of field $\mathbf{E}$ and current density $\mathbf{j}$:

$$\mathbf{j} = \sigma\mathbf{E} \tag{5.3}$$

In these relations ϵ, μ, and σ are scalars, and this simplifies the dynamic correlation of electric and magnetic fields. In their most general form, the fields are determined by Maxwell's equations as follows:

$$\operatorname{curl} \mathbf{H} = \frac{1}{c}\frac{\partial \mathbf{D}}{\partial t} + \frac{4\pi}{c}\mathbf{j} \tag{5.4}$$

$$\operatorname{curl} \mathbf{E} = -\frac{1}{c}\frac{\partial \mathbf{B}}{\partial t} \tag{5.5}$$

$$\operatorname{div} \mathbf{D} = 4\pi\rho \tag{5.6}$$

$$\operatorname{div} \mathbf{B} = 0 \tag{5.7}$$

In Eqs. (5.1)–(5.7), the Gaussian system of units is used: the magnetic quantities $\mathbf{B}$ and $\mathbf{H}$ are measured in electromagnetic units (defined by the force in dynes between unit currents at unit distance); the electric quantities are based on electrostatic units (defined by the Coulomb force of 1 dyne between unit charges at 1 cm distance). Both systems are linked in Maxwell's equations through the conversion factor c, which will turn out to be equal to the speed of light. Finally, ρ is the charge density in electrostatic units.

Maxwell's equations are simplified in applications of this chapter by the assumption that the propagation media are (1) isotropic, (2) at rest, and (3) free of static charges, div $\mathbf{E} = 0$.

5.1.1 PROPAGATION IN DIELECTRIC MEDIA

We shall first consider media such as vacuum and insulators, in which the conductivity is zero. In addition, we apply the conditions stated in the preceding paragraph with Eqs. (5.1)–(5.7). We have then

$$\text{curl } \mathbf{H} = \frac{\epsilon}{c} \frac{\partial \mathbf{E}}{\partial t} \tag{5.8}$$

$$\text{curl } \mathbf{E} = -\frac{\mu}{c} \frac{\partial \mathbf{H}}{\partial t} \tag{5.9}$$

$$\text{div } \mathbf{E} = 0 \tag{5.10}$$

$$\text{div } \mathbf{H} = 0 \tag{5.11}$$

These equations can be converted into expressions containing only $\mathbf{E}$ or only $\mathbf{H}$, e.g., by applying the operator curl to Eq. (5.8) and substituting Eq. (5.9) in the resulting expression. Thus

$$\text{curl curl } \mathbf{H} = \frac{\epsilon}{c} \text{curl } \frac{\partial \mathbf{E}}{\partial t} = -\frac{\mu\epsilon}{c^2} \frac{\partial^2 \mathbf{H}}{\partial t^2} \tag{5.12}$$

Because of Eq. (5.11) and the vector rule

$$\text{curl curl } \mathbf{H} = \text{grad div } \mathbf{H} - \nabla^2 \mathbf{H} \tag{5.13}$$

Eq. (5.12) becomes

$$\nabla^2 \mathbf{H} = \frac{\mu\epsilon}{c^2} \frac{\partial^2 \mathbf{H}}{\partial t^2} \tag{5.14}$$

A similar equation holds for the electric field:

$$\nabla^2 \mathbf{E} = \frac{\mu\epsilon}{c^2} \frac{\partial^2 \mathbf{E}}{\partial t^2} \tag{5.15}$$

Thus we have arrived at differential equations which have the form of those for elastic waves and which were discussed in Chapter 3, Eq. (3.47). We are, however, now dealing with a medium of infinite extent and, therefore, do not anticipate a pattern of standing waves. We shall treat this problem in component form (see Becker and Sauter[1]). To solve Eqs. (5.14) and (5.15) we assume the existence of a particular case, namely, a plane wave propagating in the z-direction. This means, by definition, that the locus of all points with the same vectors $\mathbf{E}$ and $\mathbf{H}$ is a family of planes which are normal to the z-direction and follow each other at an interval of the wavelength λ. Therefore, the derivatives of the fields with respect to x and y must be zero so that also z-components of curl $\mathbf{E}$ and curl $\mathbf{H}$ vanish [cf. Eqs. (5.8) and (5.9)], to wit:

$$\frac{\epsilon}{c}\frac{\partial E_z}{\partial t} = \left(\frac{\partial H_y}{\partial x} - \frac{\partial H_x}{\partial y}\right) = 0 \tag{5.16}$$

$$-\frac{\mu}{c}\frac{\partial H_z}{\partial t} = \left(\frac{\partial E_y}{\partial x} - \frac{\partial E_x}{\partial y}\right) = 0 \tag{5.17}$$

Since also, according to Eqs. (5.10) and (5.11),

$$\frac{\partial E_z}{\partial z} = 0 \tag{5.18}$$

$$\frac{\partial H_z}{\partial z} = 0 \tag{5.19}$$

the longitudinal components E_z and H_z of the electric and magnetic field, respectively, are constant in time and space. The fields propagating as waves are, therefore, restricted to the directions normal to that of propagation; electromagnetic waves are transverse.

We now proceed to write Eqs. (5.14) and (5.15) in component form, allowing for the fact that according to the preceding remarks

$$\frac{\partial^2 E_x}{\partial x^2} = \frac{\partial^2 E_x}{\partial y^2} = \frac{\partial^2 E_y}{\partial y^2} = \frac{\partial^2 E_y}{\partial x^2} = 0, \quad \text{etc.} \tag{5.20}$$

Thus there remains

$$\frac{\epsilon\mu}{c^2}\frac{\partial^2 E_x}{\partial t^2} = \frac{\partial^2 E_x}{\partial z^2} \tag{5.21}$$

$$\frac{\epsilon\mu}{c^2}\frac{\partial^2 H_y}{\partial t^2} = \frac{\partial^2 H_y}{\partial z^2} \tag{5.22}$$

with similar partial differential equations for the y-component of the electric field and the x-component of the magnetic field. The general solutions of Eqs. (5.21) and (5.22) are, respectively,

$$E_x = \frac{1}{\sqrt{\epsilon}} \{f(z - vt) + g(z + vt)\} \tag{5.23}$$

$$H_y = \frac{1}{\sqrt{\mu}} \{f(z - vt) + g(z + vt)\} \tag{5.24}$$

where f and g represent arbitrary functions of the variable $(z - vt)$ or $(z + vt)$, respectively. These functions, once determined for $t = 0$, represent wave-trains which move along the z-axis, or $(-z)$-axis, with a velocity v which, because of Eqs. (5.21) and (5.22), amounts to

$$v = \frac{c}{\sqrt{\epsilon\mu}} \tag{5.25}$$

For vacuum, in electrostatic units, $\epsilon = \mu = 1$, so that

$$v = c = 3 \times 10^{10} \quad \text{cm/sec} \tag{5.26}$$

From the historical point of view, it is of interest to note that the constant c, which in Eq. (5.26) represents the speed of electromagnetic waves in vacuum, made its appearance in Eqs. (5.4) and (5.5) merely as a conversion constant between electrostatic and electromagnetic units. As such, it turned out to be equal to the speed of light. This fact, together with the transverse nature of the wave as shown in Eqs. (5.18) and (5.19), led to the electromagnetic wave theory of light.

In dielectrics, the reduction of the velocity of propagation with respect to that in vacuum according to Eq. (5.25) gives rise to an index of refraction n, viz.,

$$n = \frac{c}{v} = \sqrt{\epsilon\mu} \tag{5.27}$$

which becomes the Maxwell relation

$$n^2 = \epsilon \tag{5.28}$$

since in general the permeability can be set equal to unity.

5.1.1.1 *Relationship of Vector Waves in Dielectric Media*

We shall now add certain alternative descriptions of wave propagation in dielectric media. First, a formulation of the plane wave equations

(5.23) and (5.24) in vector form is possible on the basis of the relationships indicated in Fig. 5.2. If **s** is the unit vector for the direction of
propagation of the plane wave, all points P with position vector **r** such
that

$$\mathbf{r} \cdot \mathbf{s} = \text{constant} \tag{5.29}$$

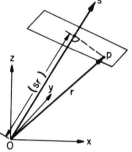

FIG. 5.2.   Vector relationships for plane wave propagation.

lie on a plane perpendicular to **s**. Hence by definition, all these points
have the same phase, and instead of Eq. (5.24) we can write

$$\mathbf{E} = \frac{1}{\sqrt{\epsilon}} \left\{ \mathbf{f}\left(\mathbf{r}\cdot\mathbf{s} - \frac{ct}{n}\right) + \mathbf{g}\left(\mathbf{r}\cdot\mathbf{s} + \frac{ct}{n}\right) \right\} \tag{5.30}$$

where we have replaced v by c/n. A corresponding expression exists for
the magnetic field vector **H**.

Any propagating electromagnetic disturbance can be represented as
a superposition of strictly monochromatic wavetrains. Suppose that the
electric field at the origin, $\mathbf{r} = 0$, varies as

$$\mathbf{E} = \mathbf{E}_0 e^{i\omega t}$$

The phase at a point P is retarded with respect to that at the origin
according to a time delay of $(\mathbf{r} \cdot \mathbf{s})/v$. The electric field changes thus
with time and position as

$$\mathbf{E} = \mathbf{E}_0 \exp\left[i\omega\left\{t - \frac{n}{c}(\mathbf{r}\cdot\mathbf{s})\right\}\right], \qquad \mathbf{H} = \mathbf{H}_0 \exp\left[i\omega\left\{t - \frac{n}{c}(\mathbf{r}\cdot\mathbf{s})\right\}\right] \tag{5.31}$$

The wavelength, which in vacuum has the value λ_0, is reduced in the
medium to

$$\lambda = \frac{v}{\nu} = \frac{c}{n\nu} = \frac{\lambda_0}{n} \tag{5.32}$$

The vector form of the wave equation (5.31) makes it immediately possible to establish the relationship between the electric and magnetic fields **E** and **H** (see Becker and Sauter[1]). For this purpose, we introduce Eq. (5.31) into Maxwell's equations (5.8) and (5.9) making use of the operator identities

$$\nabla \equiv -\frac{i\omega n}{c}\,\mathbf{s} \qquad (5.33)$$

and

$$\frac{\partial}{\partial t} \equiv i\omega \qquad (5.34)$$

We have then

$$-\frac{i\omega n}{c}\,\mathbf{s} \times \mathbf{H} = \frac{i\epsilon\omega}{c}\,\mathbf{E}, \qquad \mathbf{s}\cdot\mathbf{H} = 0 \qquad (5.35)$$

and

$$-\frac{i\omega n}{c}\,\mathbf{s} \times \mathbf{E} = -\frac{i\mu\omega}{c}\,\mathbf{H}, \qquad \mathbf{s}\cdot\mathbf{E} = 0 \qquad (5.36)$$

The vacuum case, $\epsilon = \mu = n = 1$, is particularly simple:

$$\mathbf{s} \times \mathbf{H} = -\mathbf{E}, \qquad \mathbf{s}\cdot\mathbf{H} = 0 \qquad (5.37)$$

$$\mathbf{s} \times \mathbf{E} = \mathbf{H}, \qquad \mathbf{s}\cdot\mathbf{E} = 0 \qquad (5.38)$$

First, a characteristic of all electromagnetic radiation is made particularly apparent in these equations, viz., that **E**, **H**, and the unit vector **s** of the propagation direction are oriented at right angles with respect to each other in the sense shown in Fig. 3.3. Furthermore, in vacuo, the electric and magnetic fields have not only the same phase but also the same magnitude as can be seen from Eqs. (5.37) and (5.38).

Whereas in an individual wavetrain, electric and magnetic fields have a fixed relationship to each other, ordinary light sources emit wavetrains in which the field vectors are randomly directed. In the special case, that the **E**-vectors vibrate in a single direction we speak of *linear polarization* (or *plane* polarization, viz., in a plane which contains the **H**-vector, but no component of the **E**-vector).

5.1.2 PROPAGATION IN CONDUCTORS

The presence of free electric charges, which characterizes a conducting medium, alters also profoundly the behavior of light propagation from that in dielectrics. In the idealized case that the charge carriers are completely free, Newton's equation of motion does not contain a damping term [cf. Eq. (1.2)], and thus an electron responds to a periodic field classically with a periodic motion of an amplitude which is constant in time [cf. Eq. (1.3)]. In the stationary case, then, the electron does not absorb energy but is the source of a re-emitted electromagnetic wave. It is thus plausible that a medium of infinite conductivity is completely reflecting.

Real media have, of course, a finite conductivity σ; a damping term has to be added to the Newtonian equation of motion; and the charge carriers in the stationary stage will partially reflect, partially absorb the incident electromagnetic energy at a ratio depending on the amount of σ. This is the case even for superconductors characterized by infinite conductivity at zero frequency. Thus the optical behavior of metals will be determined by three basic parameters, viz., the dielectric constant ϵ, the permeability μ, and the conductivity σ, all of which must be expected to vary with frequency.

Except for the addition of the terms containing the conductivity σ, the procedure for determining the laws of wave propagation in metals is quite analogous to the method adopted for dielectrics. Maxwell's equations for conductive media, viz., Eqs. (5.4) and (5.5), in combination with Eq. (5.30) apply here, viz.,

$$\text{curl } \mathbf{H} = \frac{\epsilon}{c} \dot{\mathbf{E}} + \frac{4\pi\sigma\mathbf{E}}{c} \tag{5.39}$$

and

$$\text{curl } \mathbf{E} = -\frac{\mu}{c} \dot{\mathbf{H}} \tag{5.40}$$

Furthermore, we can set div $\mathbf{D} = 0$, since it can be shown that any charges induced by the fields are neutralized in a relaxation time which is negligibly small compared to the period of oscillation for optical waves. The conversion of Eqs. (5.39) and (5.40) into expressions containing only either $\mathbf{E}$ or $\mathbf{H}$ follows the method already used to obtain Eq. (5.12), viz., by taking the curl of one equation and substituting the time derivative of the other equation. Then, because of the vector rule (5.13) and div $\mathbf{H} = $ div $\mathbf{E} = 0$,

$$\nabla^2\mathbf{E} = \frac{\epsilon\mu}{c^2} \ddot{\mathbf{E}} + \frac{4\pi\mu\sigma}{c^2} \dot{\mathbf{E}} \tag{5.41}$$

and

$$\nabla^2\mathbf{H} = \frac{\epsilon\mu}{c^2} \ddot{\mathbf{H}} + \frac{4\pi\mu\sigma}{c^2} \dot{\mathbf{H}} \tag{5.42}$$

The terms containing σ appear as coefficients for the first derivatives of the electric and magnetic fields and represent damping or attenuating contributions in analogy to the corresponding terms in the damped oscillator equation (see Section 1.2.2).

The solution for Eqs. (5.41) and (5.42) can be derived in close analogy to the dielectric case. We assume a plane wave at a single frequency, i.e., $\mathbf{E} = \mathbf{E}_0 e^{i\omega t}$ for all points at which $\mathbf{r} \cdot \mathbf{s} = 0$. We then assume a plane wave equation of the form (5.31). However, we have now to postulate a complex index of refraction $\hat{n}$, viz.,

$$\hat{n} = n - i\kappa \tag{5.43}$$

through which we may anticipate a term of attenuation as the wave proceeds through the medium*. In other words, we now write in analogy to Eq. (5.31):

$$\mathbf{E} = \mathbf{E}_0 \exp\left[i\omega\left\{t - \frac{\hat{n}}{c}(\mathbf{r} \cdot \mathbf{s})\right\}\right] = \mathbf{E}_0 \exp\left[i\omega\left\{t - \frac{n}{c}(\mathbf{r} \cdot \mathbf{s})\right\}\right] \exp\left[-\frac{\omega}{c}\kappa(\mathbf{r} \cdot \mathbf{s})\right] \tag{5.44}$$

If the wavefronts are normal to the z-axis, $\mathbf{r} \cdot \mathbf{s} = z$, Eq. (5.20) is valid, and Eq. (5.44) becomes

$$\mathbf{E} = \mathbf{E}_0 \exp\left[i\omega\left(t - \frac{\hat{n}}{c}z\right)\right] = \mathbf{E}_0 \exp\left[i\omega\left(t - \frac{n}{c}z\right)\right] \exp\left[\frac{-\omega}{c}\kappa z\right] \tag{5.45}$$

If we now substitute Eq. (5.45) in Eq. (5.41), we obtain the following correlation between the optical indices n and κ and the electrical parameters:

$$-\frac{\omega^2}{c^2}(n - i\kappa)^2 = -\frac{\epsilon\mu}{c^2}\omega^2 + i\frac{4\pi\mu\sigma}{c^2}\omega \tag{5.46}$$

Separation into real and imaginary components yields

$$n^2 - \kappa^2 = \epsilon\mu \tag{5.47}$$

$$2n\kappa = \frac{4\pi\mu\sigma}{\omega} \tag{5.48}$$

---

* Equivalent definitions used in the literature are $n = n + i\kappa$ or $n = n(1 + i\kappa)$ with corresponding changes in the solutions for κ.

From these last two equations follow explicit solutions for the components of the complex index of refraction:

$$n^2 = \frac{\mu}{2} \left\{ \sqrt{\epsilon^2 + \left(\frac{4\pi\sigma}{\omega}\right)^2} + \epsilon \right\}$$ (5.49)

$$\kappa^2 = \frac{\mu}{2} \left\{ \sqrt{\epsilon^2 + \left(\frac{4\pi\sigma}{\omega}\right)^2} - \epsilon \right\}$$ (5.50)

Inspection of Eq. (5.44) or (5.45) shows that the significance of n, given by Eq. (5.49) in terms of σ, ϵ, and μ, is the same as in the dielectric case. In other words, n is the factor by which the speed of light is reduced in the medium compared with its value in vacuo, and Eq. (5.49) goes over into Eq. (5.27) for the limit of vanishing conductivity σ. It also follows from the meaning of Eq. (5.44) or (5.45) that κ assumes the role of an *index of attenuation* or *extinction coefficient*. More specifically, the amplitude of the electromagnetic wave decreases by a factor A of

$$A = \frac{E(t, z + \lambda)}{E(t, z)} = \exp\left[-\frac{\omega}{c}\kappa\lambda\right] = \exp\left[-2\pi\frac{\kappa}{n}\right]$$ (5.51)

as the wave progresses by $\lambda = \lambda_0/n$ in the z-direction. The relationship of κ to the *absorption coefficient* k follows from the fact that the latter refers to the attenuation of radiation intensity $I(z)$ which is proportional to the square of the vectors $\mathbf{E}$ and $\mathbf{H}$. Thus

$$\frac{I(z)}{I(0)} = \exp\left[-kz\right] = \exp\left[-\frac{2\omega}{c}\kappa z\right]$$ (5.52)

and

$$\kappa = \frac{kc}{2\omega} = \frac{kc}{4\pi\nu} = \frac{k}{4\pi}\lambda_0$$ (5.53)

or

$$k = \frac{4\pi\kappa}{n\lambda} = \frac{4\pi\kappa}{\lambda_0}$$ (5.54)

where λ_0 and λ are the wavelengths in vacuo and in the medium, respectively.

5.1.2.1 *Limiting Cases for Large and Small Conductivities*

Whereas Eqs. (5.34)–(5.54) represent the general classical relationships for the propagation of electromagnetic radiation in homogeneous, isotropic media at rest with conductivities between zero and infinity, simplifications occur for the important cases in which σ is either very small or very large. In the first instance, for $\sigma = 0$, the relationships simply revert to those given in Section 5.1.1 for dielectric media. The second limiting case, for large σ, represents the optics of metals. The magnitude* of $4\pi\sigma/\omega$ in Eqs. (5.49) and (5.50) is for metals usually considerably larger than the magnitude of the dielectric constant ϵ. This means that attenuation of the progressing wave is high in metals. The coefficient ϵ, of course, cannot be measured with the direct current methods used for dielectrics. However, it follows from indirect optical measurements as well as from theory that ϵ is in metals of the same magnitude as in insulators.

If

$$\epsilon_0{}^2 \ll \left(\frac{4\pi\sigma}{\omega}\right)^2 \tag{5.55}$$

it can be omitted under the square root in Eqs. (5.49) and (5.50), and the following simplified relationships are obtained:

$$n = \kappa = \sqrt{\frac{2\pi\mu\sigma}{\omega}} = \sqrt{\frac{\mu\sigma}{\nu}} = \sqrt{\mu\sigma\tau} \tag{5.56}$$

where $\tau = 1/\nu$ is the period of oscillation.

The reciprocal of the absorption coefficient k, as given by Eq. (5.54), has the significance of a penetration depth or *skin depth* d_1 over which the radiation intensity is reduced by the factor e,

$$d_1 = \frac{\lambda_0}{4\pi\kappa} = \frac{\lambda_0}{4\pi}\sqrt{\frac{\nu}{\mu\sigma}} = \frac{1}{4\pi}\sqrt{\frac{c\lambda_0}{\mu\sigma}} \tag{5.57}$$

For the amplitude, the corresponding distance of attenuation, d_2, is twice the value of d_1 [cf. Eq. (5.52)]. Table 5.1 presents the penetration depths d_1 and d_2 for practically important magnitudes of conductivity and wavelength for metals with $\mu = 1$. Relationships (5.55) and (5.56) are valid for the range of λ_0- and σ-values considered in the table, but the data for σ (and also ϵ and μ), cannot be equated necessarily to those known from direct current measurements, at least not for wavelengths smaller than those of the near infrared. The examples cited in Table 5.1 for conductivities of 10^6, 10^5, 10^4 ohm^{-1}cm^{-1} correspond, respec-

---

* σ, in electrostatic units, viz., [sec^{-1}], can be obtained from practical units by multiplying σ in [ohm^{-1} cm^{-1}] by 9×10^{11}.

TABLE 5.1

SKIN DEPTH d_1, AMPLITUDE PENETRATION d_2 AND REFLECTIVITY R FOR REPRESENTATIVE
VALUES OF CONDUCTIVITY σ AND WAVELENGTH λ_0 WITH $\mu = 1$ IN THE FIRST APPROXIMA-
TION OF CLASSICAL THEORY

σ	λ_0	d_1 [cm]	d_2 [cm]	$1 - R$
10^6 [ohm^{-1} cm^{-1}]	1 μ	1.45×10^{-7}	2.9×10^{-7}	3.65×10^{-2}
(9×10^{17} sec^{-1})	100 μ	1.45×10^{-6}	2.9×10^{-6}	3.65×10^{-3}
	1 cm	1.45×10^{-5}	2.9×10^{-5}	3.65×10^{-4}
	1 m	1.45×10^{-4}	2.9×10^{-4}	3.65×10^{-5}
	10 km	1.45×10^{-2}	2.9×10^{-2}	3.65×10^{-7}
10^5 [ohm^{-1} cm^{-1}]	1 μ	4.58×10^{-7}	9.16×10^{-7}	1.15×10^{-1}
(9×10^{16} sec^{-1})	100 μ	4.58×10^{-6}	9.16×10^{-6}	1.15×10^{-2}
	1 cm	4.58×10^{-5}	9.16×10^{-5}	1.15×10^{-3}
	1 m	4.58×10^{-4}	9.16×10^{-4}	1.15×10^{-4}
	10 km	4.58×10^{-2}	9.16×10^{-2}	1.15×10^{-6}
10^4 [ohm^{-1} cm^{-1}]	1 μ	1.45×10^{-6}	2.90×10^{-6}	3.65×10^{-1}
(9×10^{15} sec^{-1})	100 μ	1.45×10^{-5}	2.90×10^{-5}	3.65×10^{-2}
	1 cm	1.45×10^{-4}	2.90×10^{-4}	3.65×10^{-3}
	1 m	1.45×10^{-3}	2.90×10^{-3}	3.65×10^{-4}
	10 km	1.45×10^{-1}	2.90×10^{-1}	3.65×10^{-5}

tively, to the direct current values of copper at $-84\,°C$, iron at $20\,°C$, and mercury at $67\,°C$. To what extent these conductivity values are valid at optical frequencies, is revealed by the optical measurements themselves. It will be shown later that, owing to the inertial effect of the electrons, the conductivities fall off sharply at frequencies above those corresponding to the near infrared region.

It is seen that as the conductivity tends to infinity, the penetration depth vanishes. Of course, the relationships derived so far in this chapter refer to propagation within a medium and say nothing about the behavior at boundaries between media. The meaning of a vanishing penetration depth is that of complete reflection as will be shown in the next section. In fact, Table 5.1 demonstrates in the last three columns how the reflectivity approaches unity as d_1/λ_0 approaches zero.

The other limiting case, namely the inverse of the inequality (5.55),

$$\left(\frac{4\pi\sigma}{\omega}\right)^2 \ll \epsilon^2 \tag{5.58}$$

leads to simple expressions by power expansions of Eqs. (5.49) and (5.50):

$$n^2 = \frac{\mu\epsilon}{2}\left\{2 + \tfrac{1}{2}\left(\frac{4\pi\sigma}{\epsilon\omega}\right)^2 \mp \cdots\right\} \approx \mu\epsilon \tag{5.59}$$

and

$$\kappa^2 = \frac{\mu\epsilon}{2} \left\{ \frac{1}{2} \left(\frac{4\pi\sigma}{\epsilon\omega} \right)^2 - \frac{1}{8} \left(\frac{4\pi\sigma}{\epsilon\omega} \right)^4 + \cdots \right\} \approx \frac{\mu}{\epsilon} \left(\frac{2\pi\sigma}{\omega} \right)^2 \tag{5.60}$$

Under these conditions, then the optical behavior is similar to that in dielectrics, characterized by an index of refraction, $n = (\mu\epsilon)^{1/2}$, and a small coefficient of absorption k which, according to Eqs. (5.54) and (5.60), is given by

$$k = \frac{4\pi}{\lambda_0} \kappa = 4\pi \sqrt{\frac{\mu}{\epsilon}} \cdot \frac{\sigma}{c} \tag{5.61}$$

Therefore in the range in which $2\sigma \ll \epsilon\nu$, the penetration depth is independent of frequency:

$$d_1' = \frac{1}{k} = \frac{1}{4\pi} \sqrt{\frac{\epsilon}{\mu}} \cdot \frac{c}{\sigma} \tag{5.62}$$

For example, the limiting condition (5.58) is obtained for all of the optical range with a σ of 10^{10} sec^{-1} (corresponding to a conductivity of 0.11 ohm^{-1} cm^{-1} which is typical, e.g., of the contribution by free carriers in semiconductors). Thus in germanium ($\epsilon = 16$) the penetration depth is in the order of 1 cm for this conductivity. It must be remembered, however, that ϵ and σ change strongly in the frequency regions of resonances or energy gaps and that the inertia of the electrons affects the conductivity for wavelengths shorter than about 10 μ.

5.1.2.2 *Relationship of Vector Waves in Conductors*

To determine how the electric and magnetic field vectors are related to each other in propagating through conducting media, we repeat the procedure of Section 5.1.1.1. Thus we assume waves of the form

$$\mathbf{E} = \mathbf{E}_0 \exp\left[i\omega \left\{ t - \frac{\hat{n}}{c}(\mathbf{r} \cdot \mathbf{s}) \right\} \right], \qquad \mathbf{H} = H_0 \exp\left[i\omega \left\{ t - \frac{\hat{n}}{c}(\mathbf{r} \cdot \mathbf{s}) \right\} \right] \tag{5.63}$$

Introduction of $\mathbf{E}$ and $\mathbf{H}$ into Maxwell's equations (5.39) and (5.40) yields with $\nabla \equiv -(i\omega\hat{n}/c)\mathbf{s}$ and $\partial/\partial t \equiv i\omega$:

$$-\frac{i\omega\hat{n}}{c} \mathbf{s} \times \mathbf{H} = \frac{i\omega}{c} \left(\epsilon + \frac{4\pi\sigma}{i\omega} \right) \mathbf{E}, \qquad \mathbf{s} \cdot \mathbf{H} = 0 \tag{5.64}$$

and

$$-\frac{i\omega\hat{n}}{c} \mathbf{s} \times \mathbf{E} = -\frac{i\mu\omega}{c} \mathbf{H}, \qquad \mathbf{s} \cdot \mathbf{E} = 0 \tag{5.65}$$

These two equations yield for the ratio of the absolute field values:

$$\frac{|\mathbf{H}|}{|\mathbf{E}|} = \sqrt{\frac{\epsilon - i4\pi\sigma/\omega}{\mu}} \tag{5.66}$$

The expression can be simplified by means of Eq. (5.43) which can be written as

$$\hat{n} = n - i\kappa = \sqrt{\epsilon\mu - i\frac{4\pi\mu\sigma}{\omega}} \tag{5.67}$$

In combination with Eq. (5.66), this yields

$$|\mathbf{H}| = \frac{\hat{n}}{\mu}|\mathbf{E}| \tag{5.68}$$

The implication of Eq. (5.68) is that $|\mathbf{H}|$ and $|\mathbf{E}|$, which are equal in vacuo, assume a complex ratio, viz., $\hat{n}/\mu$, with respect to each other in conductors; they differ not only in magnitude, but in phase as well. As in vacuo, $\mathbf{E}$, $\mathbf{H}$, and $\mathbf{s}$, according to Eqs. (5.64) and (5.65) form a rectangular coordinate system, but the relationship of the field amplitudes is altered in amount and timing (see Fig. 5.3). Since the mean energy of

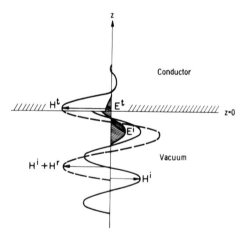

Fig. 5.3. Reflection and transmission of electromagnetic wave at normal incidence on vacuum-metal boundary. The magnetic field is in the plane of the drawing, the electric field wave, shown shaded, is perpendicular to it. The sum of the magnetic field and its reflection at a particular phase is shown by the dashed trace which maintains continuity at the boundary with the penetrating wave H^t in the conductor. H^t experiences a phase shift and rapid attenuation. Because of the negative phase relationship of the electric wave after reflection and the continuity requirement at the boundary, the penetrating electric field is much smaller than the corresponding magnetic field.

electric and magnetic fields per unit volume is, respectively, $\epsilon E^2/8\pi$ and $\mu H^2/8\pi$, their ratio has the value

$$\frac{\mu \overline{H^2}}{\epsilon \overline{E^2}} = \sqrt{1 + \left(\frac{2\sigma}{\epsilon \nu}\right)^2} \tag{5.69}$$

Thus, whereas the electric and magnetic field energies are equal in vacuo and dielectric media, they differ in conductors in that the magnetic field energy is vastly preponderant in metals. Qualitatively, this can be understood on the basis of the reflection phenomena of the $\mathbf{H}$ and $\mathbf{E}$ waves at the metal boundary as indicated in Fig. 5.3.

5.1.3 ELECTROMAGNETIC WAVES AT BOUNDARIES

The phenomena of propagation at the interface between two media involve reflection and refraction. These processes, of course, are determined by the relationship of the fields on the two sides of the boundary. We may assume here the absence of surface currents (and also charges). Then an important rule is valid and pertinent: the tangential electric and magnetic fields remain continuous at the interface, i.e., they do not experience a sudden change at that plane. It is shown in detail in the theory of electromagnetism that if that were not the case, the curl over the area of any loop passing through two media would remain finite for vanishing area, even though the surface integrals of the right sides in Eqs. (5.39) and (5.40) would vanish.

5.1.3.1 *Reflection at Normal Incidence*

We now assume that a plane wave ($\mathbf{E}^i$, $\mathbf{H}^i$) is propagating in the z-direction in vacuo and that it is incident perpendicularly on a homogeneous, isotropic conductor at a boundary defined by $z = 0$ (see Fig. 5.3). It follows from symmetry considerations that also all subsequent propagation is directed along either the $(+z)$ or $(-z)$ axis. Thus if the incident wave has the form

$$\mathbf{E}^i = \mathbf{E}_0{}^i \exp\left[i\omega\left(t - \frac{z}{c}\right)\right], \qquad \mathbf{H}^i = \mathbf{H}_0{}^i \exp\left[i\omega\left(t - \frac{z}{c}\right)\right], \tag{5.70}$$

there results a reflected wave (index r) and a transmitted wave (index t) which are written, respectively, as

$$\mathbf{E}^r = -\mathbf{E}_0{}^r \exp\left[i\omega\left(t + \frac{z}{c}\right)\right], \qquad \mathbf{H}^r = \mathbf{H}_0{}^r \exp\left[i\omega\left(t + \frac{z}{c}\right)\right] \tag{5.71}$$

$$\mathbf{E}^t = \mathbf{E}_0{}^t \exp\left[i\omega\left(t - \hat{n}\frac{z}{c}\right)\right], \qquad \mathbf{H}^t = \mathbf{H}_0{}^t \exp\left[i\omega\left(t - \hat{n}\frac{z}{c}\right)\right] \tag{5.72}$$

Here $\hat{n} = n - i\kappa$ is the complex index of refraction in the conductive medium ($\kappa = 0$ for dielectrics). Altogether, there are three pairs of **E** and **H** waves, and for each pair the amplitude ratio is determined, viz., for vacuo by the equality of $|\,\mathbf{H}\,|$ and $|\,\mathbf{E}\,|$ and for the conductor by the factor $\hat{n}/\mu$. Furthermore, the relationship between the amplitudes of the different pairs is uniquely determined by the continuity conditions at the boundary, $z = 0$, and by the requirement that each **E** and **H** pair must form a right-handed coordinate system with the propagation direction along the positive or negative z-axis.

The continuity conditions reduce to a statement about the amplitude values since, at $z = 0$, the exponential factor can be cancelled out. We have thus the relation for the E-vector amplitudes

$$E_0{}^i - E_0{}^r = E_0{}^t \tag{5.71}$$

and for the H-vectors (oriented perpendicular to E in the plane $z = 0$) the magnitudes

$$H_0{}^i + H_0{}^r = H_0{}^t \tag{5.72}$$

Because of the relationships between the field amplitudes [cf. Eqs. (5.37), (5.38), and (5.68)], Eq. (5.72) can be expressed in terms of E-values:

$$E_0{}^i + E_0{}^r = \frac{\hat{n}}{\mu} E_0{}^t \tag{5.73}$$

This permits at once the correlation of the transmitted and reflected amplitude to that of the incident wave. By addition of Eqs. (5.71) and (5.73),

$$E_0{}^t = \frac{2\mu}{\mu + \hat{n}} E_0{}^i \tag{5.74}$$

and by substitution in Eq. (5.73):

$$E_0{}^r = \frac{\mu - \hat{n}}{\mu + \hat{n}} E_0{}^i \tag{5.75}$$

We define as *reflectivity* or *reflection coefficient* the ratio of reflected to incident light intensity, i.e., the ratio of the corresponding wave amplitudes squared. It follows then from Eq. (5.75) that

$$R = \left|\,\frac{E_0{}^r}{E_0{}^i}\,\right|^2 = \left|\,\frac{H_0{}^r}{H_0{}^i}\,\right|^2 = \left|\,\frac{\hat{n} - \mu}{\hat{n} + \mu}\,\right|^2 \tag{5.76}$$

Adding the squares of the real and imaginary components of the term on the right (i.e., forming the product of the conjugate complex numbers) yields for the reflectivity at normal incidence

$$R = \frac{(n - \mu)^2 + \kappa^2}{(n + \mu)^2 + \kappa^2} \tag{5.77}$$

The values of n and κ can be reduced further by means of Eqs. (5.49) and (5.50) so that the reflectivity is given ultimately in terms of the electrical parameters ϵ, μ, and σ. In dielectric materials with $\mu = 1$ and $\kappa = 0$, Eq. (5.77) reduces to

$$R = \frac{(n - 1)^2}{(n + 1)^2} \tag{5.78}$$

For optical frequencies we can set $\mu = 1$ also in metals. If the inequality (5.55) is valid, i.e. if $n = \kappa = \sqrt{\sigma/\nu}$ and also $\sigma/\nu \gg 1$, the expansion of Eq. (5.77) yields

$$R = \frac{2\dfrac{\sigma}{\nu} - 2\sqrt{\dfrac{\sigma}{\nu}} + 1}{2\dfrac{\sigma}{\nu} + 2\sqrt{\dfrac{\sigma}{\nu}} + 1} = 1 - 2\sqrt{\dfrac{\nu}{\sigma}} + \cdots \tag{5.79}$$

It follows from Eqs. (5.76) and (5.79) that the reflectivity R can approach 100% for a variety of circumstances. Most important among these is the condition—approximated by metals or dielectrics at resonance points to be discussed later—that $\hat{n}$ is purely imaginary. In this case $|\hat{n} - \mu| = |\hat{n} + \mu|$ and $R = 1$. Phenomenologically this can be understood through Eq. (5.1) as follows: $\hat{\epsilon} = \hat{n}^2$ is real and negative; therefore **D**, the net field in the medium is in opposition or antiphase to the incident wave **E**. According to Eq. (5.79), complete reflectivity is approached in metals for low frequencies or high conductivities.

Thus large values of the extinction coefficient κ in metals imply almost vanishing penetration depths together with almost complete reflectivity (cf. Table 5.1). The limitations of Eq. (5.79) will be discussed at the end of the next section where theoretical values for the reflectivity are compared with those obtained by experimental methods in Table 5.2.

5.1.3.2 *The Laws of Reflection and Refraction*

The boundary phenomena which we have considered in the preceding section were limited to normal incidence of the electromagnetic wave on the interface between two media. If the propagation vector $\mathbf{s}^i$ of the wave forms an angle of incidence θ^i with the normal, there results a disturbance of the symmetry which had simplified the relationships of angles and intensities. At oblique incidence we must expect a more or less complex dependence of these quantities on the angle θ^i. This subject matter is treated in detail in the textbooks on classical optics. We propose to present here only an abbreviated version of some of the more important results following the procedure of Born and Wolf.[2]

(a) *Oblique Incidence on Dielectric Surface.* Suppose an electromagnetic wave is incident at an angle θ^i on an interface given by $z = 0$ between two dielectric media with the indices n_1 and n_2 (see Fig. 5.4). Then the continuity conditions of the fields discussed in the preceding section for normal incidence have to be generalized. The incident, reflected, and transmitted wave equations are, respectively,

$$\mathbf{E}^i = \mathbf{E}_0^{\ i} \exp\left[i\omega\left\{t - \frac{n_1}{c}(\mathbf{r}\cdot\mathbf{s}^i)\right\}\right], \qquad \mathbf{E}^r = \mathbf{E}_0^{\ r} \exp\left[i\omega\left\{t - \frac{n_1}{c}(\mathbf{r}\cdot\mathbf{s}^r)\right\}\right]$$

$$\mathbf{E}^t = \mathbf{E}_0^{\ t} \exp\left[i\omega\left\{t - \frac{n_2}{c}(\mathbf{r}\cdot\mathbf{s}^t)\right\}\right]$$

(5.80)

In particular, the continuity conditions require that the arguments of the three functions in Eq. (5.80) are the same for any point $r(x, y, 0)$ on the interface of the medium.

For in order that at all times t

$$\mathbf{E}^i(t, \mathbf{r}) + \mathbf{E}^r(t, \mathbf{r}) = \mathbf{E}^t(t, \mathbf{r})$$

(5.81)

it is necessary that

$$t - \frac{n_1}{c}\mathbf{r}\cdot\mathbf{s}^i = t - \frac{n_1}{c}\mathbf{r}\cdot\mathbf{s}^r = t - \frac{n_2}{c}\mathbf{r}\cdot\mathbf{s}^t$$

(5.82)

These last three relationships represent at once statements regarding the direction of the reflected and refracted wavefronts. We define as the *plane of incidence* that which contains the normal on the surface $z = 0$ and the unit vector $\mathbf{s}^i$ and locate it in the $x - z$ plane so that $s_y^{\ i} = 0$. It follows then from the presentation of Eq. (5.82) in component form that $s_y^{\ r} = 0$ and with the aid of Fig. 5.4 that

$$\theta^i = \theta^r$$

(5.83)

Thus the reflected beam lies in the plane of incidence and the angle of reflection equals that of incidence.

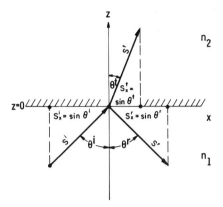

FIG. 5.4. Continuity requirements at oblique incidence on boundary between media of indices n_1 and n_2. The laws of reflection and refraction result geometrically from the continuity conditions.

Furthermore, we find for the refracted beam with Eq. (5.82) and Fig. 5.4 that

$$\mathbf{s}_y{}^t = \mathbf{s}_y{}^i = 0 \qquad (5.84)$$

and

$$n_1 \sin \theta^i = n_2 \sin \theta^t \qquad (5.85)$$

This last relationship is known as Snell's law and can be written as

$$\frac{\sin \theta^i}{\sin \theta^t} = \frac{n_2}{n_1} = \sqrt{\frac{\epsilon_2 \mu_2}{\epsilon_1 \mu_1}} = \frac{v_1}{v_2} = n \qquad (5.86)$$

where $n = n_2/n_1$, the refractive index from medium 1 to medium 2, has been introduced. It is equal to the inverse ratio of the propagation velocities in the two media. The case of $n < 1$ corresponds to lower optical density of the second medium. If $\theta^t = 90°$, $\sin \theta^i = n$; for $\theta^i > \sin^{-1} n$, *total reflection* takes place.

The computation of the reflected intensity at oblique incidence is considerably more complex than that at normal incidence in which the field vectors are perpendicular to the plane of incidence (or tangential to the boundary surface). The procedure, which we can only indicate here, consists of resolving the incident fields into components which are parallel and perpendicular to the plane of incidence. The application of the boundary conditions is then possible separately for the two component waves of each field. The conditions of vector addition depend, of course, on the orientation of the incident vector and, as a result, the reflected and transmitted field amplitudes are different for vibrations parallel and perpendicular to the plane of incidence; in other words, the reflected and transmitted beams are polarized. In particular, the reflectivities referred to the radiation intensities (square of the amplitudes), i.e., $R_\parallel$ for the square of the electric vector parallel to the plane of incidence and $R_\perp$ for perpendicular electric fields are, respectively,

$$R_\parallel = \frac{\tan^2(\theta^i - \theta^t)}{\tan^2(\theta^i + \theta^t)} \qquad (5.87)$$

$$R_\perp = \frac{\sin^2(\theta^i - \theta^t)}{\sin^2(\theta^i + \theta^t)} \qquad (5.88)$$

The relative contribution of the two components depends on the angle α^i which the electric vector $\mathbf{E}^i$ forms with plane of incidence such that the total reflectivity is given by

$$R = R_\parallel \cos^2 \alpha^i + R_\perp \sin^2 \alpha^i \qquad (5.89)$$

These so-called *Fresnel equations* (5.87)–(5.89) must contain the case of normal incidence, and this can be verified by setting $\alpha^i = \pi/2$ and by expansion and transition of θ^i and θ^t to the limit of zero. The complete plot of $R_\parallel$ and $R_\perp$ versus θ^i is shown in Fig. 5.5. Of particular importance is *Brewster's angle* $\theta_B{}^i$ of incidence at which $\theta_B{}^i + \theta_B{}^t = \pi/2$ so that $R_\parallel = 0$, and the reflected beam is plane-polarized, the E-field vibrating perpendicular to the plane of incidence. Since the significance of the complementary relationship $\theta_B{}^i + \theta_B{}^t = 90°$ lies in the fact that the reflected beam is at right angles to the refracted beam, the dipole radiation of the excited electrons parallel to the plane of incidence, must, of course, be zero. The polarization measurement of Brewster's angle leads to a particularly simple determination of the index of refraction n inasmuch as

$$\tan \theta_B{}^i = \frac{\sin \theta_B{}^i}{\sin \left(\dfrac{\pi}{2} - \theta_B{}^i\right)} = n \qquad (5.90)$$

The measurement of n by means of the polarizing angle of the reflected beam is especially important for the cases in which the refracted beam cannot be observed.

(b) *Oblique Incidence on Metallic Surfaces.* It will be obvious that the considerations in the preceding section of continuity and of the resulting propagation phenomena are immediately applicable to the oblique in-

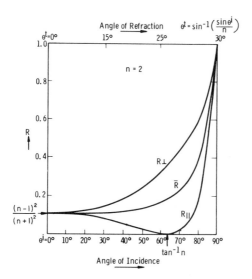

FIG. 5.5. Reflectivity as a function of angle of incidence for $n = 2$. The two curves for $R_\perp$ and $R_\parallel$, respectively, correspond to the relative reflected intensity of components vibrating perpendicularly and parallel to the plane of incidence. All other vector orientations of the incident wave have net reflectivities between the two curves as upper and lower limits. Ordinary light (black body radiation) contains random orientations of the electric vector such that $\bar{R} = \frac{1}{2}(R_\parallel + R_\perp)$. The curve for $\bar{R}$ thus represents the net reflectivity for light from ordinary sources. At $\theta^i = \tan^{-1} n$ (Brewster angle) $R_\parallel = 0$, and the reflected beam is plane-polarized in the plane of incidence.

cidence on metallic surfaces except for the substitution of a complex index $\hat{n}$ instead of the real $n = n_2/n_1$. With this proviso, Eqs. (5.80)–(5.88) are also valid for reflection and refraction at the boundary between a dielectric and a metal.

Thus, it follows, as before, that the reflected beam lies in the plane of incidence and that the angle of reflection equals that of incidence. Again, the law of refraction has the form

$$\sin \theta^t = \frac{1}{\hat{n}} \sin \theta^i = \frac{\sin \theta^i}{n - i\kappa} \tag{5.91}$$

However, $\sin \theta^t$ is complex here since $\hat{n}$ is complex, and therefore it cannot be interpreted simply as an angle of refraction. Indeed, the real angle of refraction has a more complicated dependence on θ^i than given by Eq. (5.86).

Similarly, corresponding to Eqs. (5.87) and (5.88), we have for the complex *amplitudes* of the reflected electric field components vibrating parallel and perpendicular to the plane of incidence, respectively:

$$E_{\parallel}{}^{r} = \frac{\tan(\theta^{i} - \theta^{t})}{\tan(\theta^{i} + \theta^{t})} E_{\parallel}{}^{i} \tag{5.92}$$

$$E_{\perp}{}^{r} = -\frac{\sin(\theta^{i} - \theta^{t})}{\sin(\theta^{i} + \theta^{t})} E_{\perp}{}^{i} \tag{5.93}$$

where we have omitted the index 0 for the amplitude designation. Corresponding relationships obtain for the magnetic fields.

The meaning of complex trigonometric functions is that of phase and amplitude changes after reflection. In other words, Eqs. (5.92) and (5.93) can be written in the form

$$\frac{E_{\parallel}{}^{r}}{E_{\parallel}{}^{i}} = \rho_{\parallel} e^{i\phi_{\parallel}} \tag{5.94}$$

and

$$\frac{E_{\perp}{}^{r}}{E_{\perp}{}^{i}} = \rho_{\perp} e^{i\phi_{\perp}} \tag{5.95}$$

Then, too, the relationship between the reflected field components, which are, respectively, parallel and normal to the plane of incidence, has changed in two respects over that of the incident beam:

First the components vibrate with a phase difference

$$\Delta = \phi_{\perp} - \phi_{\parallel} \tag{5.96}$$

Secondly, the amplitude ratio has been distorted by a factor of

$$P = \frac{\rho_{\perp}}{\rho_{\parallel}} = \tan \psi \tag{5.97}$$

where ψ, the angle of amplitude ratio distortion has been introduced for mathematical convenience. The time relationship of the two components can be expressed in the form

$$\begin{aligned} E_{t\parallel}^{r} &= \rho_{\parallel} E_{\parallel}{}^{i} \cos \omega t, \\ E_{t\perp}^{r} &= \rho_{\perp} E_{\perp}{}^{i} \cos(\omega t + \Delta) \end{aligned} \tag{5.98}$$

Equation (5.98) indicates in parameter presentation that the field vector describes an ellipse in the plane normal to the propagation direction of the beam (see Fig. 5.6). Thus a linearly polarized electromagnetic

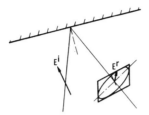

F IG. 5.6. Elliptic polarization. The linearly polarized wave E^i experiences different phase shifts for the parallel and perpendicular components after reflection so that elliptcial polarization results.

wave is reflected *elliptically polarized* such that, according to Eqs. (5.94) and (5.95), the ratio of the electric field components is

$$\frac{E_\perp{}^r}{E_\parallel{}^r} = -\frac{\cos(\theta^i - \theta^t)}{\cos(\theta^i + \theta^t)}\frac{E_\perp{}^i}{E_\parallel{}^i} = \frac{\rho_\perp}{\rho_\parallel}\cdot e^{i\Delta}\cdot\frac{E_\perp{}^i}{E_\parallel{}^i} = \tan\psi\cdot e^{i\Delta}\frac{E_\perp{}^i}{E_\parallel{}^i} \qquad (5.99)$$

with corresponding expressions for the magnetic field components. In the special cases that Δ equals 0 or π, the reflected beam is itself linearly polarized, such as for normal and grazing incidence. Another limiting case of elliptical vibration is that for $\Delta = \pi/2$ and $(\tan\psi)\cdot E_\perp{}^i/E_\parallel{}^i = 1$ which results in *circular polarization*.

The phase and amplitude relationships (5.96) and (5.97) are implicitly dependent on n and κ, aside from the angle of incidence θ^i, as follows from Eqs. (5.91)–(5.93). Therefore the measurement of two polarization parameters in the beam reflected by a metal can serve for the determination of the indices n and κ, just as the measurement of the Brewster angle can serve for the determination of n in dielectrics (in which κ is zero). We cite here without derivation two useful equations which permit the determination of n and κ from easily measurable polarization quantities:

$$n \sim -\frac{\sin\theta^i\tan\theta^i\cos 2\Psi}{1 + \sin 2\Psi\cos\Delta} \qquad (5.100)$$

and

$$\kappa \sim \frac{\sin\theta^i\tan\theta^i\sin 2\Psi\sin\Delta}{1 + \sin 2\Psi\cos\Delta} \qquad (5.101)$$

These formulae assume particularly simple forms for the *principal angle* of incidence $\overline{\theta^i}$, for which $\Delta = \pi/2$ and which corresponds to Brewster's angle for dielectric boundaries:

$$n \sim -\sin\overline{\theta^i}\tan\overline{\theta^i}\cos 2\Psi \qquad (5.102)$$

and

$$\kappa \sim -\sin\overline{\theta^i}\tan\overline{\theta^i}\sin 2\Psi \qquad (5.103)$$

TABLE 5.2

Optical Indices and Reflectivity of Metals

I	II λ_0	III[a] n	IV[a] κ	V[a] R Comput. in % from exp. n, κ	VI[b] R Direct exp. value %	VII R From Eq. (5.79) with dc values of σ, $\mu = 1$, %
Metal	[μ]					
Al	0.492	0.68	4.80	90 (e) (VM)		91.2
	0.950	1.75	8.50		90 (e)	93.6
	2.0	2.30	16.5	98.4 (e) (VM)	95 (e)	95.6
	5.0	8.19	36.8	98.8 (e) (VM)	96 (p)	97.3
	12.0	33.6	76.4	99.1 (e) (VM)	98 (p)	98.2
Ag	0.226	1.406	1.107	19.8 (p)	20	
	0.316	1.127	0.427	4.2 (p)	5	
	0.492	0.123	2.72	94.3 (e) (VM)		93.3
	1.00	0.24	6.96	98.1 (c)	97.0 (p)	95.3
	2.00	0.68	13.7	98.5 (c)	97.8 (p)	96.7
	4.37	4.34	32.6	98.6 (c)	98.5 (p)	97.7
	10.0				99	98.5
Au	0.508	0.908	2.075	54.3 (el)	50 (p)	92.0
	1.00	0.194	5.57	97.5 (e)		94.3
	2.00	0.56	15.4	99.1 (e)	96.5 (s)	96.0
	3.14	0.80	18.9	99.2 (el)		96.8
	4.83	1.83	33.0	99.3 (el)	97.2	97.4
	9.0				98.0 (el)	98.1
Ni	0.500	1.54	2.975	59.7 (p)	60.8 (el)	87.2
	1.00	2.63	5.26	74.1 (p)	72.0 (el)	90.9
	2.00	3.70	8.54	84.4 (p)	83.5 (el)	93.6
	9.0				95.6 (el)	97.0
Cu	0.500	1.098	2.341	55.5 (p)	43.7 (p)	93.3
		1.500	2.370	49.0 (el)		
	1.00	0.224	6.14	97.5 (e)	90.1 (p)	95.2
	2.00	0.25	11.7	97.1 (e)	95.5 (p)	96.6
	4.20	1.92	22.8	98.7 (s)	97.3 (p)	97.4
	9.0				98.4 (p)	98.4
Fe	0.257	1.006	0.880	16.2 (s)		
	0.508	1.382	1.495	30.2 (s)	55 (p)	
	1.00				65	88.5

Table 5.2 presents the optical indices n and κ determined from polarization measurements for certain metals of interest at various wavelengths λ. The reflectivity R can be computed from n and κ according to Eq. (5.77), where the permeability μ is set equal to unity at the frequencies concerned, and these values are listed under column V. Values of the reflectivity R measured directly at normal incidence are shown for comparison in column VI. Differences between the values of these two columns are mostly small and, as a rule, can be ascribed to variations of the surface and preparation conditions. Column VII contains theoretical values of R on the basis of Eq. (5.79), where the conductivity σ known from direct current measurements has been introduced. It is seen that agreement is satisfactory at long wavelengths, and that in many cases it is only in the near infrared or the visible region that the electronic mass effect leads to major discrepancies. It can be shown that at these frequencies the electron oscillation amplitude is smaller than the mean free path so that a decrease of conductivity and reflectivity must be expected on the basis of such simple arguments.

5.2 Optical Properties of the Materials

It was shown in the preceding treatment that virtually all macroscopic phenomena of light propagation, such as reflection, refraction, absorption, and polarization, are derivable from Maxwell's equations. In particular, the optics of insulators and metals can be linked to optical indices, viz., n and κ, and these in turn are given by certain electrical parameters such as ϵ and σ in the cases considered so far. The electromagnetic wave theory, as presented in the preceding section, deals with the indices and parameters as given (and fixed) constants and has little to say about their wavelength dependence. Thus dielectrics were treated as elastic media without regard to their regions of resonance and metals were discussed only in their regions of reflectivity. Actually, there exist drastic changes of the characteristics for various wavelengths. A survey

---

[a] From H. Schopper, *in* Landolt-Boernstein, "Zahlenwerte und Funktionen," 6th ed., Vol. II, Part 8. Springer, Berlin, 1962.

[b] From "Handbook of Chemistry and Physics," 44th ed. Chem. Rubber Publ. Co., Cleveland, Ohio, 1962-1963.

[c] (c)   chemically deposited   (e)   evaporated
   (el)   electrolytically formed   (p)   polished in bulk
   (s)   sputtered   (VM)   vacuum measurement

TABLE 5.3

PROPERTIES OF LIGHT PROPAGATION TYPICAL FOR A MAJORITY
OF DIFFERENT MEDIA AND WAVELENGTH REGIONS

Materials	Far infrared $\lambda > 10\mu$	Near infrared and visible region $3\mu > \lambda > 0.4\mu$	Ultraviolet $\lambda < 0.3\mu$
Metals	Reflectivity near 100%, high κ	Good to intermediate reflectivity	Low reflectivity, transparency for alkali metals
Semiconductors	Transparent	Partially reflecting and absorbing	Low reflectivity, high absorption
Insulators	Reststrahlen region of high absorption and reflection	Transparent	Onset of fundamental absorption band
Gases (in ground state)			
Homopolar	Transparent	Transparent	Absorption
Heteropolar	Absorption in extended regions	Transparent	Absorption

of optical properties of the materials throughout the optical spectrum is given in Table 5.3. Obviously, the varied behavior of the optical indices can be explained only by detailed atomistic models. This microscopic aspect of optical phenomena involves a distinctly different approach, namely, that which derives the optical properties from the behavior of free and bound charges.

5.2.1 GENERAL RELATIONSHIPS

The task of the atomistic theory then is to postulate suitable models for the structure of matter and explore the response of electric charges under the various bonding conditions to the incident electromagnetic wave. As pointed out before, the classical approach goes quite far in describing all pertinent phenomena.

The elementary models discussed in Chapter 1 are now the basis for the postulated structures: the free electron picture for metal optics, bound charges in dielectrics, and both in semiconductors. A realistic description must also include damping forces which in effect are provided, for instance, by collisions of electrons in metals with phonons and impurity atoms or by various lossy mechanisms in the case of the oscillators in dielectrics. Nevertheless, many optical properties of the various materials can be explained on the basis of ideal, undamped charged particles, and the quantitative descriptions on this premise are often remarkably accurate.

The computation of the response of the charges to the incident radiation amounts to setting up a driving force $eE_0e^{i\omega t}$ in the Newtonian equation of motion which depending on the structural model may, or may not, contain a restoring force and a damping term. The direct solution of the equation of motion, $x(t)$, will be real, if the damping term is absent, and complex, if it is present.

It will be seen that the displacement $x(t)$ of a charge e represents a time-variant electric moment $X = e \cdot x(t)$ which is induced by the incident wave. In addition, there may be permanent dipoles (polar molecules) which we shall ignore here. It there are N charges (usually electrons) per cubic centimeter, there results a total electric moment per unit volume, the so-called *polarization* P:

$$\mathbf{P} = N \cdot \mathbf{X} = N \cdot e \cdot \mathbf{x}(t) \tag{5.104}$$

It is shown in electrostatic field theory that as a result of Eq. (5.6), the electric displacement $\mathbf{D} = \epsilon\mathbf{E}$ is given by

$$\mathbf{D} = \epsilon\mathbf{E} = \mathbf{E} + 4\pi\mathbf{P} \tag{5.105}$$

Therefore we have for the complex dielectric constant, if $\mu = 1$:

$$\hat{\epsilon} = 1 + \frac{4\pi\,|\,\mathbf{P}\,|}{|\,\mathbf{E}\,|} = (\hat{n})^2 = (n - i\kappa)^2 \tag{5.106}$$

where the relationship between $\hat{\epsilon}$ and $\hat{n}$ follows from Maxwell's equations in analogy to Eq. (5.28) and the real and imaginary components of $\hat{n}$ are defined by Eq. (5.43).

Furthermore, if the density of the charges is small enough so that they do not affect each other (and the microscopic field is indeed $\mathbf{E}$), the solution of the Newtonian equation yields the general result

$$e\mathbf{x}(t) = \mathbf{X}(t) = \alpha\mathbf{E} = \alpha\mathbf{E}_0\,e^{i\omega t} \tag{5.106a}$$

where the *polarizability* α is to be computed for each model. Hence, in combination with Eqs. (5.104) and (5. 106) it follows that

$$(\hat{n})^2 = (n^2 - \kappa^2) - i2n\kappa = 1 + 4\pi\alpha N \tag{5.107}$$

in which the left side becomes $\epsilon = n^2$ if $\kappa = 0$.

This, in a general form, represents the solution to the problem of reducing the optical indices to atomic units inasmuch as α will be given explicitly in terms of fundamental constants through the Newtonian equation in a manner depending on the bonding forces. Although detailed applications of this procedure to the various media will be presented in Sections 5.2.2 and 5.2.3, certain common features deserve mention at this point.

First, the physical connection between reflection, refraction, and absorption, which is phenomenologically most impressive at oscillator resonance conditions, is much clearer in the atomistic interpretation than through the derivation by Maxwell's equations. The picture of the induced oscillating dipole explains at once that re-emission, phase change, and power dissipation must occur together so that a medium, transparent up to a certain frequency limit, will exhibit above such limit the phenomena of reflection, refraction, polarization, and absorption to an enhanced degree.

Another general feature concerns the dependence of the optical indices on the concentration N of the active charges [cf. Eq. (5.107)]. For low densities, the relationship between n, κ and N is, in fact, approximately linear in some cases as will be demonstrated later. Under these conditions it is meaningful to express, for instance, the absorption coefficient in terms of an effective cross section q. In other words, the fractional loss of intensity dI/I for a beam propagating in the z-direction through a lossy medium can be written in a form (referred to as *Beer's law*):

$$\frac{dI}{I} = -k_\nu \, dz = -Nq \, dz \qquad (5.108)$$

the integral of which is

$$I(z) = I_0 e^{-Nqz} \qquad (5.109)$$

Interpretations of k_ν in terms of a cross section have been discussed before, as for example in the case of a resonance line in natural broadening [see Eq. (3.102)] where q is in the order of the square of the wavelength.

Equations (5.108) and (5.109) refer to a single frequency indicated by the index ν and imply, furthermore, that k_ν is indeed independent of the path length. In practical situations, however, finite spectral widths are encountered in which $k(\nu)$ may vary strongly. Consider, e.g., absorption in the region of narrow spectral lines. The absorption coefficient $k(\nu)$ has a maximum at the peak of the spectral line. Hence, as the beam progresses along the direction z, the intensity around the spectral region near k_{max} will decrease rather quickly towards zero, whereas the intensity for wavelengths farther away from the center of the absorption line is much less diminished. The result is that the absorption coefficient for the total band $\nu_2 - \nu_1$, i.e., the fraction of intensity absorbed per unit length, is not constant along the propagation direction.

Thus, we have to integrate over the various wavelength regions in order to obtain the variation of intensity with path length:

$$I(z) = \int_{\nu_1}^{\nu_2} I_0(\nu) e^{-k_\nu(\nu)z} \, d\nu \tag{5.110}$$

In general, this expression can not be evaluated as simply as Eq. (5.109). In the case of absorption lines, $k_\nu(\nu)$ can have any of the profiles discussed under Section 3.4. An important special case is the condition that the observed absorption line or the spectral region admitted by the slit of the spectrograph consists of many unresolved lines. Assuming that $k(\nu)$ in these components has a Lorentzian shape, the intensity absorbed by each can be found by an approximate solution of Eq. (5.110) in the form

$$\frac{I_0 - I(z)}{I_0} = 2\Delta\nu_h(\pi k_0 z)^{\frac{1}{2}} \tag{5.111}$$

where it has been assumed that $\Delta\nu_h$, the half-width of the individual lines, is small compared to the interval between them, k_0 is the peak absorption and I_0 constant over the frequency range considered. The total absorption here is seen to increase approximately with the square root of the path length as was first found experimentally by Strong[3] (see also Elsasser[4]).

5.2.2 Dispersion Theory of Dielectric Media

We now apply the procedure outlined in the beginning of the preceding section to the determination of n and κ in dielectric media with power dissipation. The model to be applied here is that of the damped forced oscillator, viz. the electron or ion bound to a fixed atom site. The differential equation for forced oscillation then contains a frictional term $R\dot{x}$:

$$m\ddot{x} + R\dot{x} + Gx = eE_0 e^{i\omega t} \tag{5.112}$$

where e, m, and G are charge, mass, and restoring force constant, respectively, of either an electron or an ion. However, it is also possible to interpret these constants as distributed over several mass points or over the entire lattice.

The solution, as already discussed in Section 3.1.4, is completely determined by the sum of the general integral for Eq. (3.25), i.e., for zero driving force, and any particular integral of the Eq. (5.112). The first of these integrals defines the initial conditions and describes a transient oscillation, the amplitude building up or decaying with a time constant of $\gamma/2 = R/2m$. This has been discussed in Sections 1.2.2 and 3.1.4, and we shall not include the transient term in the subsequent equations since for phenomena of propagation only the second integral is important which represents the stationary solution.*

Setting $x = x_0 e^{i\omega t}$, $R/m = \gamma$, and $G/m = \omega_0{}^2$, we obtain by substitution in Eq. (5.112)

$$\mathbf{x} = \frac{e\mathbf{E}_0}{m} \frac{1}{\omega_0{}^2 - \omega^2 + i\gamma\omega} e^{i\omega t} \tag{5.118}$$

It is seen that the polarizability, $\alpha = ex/E$, is inversely proportional to the mass m. Hence electron polarizabilities are important throughout the optical spectrum and they determine the index of refraction in the transparent regions. Ionic polarizabilities are dominant only in the region of resonance.

5.2.2.1 Gases

If it is assumed first that the atoms are sufficiently far apart as in a low pressure gas, their mutual interaction can be ignored. We deal then

---

* The behavior of the oscillating mass point is seen better from a presentation in real numbers rather than the complex form for $n - i\kappa$. If we set [cf. Eq. (3.27)]

$$\mathbf{x} = \mathbf{x}_0 e^{i(\omega t - \beta)}, \tag{5.113}$$

one obtains from Eq. (5.112)

$$\mathbf{x}_0 \left(-\omega^2 + \frac{G}{m} + i\frac{R}{m}\omega\right) = \frac{e\mathbf{E}}{m} e^{i\beta} = \frac{e\mathbf{E}}{m}(\cos\beta + i\sin\beta) \tag{5.114}$$

From this equation, amplitude $\mathbf{x}_0$ and phase angle β can be determined, the former by comparing the absolute amounts, the latter by the ratio of complex and real amounts:

$$\mathbf{x}_0{}^2[(\omega_0{}^2 - \omega^2)^2 + \gamma^2\omega^2] = \left(\frac{e\mathbf{E}}{m}\right)^2 \tag{5.115}$$

or

$$\mathbf{x}_0 = \frac{e\mathbf{E}_0}{m} \frac{1}{\sqrt{(\omega_0{}^2 - \omega^2)^2 + \gamma^2\omega^2}} \tag{5.116}$$

and

$$\beta = \tan^{-1}\frac{\gamma\omega}{(\omega_0{}^2 - \omega^2)} \tag{5.117}$$

where again we have set γ for R/m and $\omega_0{}^2$ for G/m. Equation (5.116) describes absorption curves (as shown in Fig. 3.4). Equation (5.117) states that as ω changes from zero to infinity, the phase delay varies from 0 to $-\pi$. At resonance, $\omega = \omega_0$, the charge oscillates 90° out of phase with the driving field, and at this point, for simple cases as in Section 5.2.2.1, $n = 1$ or $v = c$.

with the electrons as oscillating mass points, the polarization and index of refraction being given by Eqs. (5.106) and (5.107), respectively. According to Eqs. (5.106) and (5.118), the polarizability α is

$$\alpha = \frac{e^2}{m} \frac{1}{\omega_0^2 - \omega^2 + i\gamma\omega} \tag{5.119}$$

Introduction of the expression into Eq. (5.107) yields after separation into real and imaginary parts $\mathcal{R}$ and $\mathcal{I}$:

$$\hat{n}^2 = (n^2 - \kappa^2) - i(2n\kappa) = \mathcal{R} + i\,\mathcal{I} = \left(1 + \frac{4\pi Ne^2}{m} \frac{\omega_0^2 - \omega^2}{(\omega_0^2 - \omega^2)^2 + \gamma^2\omega^2}\right)$$
$$-i\left(\frac{4\pi Ne^2}{m} \cdot \frac{\omega\gamma}{(\omega_0^2 - \omega^2)^2 + \gamma^2\omega^2}\right) \tag{5.120}$$

From this follow explicit relationships for the optical indices:

$$n^2 = \tfrac{1}{2}\{\sqrt{\mathcal{R}^2 + \mathcal{I}^2} + \mathcal{R}\} \tag{5.121}$$

$$\kappa^2 = \tfrac{1}{2}\{\sqrt{\mathcal{R}^2 + \mathcal{I}^2} - \mathcal{R}\} \tag{5.122}$$

where $\mathcal{R}$ and $\mathcal{I}$ are given in terms of atomic constants by Eq. (5.120).

In the case of gases at not too high densities, certain inequalities and approximations are valid, viz.:

$$\kappa \ll 1, \qquad |n - 1| \ll 1, \qquad n^2 - 1 \cong 2(n - 1) \tag{5.123}$$

With these simplifications, the following equations for n and κ result from Eq. (5.120):

$$n = 1 + \frac{2\pi Ne^2}{m} \frac{\omega_0^2 - \omega^2}{(\omega_0^2 - \omega^2)^2 + \gamma^2\omega^2} \tag{5.124}$$

$$\kappa = \frac{2\pi Ne^2}{m} \cdot \frac{\gamma\omega}{(\omega_0^2 - \omega^2)^2 + \gamma^2\omega^2} \tag{5.125}$$

The variation of κ and n in the neighborhood of the resonance frequency $\omega = \omega_0$ is shown in Figs. 5.7a and 5.7b. It is seen that outside this region the index of refraction n increases as the frequency increases. The corresponding negative coefficient in wavelength, $dn/d\lambda$, is called *normal dispersion*. In the environment of the resonance point, however, the dispersion changes sign; it is called *anomalous*, and it is in this region of anomalous dispersion that absorption, too, is appreciable. In the idealized case that $\gamma = 0$, i.e., for oscillation without damping, the dispersion,

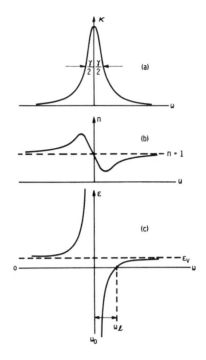

Fig. 5.7. Theoretical dispersion curves near resonance: (a) extinction coefficient for low pressure gas; (b) index n in the presence of damping; (c) dielectric constant $\epsilon = n^2$ for solids in the absence of damping. Reflectivity is total in the wavelength interval for which $\epsilon < 0$. As indicated by the arrowed line, this domain lies between the resonant frequency ω_0 and the value $\omega_l = \omega_0(\epsilon_0/\epsilon_v)^{1/2}$ (where ω_l is also the fundamental longitudinal optical mode).

instead of passing through a maximum and a minimum, tends to $+\infty$ or $-\infty$ as ω_0 is approached from smaller or larger frequencies, respectively. Figure 5.7c shows the curve $\epsilon = n^2$ for the case of $\gamma = 0$. The interval in which $\epsilon < 0$ is characterized by total reflection (see Section 5.2.2.3).

Qualitatively, Eqs. (5.124) and (5.125) agree with measured profiles of absorption lines in the spectrum of gases. Quantitatively, however, one finds in general that the effective number of oscillators for any given resonance point is smaller than the number of atoms N per unit volume even for excitations from the ground state. The reason for this is obvious from the discussion of Section 3.3.3.1. Atoms are not simple oscillators but have in principle an unlimited number of resonances among which the classical value for the index $n - 1$ (per electron) is subdivided so that each vibration has a characteristic oscillator strength f_i. We can take this situation into account by writing

$$n - 1 = \frac{2\pi Ne^2}{m} \sum_i \frac{f_i(\omega_i^2 - \omega^2)}{(\omega_i^2 - \omega^2) + \gamma^2\omega^2} \qquad (5.126)$$

Furthermore, however, if excited states are present, they can contribute to negative absorption (stimulated emission) and dispersion so that the number of oscillators is reduced by an additional factor of $(1 - g_m N_n / g_n N_m)$ according to Section 3.3.3.1. We have then for an individual line

$$n - 1 = \frac{2\pi N_m e^2}{m} \frac{f_{mn}(\omega_{mn}^2 - \omega^2)}{(\omega_{mn}^2 - \omega^2)^2 + \gamma^2\omega^2} \left(1 - \frac{g_m}{g_n} \frac{N_n}{N_m}\right) \qquad (5.127)$$

Outside the immediate neighborhood of the resonance point $\omega = \omega_{mn}$, one can ignore the damping term γ, and Eq. (5.127) passes over into Eq. (3.116).

5.2.2.2 *Dense Media; Lorentz-Lorenz Equation*

The interatomic distance in solids and liquids (or gases at about 10^3 atmospheres) is in the order of 10 times smaller than that in gases at atmospheric pressure. This factor is not only decisive for the degree of coupling between the oscillators but it implies a density larger by 10^3. As a result, the optical properties of dense media differ from those in normal gases in two major respects: (1) the field $\mathbf{E}$ impressed from the outside is modified in microscopic regions by the surrounding induced dipoles; (2) the values of $n - 1$ and κ can no longer be considered small compared with unity, particularly in the region of anomalous dispersion where reflectivity and absorption assume high values.

It is known from electrostatic theory that the field produced at a point by the polarization $\mathbf{P}$ of a surrounding medium equals $4\pi\mathbf{P}/3$. Therefore Eq. (5.106a) for the relationship between the induced moment and the microscopic field $\mathbf{E}$ has to be modified as follows:

$$e\mathbf{x}(t) = \mathbf{X}(t) = \alpha\left(\mathbf{E} + \frac{4\pi}{3}\mathbf{P}\right) \qquad (5.128)$$

Combined with Eq. (5.104) this becomes

$$\mathbf{P} = N\alpha\left(\mathbf{E} + \frac{4\pi}{3}\mathbf{P}\right) \qquad (5.129)$$

Solved for $N\alpha$, Eq. (5.129) yields the relation called, after its discoverers, the Lorentz[5]-Lorenz[6] formula:

$$\frac{4\pi}{3}N\alpha = \frac{4\pi(P/E)}{3 + 4\pi(P/E)} = \frac{\hat{n}^2 - 1}{\hat{n}^2 + 2} \tag{5.130}$$

The term on the left-hand side of this equation may contain additive contributions from different kinds of oscillators [cf. Eq. (5.126)], i.e., for high frequencies various electron resonances, for the infrared, in addition, various ion resonances. In this connection, it is often desirable to refer the expression (5.130) to a mole of a substance. One cubic centimeter contains a number N of molecules which is proportional to Avogadro's number N_m and the density ρ and indirectly proportional to the molecular weight M':

$$N = \frac{N_m}{M'}\rho \tag{5.131}$$

For frequencies far enough away from any resonance points we can ignore the damping factor γ and hence the imaginary portion of $\hat{n}$. Under these conditions we define as *molar refractivity* a term containing the product of the polarizability and Avogadro's number, viz.,

$$\frac{4\pi}{3}N_m\alpha = \frac{M'}{\rho}\frac{n^2 - 1}{n^2 + 2} \tag{5.132}$$

or

$$\frac{M'}{\rho}\frac{n^2 - 1}{n^2 + 2} = \frac{4\pi}{3}\frac{N_m e^2}{m}\sum_i \frac{f_i}{\omega_0^2 - \omega^2} \tag{5.133}$$

where we have used Eq. (5.119) for $\gamma = 0$ in a summation over several oscillators.

The connection between anomalous dispersion and absorption is recognized by solving Eq. (5.129) for P and n^2:

$$P = E\frac{N\alpha}{1 - (4\pi/3)N\alpha} \tag{5.134}$$

hence, with Eq. (5.106) and (5.119)

$$\hat{n}^2 - 1 = 4\pi\frac{P}{E} = \frac{4\pi N\alpha}{1 - (4\pi/3)N\alpha}$$

$$= \frac{4\pi Ne^2/m}{(\omega_0^2 - \omega^2 + i\gamma\omega)\left(1 - \frac{4\pi}{3}N\frac{(e^2/m)}{(\omega_0^2 - \omega^2 + i\gamma\omega)}\right)} \tag{5.135}$$

We have thus

$$\hat{n}^2 = (n - i\kappa)^2 = 1 + \frac{4\pi N e^2}{m} \frac{1}{\omega_1{}^2 - \omega^2 + i\gamma\omega} \qquad (5.136)$$

In this equation, the resonant frequency ω_1, given by

$$\omega_1{}^2 = \omega_0{}^2 - \frac{4\pi}{3} \frac{e^2}{m} N = \frac{G}{m} - \frac{4\pi}{3} \frac{e^2}{m} N \qquad (5.137)$$

has been introduced to emphasize the similarity with Eq. (5.120). Thus n^2 and κ^2 can be evaluated again by Eqs. (5.121) and (5.122) with the proviso that ω_0 is replaced by ω_1. Again, the last term of Eq. (5.136) has to be summed over several resonators in the general case. We conclude from the last three equations that the resonant frequencies of the optically denser media are shifted towards longer wavelengths from the values characteristic for the undisturbed oscillator. At these regions of resonance, which will be quite broad because of the dense oscillator packing, absorption, reflection, and the absolute value of dispersion reach relatively high values.

5.2.2.3 *Optical Modes of Lattice Vibrations*

The optical indices of crystals are of special interest among the dense media because the relationship of the dispersion spectra to vibration modes and, generally, solid state parameters can be understood relatively well. The enumeration of oscillators in a volume filled with an elastic medium has been discussed already in Section 1.2.3, and this model has been applied to the density distribution of radiation contained in a black body as well as to electron wave functions in a solid (see Section 4.6.1.3). It has been brought out that there exist $3N$ possible vibrations if the particles performing such oscillations have a finite number N. Therefore it might appear that there also exist $3N$ modes by which light can interact with the N vibrating ions of a crystal lattice so that the spectrum extends from wavelengths in the order of the specimen dimensions to that of the interatomic distance. This, however, is not the case. There are only a few modes able to interact with the electromagnetic field, and in the simplest case, namely that of the alkali halides, there is only one *fundamental optical mode*. The reason can be indicated here only qualitatively (for a complete treatment of lattice dynamics, see Born and Huang[7]). The solution for the equations of motion in the case of a lattice with two kinds of atoms can be expressed in terms of a parameter **k**, viz., the wave vector already introduced by Eq. (4.97). There exist two possibilities (see Fig. 5.8): (1) solutions in which neighboring

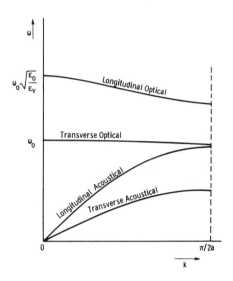

FIG. 5.8.   Example for $\omega(k)$ curves representing optical and acoustical lattice vibrations.

ions move more or less in unison so that no change of dipole moment occurs (*acoustical branches*); and (2) vibrations of neighboring ions with respect to each other (*optical branches*). The acoustical vibrations are typically that of an elastic medium; their frequency rises from $\omega = 0$, for $k = 0$, to an upper limit at $k = \pi/2a$ (the first Brillouin zone for a lattice parameter a). The optical frequencies are above those of the acoustical branches, separated from them by a forbidden gap. For large k, however, the gap may approach zero, as shown in Fig. 5.8. In either case, the vibrations may be longitudinal or transverse, i.e., oriented in the direction of wave propagation or perpendicular to it. The ratio of the fundamental ($k = 0$) longitudinal and transverse optical frequencies is given by the static and high frequency dielectric constants as $\omega_l/\omega_0 = (\epsilon_0/\epsilon_v)^{1/2}$. First-order dipole interaction, of course, is restricted to the transverse optical mode; but even there only the vibration with $k = 0$ makes a large contribution to net radiation since only then the momentum vector of light matches that of the lattice and only for that mode a nonvanishing macroscopic dipole can be sustained by the crystal. Thus the infrared absorption spectra of the alkali halides may be explained by the existence of a single oscillator, viz., the vibration of the anion lattice 180° out of phase with the cation lattice.

The optical indices thus contain contributions of only one, or at most only a few, resonances. However, because of their small mass, the electrons also contribute to the index of refraction, and we account for this in the following manner.

The electrons give rise to a dielectric constant ϵ_v at frequencies sufficiently large compared to the ionic resonance (but still smaller than corresponding to the electronic band transitions). The long-wavelength dielectric constant follows then for $\omega = 0$ from Eqs. (5.107) and (5.119):

$$\epsilon_0 = n_0^2 = \epsilon_v + \frac{4\pi N e^2}{M \omega_0^2} \tag{5.138}$$

We have thus as a variation* to Eqs. (5.120) or Eq. (5.136):

$$n^2 - \kappa^2 = \epsilon_v + \frac{(\epsilon_0 - \epsilon_v)\omega_0^2(\omega_0^2 - \omega^2)}{(\omega_0^2 - \omega^2)^2 + \gamma^2 \omega^2} \tag{5.139}$$

and

$$2n\kappa = \frac{(\epsilon_0 - \epsilon_v)\omega_0^2 \omega \gamma}{(\omega_0^2 - \omega^2)^2 + \gamma^2 \omega^2} \tag{5.140}$$

In this form, the dispersion is reduced to ϵ_0 and ϵ_v rather than the atomic constants, except for the resonant frequency ω_0.

The trend of the reflectivity R in the resonance region is of particular significance. According to Eqs. (5.76), R is given for $\mu = 1$ by

$$R = \left| \frac{\hat{n} - 1}{\hat{n} + 1} \right|^2 \tag{5.141}$$

If $\hat{n}$ is purely imaginary, then $|\hat{n} - 1| = |\hat{n} + 1|$, hence $R = 1$. If we assume the case of $\gamma = 0$ illustrated in Fig. 5.7c, we find that $\epsilon = n^2$ is real and negative in the frequency region between $\omega = \omega_0$ and $\omega = (\epsilon_0/\epsilon_v)^{1/2}\omega_0$, these limits resulting from Eq. (5.139) by setting $n^2 - \kappa^2 = \gamma = 0$. Now in this interval, in which ϵ is negative, n is purely imaginary, and the reflectivity is, therefore, total, the limits define sharply the reststrahlen region. Under these circumstances, there is no absorption, but also no transmission. The effect of a finite γ is a reduction of the peak reflectivity from 100% with subsequent penetration and absorption of the incident wave.

Similar considerations apply, of course, to reflection in the ultraviolet absorption band.

---

* Regarding the validity of these and related equations, see, e.g., Szigeti.[8]

5.2.3 EXPERIMENTAL PROCEDURES AND RESULTS OF
DIELECTRIC OPTICS

In general, two major regions of resonance should be expected in dielectric crystals: (1) optical lattice vibrations which may be visualized as positive ion groups oscillating as a whole with respect to negative ion groups; (2) electronic transitions which have been discussed in Section 4.6 on the spectra of solids. The former lie in the far infrared, typically at wavelengths longer than $10\,\mu$; the latter, for which $h\nu$ must be larger than the gap energy eE_g, lie in the ultraviolet, at least for insulators ($E_g \approx 4$–10 V). Classically, these locations can be understood on the basis of the resonance equation $\omega_0{}^2 = G/M$, where M is the mass of the ion in the first instance, that of the electron in the second. According to the results of the preceding section, the existence of these resonances should reveal itself by pronounced changes in dispersion, reflectivity, and absorptivity. The reflectivity effects are particularly apparent for the optical lattice vibrations, giving rise to the *reststrahlen* (hence also the name *reststrahlen region*), as will be discussed later.

5.2.3.1 *Absorption and Dispersion of Some Dielectric Materials*

The behavior of the absorption coefficient throughout most of the optical spectrum is shown in Fig. 5.9 for the case of diamond. The lattice vibration band of this material is of interest in two respects. First, the combination of a large force constant (large G as shown by the great hardness of diamond) and low atomic mass brings about an unusually high frequency of the optical lattice vibration, corresponding to a peak at $5\,\mu$. Secondly, the covalent bond between the diamond atoms does not give rise to a first order dipole moment. Therefore the optical modes actually observed must be due to higher order perturbations such as induced moments, and the absorption coefficient is by orders of magnitude smaller than that typical of ionic crystals. The spectrum shown in Fig. 5.9 is that of pure diamond. Throughout the spectrum additional bands may appear if (naturally occurring) impurities are added. The infrared bands in Fig. 5.9 characterize the so-called Type II diamond, in contrast to the Type I material which exhibits bands also between 8 and 10 μ with a strength proportional to the number of nitrogen impurity centers[9].

Another example for the trend of the absorption coefficient throughout the optical spectrum is presented in Fig. 5.10 for the case of quartz.

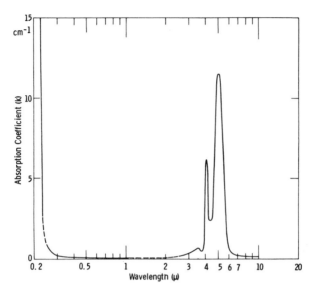

Fig. 5.9. Spectral profile of absorption coefficient in diamond between the ultraviolet and far infrared. Curve represents only typical gross outline since it is composed from data of various observers, viz., in the ultraviolet of F. Peter [*Z. Physik* **15**, 358 (1923)], in the infrared of O. Reinkober [*Ann. Physik* [4] **34**, 243 (1911)] and R. Robertson, J. J. Fox, and A. E. Martin [*Phil. Trans. Roy. Soc.* **232**, 463 (1934)]. For more recent work in the ultraviolet, see e.g. H. R. Philipp and E. A. Taft, *Phys. Rev.* **136**, A1445 (1964).

This material has certain qualities in common with diamond. It is a good insulator, and the intrinsic absorption band lies in the far ultraviolet[10,11]. Its crystal bonds are not typically ionic, but rather covalent; hence the infrared vibration spectrum will not correspond to a single oscillation, but rather show several resonances. It is also a hard material with a large modulus of elasticity, hence presumably with a large G-value. However, the structure of SiO_2 can be expected to give rise to a first order dipole moment, and the ionic masses involved are larger than the mass of carbon. Thus, in contrast to diamond, quartz shows relatively strong lattice bands in the reststrahlen region between 5 and 40 μ. The absorption coefficients in the infrared—evaluated in Fig. 5.10 from the reflection and transmission data of Spitzer and Kleinman[12]—vary from less than 10 to 10^5 cm^{-1}. Several low intensity absorption bands exist also at 38, 77, 105, and 122 μ (Barnes[13]). The correlation of the resonances to the various possible modes of vibration is more difficult than in ionic crystals, although the task has been undertaken with apparently good success by Saksena[12,14] with infrared and Raman spectra.

5. PHENOMENA OF PROPAGATION

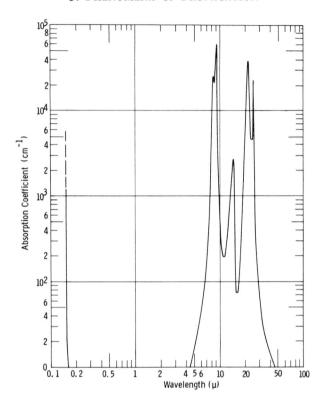

FIG. 5.10. Absorption coefficient in quartz between the ultraviolet and far infrared. Data between 5 and 37 μ have been converted from determinations of the extinction coefficient κ by W. G. Spitzer and D. A. Kleinman [*Phys. Rev.* **121**, 1324 (1961)], measured for the ordinary ray in α-quartz with plane polarized light (**E** perpendicular to the optical axis). Not shown are several resonances beyond 45 μ [R. B. Barnes, *Phys. Rev.* **39**, 562 (1932)] which have absorption coefficients smaller than 10 cm^{-1}. [Data for the electronic absorption band in the far ultraviolet are taken from A. Gilles, R. Bauplé, J. Romand, and B. Vordar, *Compt. Rend.* **229**, 876 (1949), and E. W. J. Mitchell and E. G. S. Paige, *Phil. Mag.* [8] **1**, 1085 (1956)].

The largest amount of work in this area has of course dealt with ionic crystals, particularly the alkali halides. As a group, these materials have large electronic band gaps so that their intrinsic absorption band lies in the far ultraviolet; they have often small or intermediate G-values, and their ionic masses vary from small to large values. They are capable of well-defined, strong dipole oscillations. For these reasons, their lattice absorption spectrum is intense, relatively simple, and can extend over much of the infrared region. Figure 5.11 shows the spectral behavior of the index of refraction for a group of alkali halides. It is seen that the

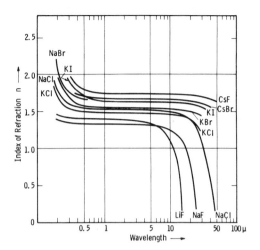

FIG. 5.11. Index of refraction between the fundamental and the lattice vibration bands for alkali halides [H. Pick *in* Landolt-Boernstein, "Zahlenwerte und Funktionen," 6th ed., Vol. II, Part 8. Springer, Berlin, 1962; also G. Joos, "Optical Properties of Solids," F.I.A.T., Review of German Science 1939-1946, Part II: The Physics of Solids (P. B. 95684)].

variation of the index n with wavelength, i.e., the dispersion, is large in regions of lattice vibrations and intrinsic transitions, whereas it is small in the spectral interval between the two. This interval is characterized by transparency as shown for some of these crystals, in addition to certain other materials of interest, in Figs. 5.12–5.16. It is also seen that the regions of anomalous dispersion nearly coincide with those of vanishing transparency. We define the *transmittance T* as the fraction of incident power transmitted through the object, just as A and R represent the

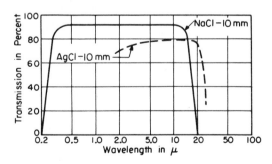

FIG. 5.12. Transmittance of sodium chloride and silver chloride (data from Servo Corporation).

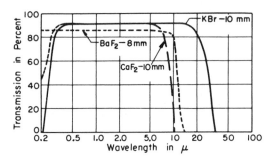

FIG. 5.13. Transmittance of potassium bromide and calcium fluoride (data from Servo Corporation) and barium fluoride (data from Harshaw Company).

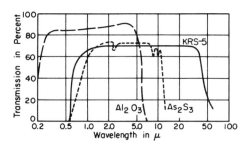

FIG. 5.14. Transmittance of thallium-bromide-iodide (KRS-5), servofrax (Servo Corporation), and sapphire (Linde Company).

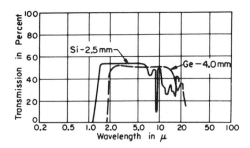

FIG. 5.15. Transmittance of silicon and germanium (data from Servo Corporation).

fractions absorbed and reflected, respectively.* Because of the obvious relationship

$$A + T + R = 1 \qquad (5.142)$$

---

* These definitions include the effects of multiple reflections within a slab of material, whereas the terms *reflectivity*, *absorptivity*, etc., refer to the elementary process (no internal reflections), hence are independent of the slab thickness.

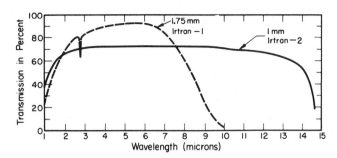

FIG. 5.16.   Transmittance of magnesium fluoride (irtran 1) and zinc sulfide (irtran 2) in the infrared (Kodak Company).

a measurement of the transmitted fraction, even though it is the simplest approach, makes a statement only about the sum of reflectivity and absorptivity. Another limitation affecting the determination of anomalous dispersion as well as of transmission in crystals is that in resonance regions reflection and absorption are very high. It is therefore often necessary to work with samples as thin as 0.1–1 μ. Crystals in these sizes are difficult to produce, especially as strain-free slivers, and evaporated layers are typically amorphous or polycrystalline. Therefore, as a rule, transmission and dispersion measurements will only indicate the outer limits of the resonance regions, and details and fine structure within the absorption intervals are not easily accessible to these methods.

Nevertheless, a number of features are recognizable and noteworthy in the dispersion curves of Fig. 5.11 and the transmission curves of Figs. 5.12–5.16. The most important fact is the qualitative agreement with Eq. (5.137), particularly to the extent to which it confirms the trend of $\omega_0{}^2$ with the force constant G and the effective mass M. In other words, the curves show the trend of the resonance regions to longer wavelengths, if under otherwise nearly equal conditions the molecular weight increases or the bonding becomes weaker.

5.2.3.2 *Practical Considerations of Transmittance and Dispersion*

The development of infrared windows, lenses, and prisms is guided, in fact, by the search for crystals or glasses with heavy ion masses and small binding forces. Examples for such selectively developed materials are shown in Figs. 5.14 and 5.15, notably thallium bromide-iodide (also called KRS-5) with a transmission up to about 40 μ, arsenic trisulfide (also called Servofrax), transparent up to 15 μ, magnesium fluoride (Irtran-1) or zinc sulphide (Irtran-2) which have selective transmission regions between about 1 and 9, or 1 and 15 μ, respectively. The fact that the intrinsic absorption edge of these materials lies also at relatively long wavelengths (i.e., in the red or near infrared rather than in the ultraviolet as for most insulators) is of course no disadvantage in many applications in which one wishes to exclude visible light anyway. The heavy ionic crystals transmit radiation between the ultraviolet and the far infrared; however, many of them also exhibit physical properties which are undesirable from the point of view of application. The alkali halides, for instances, with a few exceptions are quite water-soluble and hygroscopic. Other materials, such as silver chloride, become partially opaque after prolonged exposure to blue or ultraviolet light because of F-center generation or are too soft to maintain accurate dimensions.

Despite these and other disadvantages, materials of the types considered in Figs. 5.12–5.16 continue to be in use for lack of more perfect alternatives. For example, applications to prisms, which require transparent crystals of sufficiently high dispersion, include NaCl (below 16 μ) and KBr (below 28 μ) in the infrared, glass in the visible (between about 0.3 and 2.5 μ), quartz (0.2–3.5 μ), NaCl (above 0.18 μ), CaF$_2$ (above 0.18 μ), and LiF (above 0.12 μ). These materials are also used for lenses and windows, although high dispersion is here undesirable, particularly at large apertures, because of the resulting chromatic aberration. Furthermore, requirements for moisture resistance are higher, whereas those for transparency are lower. As a result, such materials as germanium, combining a high index of refraction with low dispersion in their transparency region of the infrared, have become of increasing importance, particularly for lenses. Infrared windows, in addition, employ crystals such as of BaF$_2$ (between 0.18 and 11 μ), Al$_2$O$_3$ (sapphire), MgF$_2$, and ZnS, aside from the materials mentioned before.

The curves for the index of refraction and for the transmittance in Figs. 5.12 to 5.16 show the effect of the optical lattice vibrations for these optically important dielectrics only in part, viz., only for the short-wavelength end of the ionic resonance region. Depending on the an-harmonicity of the binding forces, the absorption band may be more or less broad. Beyond this region, i.e., for many cases in the submillimeter part of the spectrum, the materials become again transparent. Examples, are given in Fig. 5.17 which shows the transmission in the far infrared of quartz, silicon, and polystyrene as measured by Decamps and Hadni[15]. The existence of broad reststrahlen regions lends itself to selective filters or window action in the far infrared. Thus a high pressure mercury discharge in a quartz envelope is a source of radiation above 50 μ (as well as between 0.2 to 4.5 μ). The high transparency of polystyrene in

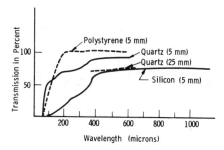

Fig. 5.17. Transmittance of materials at wavelengths above reststrahlen region [E. Decamps and A. Hadni, *Compt. Rend.* **250**, 1827 (1960)].

the far infrared is used by Yamada *et al.*[16] in combination with the reststrahlen bands of various ionic crystals. In this application, materials such as NaCl, KBr, etc., are admixed in powder form to polystyrene sheet when it is soft. Contained in (and protected by) this sheet, the salts maintain their characteristic absorption spectra (see Fig 5.18) so that various filter effects can be produced in this easily attainable manner.

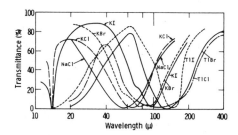

Fig. 5.18. Filter action of crystallite powders in their reststrahlen regions. The active materials are contained in transparent polystyrene sheet [Y. Yamada, A. Mitsuishi, and H. Yoshinaga, *J. Opt. Soc. Am.* **52**, 17 (1962)].

5.2.3.3 *Reflectivity Measurements and Applications*

Historically, reflection phenomena in the far infrared gave the first indications for the existence of lattice vibrations. If radiation, e.g. from a black body, is reflected repeatedly from polished surfaces of a suitable crystal material, a narrow wavelength interval is filtered out which corresponds to the reflection peak of the material (see Fig. 5.19). This method of residual rays or "reststrahlen" was applied first by H. Rubens to the production of narrow wavelength regions in the far infrared. At

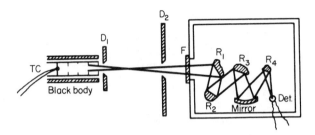

FIG. 5.19. Reststrahlen arrangement of Rubens. R_1–R_3 are polished surfaces of crystals with highly selective infrared reflectivity.

the peak, reflectivity may be as high as 97% or more, particularly at low temperatures. After 10 reflections the peak intensity is reduced by only about 30%, whereas at the half-width points of reflectivity the intensity has dropped to 2^{-10} or about 0.001 of the initial value. In this manner, a selection and narrowing of a wavelength interval occurs with the reststrahlen method.

The infrared reflectivity spectrum of a number of ionic crystals is shown in Fig. 5.20. A major peak corresponding to the optical resonance frequency is seen in all curves which also exhibit a smaller peak at the high frequency side of the region. The reflectivity curves themselves, therefore, provide a clue to the existence of other than the fundamental optical crystal vibration. The second peak has been interpreted as

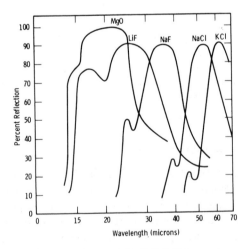

FIG. 5.20. Infrared reflection spectra of alkali halides and magnesium oxide [M. Lax and E. Burstein, *Phys. Rev.* **97**, 39 (1955)].

evidence for anharmonicity of the binding forces, which would give rise to harmonics, or as indication for a second-order dipole moment, e.g., between ionic shells and nuclei. The direct data of reflectivity reveal in fact such subtle features as the shift of the resonant frequency with isotope mass as shown by the comparison of the Li^6F and Li^7F spectrum[17] in Fig. 5.21. Nevertheless, a quantitative evaluation of the meaning

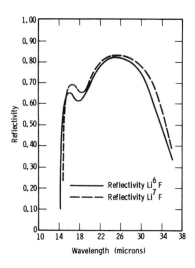

FIG. 5.21. Isotope shift of reflectivity for lithium fluoride. Measurements were made at 300°K on 10 μ thick films deposited on natural LiF substrate [M. Gottlieb, *J. Opt. Soc. Am.* **50**, 350 (1960)].

implicit in the reflectivity data requires the analysis in terms of n and κ.

The difficulties of determining the optical indices n and κ (and from these the dispersion parameters γ, ω_0, etc.) by means of transmittance or dispersion measurements in the highly absorbing reststrahlen region are avoided by the methods which derive the needed information from reflectivity data. The burden is then shifted from experimental to analytical complexity since the indices n and κ depend on the reflectivity R only implicitly and must be derived from it by approximation methods [cf. Eqs. (5.138)–(5.141)]. Nevertheless, the advent of high-speed computer techniques has vastly simplified this task, and the analytical evaluation of reflectivity measurements is now practical by means of a variety of procedures.

One of the methods already mentioned in connection with metal optics is the determination of the optical indices by polarization effects, e.g., by Eqs. (5.100) and (5.101). However, polarization techniques in the far infrared are often unsuitable, inasmuch as the conditions for sufficient intensity, choice of available materials for the purpose, etc., represent mutually conflicting requirements. For example, Nicol prisms (calcite) can be used only at wavelengths below about 2 μ. Reflection at the Brewster angle as polarizing means encounters a large penalty in intensity[18]. It is possible, of course, to polarize a beam at the Brewster angle by transmission, rather than reflection, and this method has been applied successfully to film stacks of selenium and sheets of silver chloride. Nevertheless, even this method fails at wavelengths for which the absorption coefficient of the polarizer material assumes high values.

An entirely different approach is that of *dispersion analysis* (Spitzer *et al.*[19,20]) in which the optical indices and parameters are determined by a trial-and-error method: tentative values for the parameters are "tried out" in the dispersion equations by a computer and compared with the observed reflectivity. The values are then readjusted repeatedly until satisfactory agreement is reached. In the general case of several oscillators (subscript i), Eqs. (5.139)–(5.141) are modified by a summation process:

$$n^2 - \kappa^2 = \epsilon_v + \sum_i \frac{4\pi N_i e^2 f_i}{m_i} \cdot \frac{\omega_i^2 - \omega^2}{(\omega_i^2 - \omega^2)^2 + \gamma_i^2 \omega^2} \tag{5.143}$$

$$2n\kappa = \sum_i \frac{4\pi N_i e^2 f_i}{m_i} \frac{\omega \gamma_i}{(\omega_i^2 - \omega^2)^2 + \gamma_i^2 \omega^2} \tag{5.144}$$

$$\epsilon_0 = \epsilon_v + \sum_i \frac{4\pi N_i e^2 f_i}{m_i \omega_i^2} \tag{5.145}$$

n and κ being related to the reflectivity by

$$R = \frac{(n-1)^2 + \kappa^2}{(n+1)^2 + \kappa^2} \tag{5.146}$$

Figure 5.22 represents values obtained by Spitzer and Kleinmann[12] by means of dispersion analysis for the complex case of quartz (only those for the ordinary ray are shown). The solid curve in Fig. 5.22a is the curve adjusted to fit the actually measured reflectivities indicated by the points. The match achieved establishes confidence in the pivotal values of n and κ which are shown in Figs. 5.22b and 5.22c, respectively. The data of Fig. 5.22c have been used for the infrared portion of Fig. 5.10.

A third procedure of evaluating the dispersion indices from the

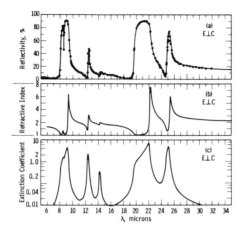

Fɪɢ. 5.22. Reflectivity, refractive index, and extinction coefficient of quartz for the ordinary ray. (a) Theoretical curve for reflectivity (solid trace) is adjusted by variation of dispersion parameters until satisfactory fit with measured values (points) is obtained; (b) and (c) traces for n and κ, respectively, as computed from the adjusted dispersion parameters. Four major and several minor resonances are seen to coincide in the three curves. [Data are taken from the work of W. G. Spitzer and D. A. Kleinmann, *Phys. Rev.* **121**, 1324 (1961).]

reflectivity data uses the *Kramers-Kronig* relationships. The complex reflectivity of the amplitude [cf. Eq. (5.94)] can be written

$$\hat{r} = \frac{n - i\kappa - 1}{n - i\kappa + 1} = re^{i\phi}, \qquad r = \sqrt{R} \tag{5.147}$$

or

$$\ln \hat{r} = \ln r + i\phi \tag{5.148}$$

It can be shown that ϕ may be expressed as the following integral:

$$\phi(\omega) = \frac{2\omega}{\pi} \int_0^{\infty} \frac{\ln r(\omega_1)}{\omega^2 - \omega_1^2} d\omega_1 \tag{5.149}$$

Therefore it is possible[21] to obtain the values $\phi(\omega)$ by the measurement of a single variable, viz., $r(\omega)$, over the frequency spectrum which may be limited, in fact, to the neighborhood of ω. Furthermore, it is at once possible to evaluate n and κ explicitly from Eq. (5.147).

The Kramers-Kronig relationships were applied by Gottlieb[21] to reflection measurements on lithium fluoride. The most pertinent results of this work concern the behavior of the imaginary part of the dielectric constant, viz., $2n\kappa$ (see Fig. 5.23). Three distinct peaks are apparent,

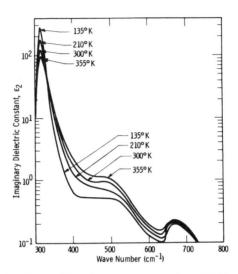

FIG. 5.23. The imaginary dielectric constant $\epsilon_2 = 2n\kappa$ of lithium fluoride as determined by M. Gottlieb [*J. Opt. Soc. Am.* **50**, 343 (1960)] from measured reflectivities.

although they differ in absolute height by orders of magnitude. The largest of these corresponds to the fundamental optical mode with Q'-values ω_0/γ decreasing from 25 at 135 °K to 13 at 355 °K as a more detailed inspection shows ($Q' = 3$ for 893 °K). It is noteworthy that the reflection spectrum itself has a width of several hundred (centimeters)$^{-1}$ (cf. also Fig. 5.20) so that its profile (as that of the transmittance which, of course, is affected by the reflectivity) is in no way representative of the spectral oscillator behavior. The second peak in the traces of Fig. 5.23 is attributed to combination modes of vibration. The third resonance, by virtue of its magnitude and position, could well constitute the second harmonic of the fundamental vibration; another interpretation, however, identifies the peak with the fundamental longitudinal optical mode. In first approximation, no light is emitted or absorbed in the axial direction of an oscillating dipole, but any dissymmetry would contribute to a radiative interaction, and the spectral position appears to agree with theoretical expectations for such a mode. Furthermore, peaks are seen for other alkali halides, too, at spectral locations which correspond to the longitudinal mode, but do not match the second harmonic[21].

Reflection measurements have also been successful in the intrinsic absorption band of various materials. The alkali halides, for instance, show in the order of 10 reflectivity peaks between 500 and 2500 Å. As in the case of reststrahlen, detailed information can be obtained from the reflectivity spectrum only after its evaluation in terms of the optical indices. Using the Kramers-Kronig relations for the analysis of their data on potassium-chloride, bromide, and iodide, Phillip and Ehrenreich[22] obtained relatively well-resolved structures for the profiles of the real and imaginary dielectric constants. They interpreted the various peaks as exciton resonances involving deeper lying bands and as plasma oscillations.

Measurements of fundamental absorption in thin films, despite their drawbacks because of uncertain reflection losses, lattice imperfections, thickness variations, etc., are sometimes at least as reliable as reflection measurements. This is the case when it is difficult to prepare and maintain the reflecting surfaces in satisfactory condition. For example, absorption measurements on the ultraviolet absorption of barium oxide films[23] are substantially confirmed by reflectivity determinations with freshly prepared single crystal surfaces. The latter change their behavior in time, however, presumably because of the formation of $Ba(OH)_2$ layers[24].

5.2.4 Free Electron Model and Optics of Metals

Metals are characterized by the fact that the valence electrons of the constituent atoms are not bound to fixed ion sites but free to move through the lattice with little or no restriction. The consequences of the free-electron structure for crystal bonding, conductivity phenomena, and energy distribution have been discussed already in Section 4.6.1. For the present subject other considerations are of importance. The absence of forces restraining the electron to certain centers is mirrored by the absence of the resonances which in the case of dielectrics led to a considerable atomistic extension of the phenomenological electromagnetic continuum theory. Thus in metals the variation of the optical indices with frequency is more gradual, and the prediction of the theory based only on Maxwell's equations could be carried successfully at least up to the frequencies of the near infrared on the basis of electrical parameters measured with direct currents.

Nevertheless, it is necessary to resort to the electron theory of metals in order to explain their optical behavior in the visible and ultraviolet region. The free electron model of metals proceeds, of course, through various stages of approximation. The very simplest of these are already able to describe in first order some of the pertinent phenomena. The Newtonian equation of motion of the completely free electron consists only of two terms as given by Eq. (1.2). If it is also necessary to account for the finite resistance of metals, a term $R\dot{x}$ for friction or damping has to be added to the equation:

$$m\ddot{x} + R\dot{x} = e\mathbf{E} = e\mathbf{E}_0 e^{i\omega t} \tag{5.150}$$

The parameter R in this model arises from collisions of the electrons with phonons and impurity atoms, the effect of which amounts to a damping force in proportion to the frequency of such encounters and thus to the velocity $\dot{x}$. The *relaxation time* between collisions according to this model is given by $\tau = \Lambda/v$, where the mean free path Λ at room temperature is typically 10^{-5} cm and the velocity v (from the Fermi-Dirac distribution) is in the order of 10^8 cm/sec. Thus $\tau \approx 10^{-13}$ sec, a time much larger than the period of oscillations for visible and ultra-violet light (e.g., 10^{-15} sec for 3000 Å), so that the electrons can describe many cycles without disturbance. For these reasons it is in first order possible to ignore R in the integration of Eq. (5.150) at the higher frequencies, and we obtain as the stationary solution

$$\mathbf{x} = -\frac{e\mathbf{E}}{m\omega^2} ; \quad \gamma = 0 \tag{5.151}$$

According to Eqs. (5.106)—(5.107), we have then for the dielectric constant the real value

$$\epsilon = n^2 = 1 + 4\pi\alpha N = 1 - \frac{4\pi Ne^2}{m\omega^2} \tag{5.152}$$

It follows that ϵ is negative for sufficiently low frequencies, i.e., below a critical value

$$\nu_c^2 = \left(\frac{c}{\lambda_c}\right)^2 = \frac{Ne^2}{\pi m} \tag{5.153}$$

The implication of a negative displacement $\mathbf{D}$, as was shown before, is that of total reflection which is the expected behavior for metals with vanishing loss factor R. However, above ν_c, the dielectric constant becomes positive and the metal has the optical properties of a transparent dielectric. In terms of wavelengths, this occurs at a critical value

$$\lambda_c = 2\pi\sqrt{\frac{mc^2}{4\pi Ne^2}} = \frac{c}{e}\sqrt{\frac{\pi m}{N}} \tag{5.154}$$

The transition from reflecting opaqueness to transparancy was discovered, although in fact not expected, by R. Wood[25] in alkali metals in the ultra-violet and subsequently explained by Zener[26] (1933) by the change of sign for the dielectric constant as shown here. The actually measured values of the critical wavelength are up to about 30% larger than the theoretical values according to Eq. (5.154), if N is set equal to the number of alkali metal atoms per unit volume. The exceptions are sodium, for which both values are 2100 Å, and potassium, for which λ_c is measured at 3150 Å and calculated at 2900 Å.

Presumably the agreement is improved, if the damping factor R is taken into account (see Born and Wolf,[2] p. 622). The stationary solution of Eq. (5.150) is

$$\mathbf{x} = -\frac{e}{m(\omega^2 + i\gamma\omega)}\mathbf{E} \qquad (5.155)$$

where $\gamma = R/m$. This approximation, which replaces Eq. (5.151), results in a modified value λ_c' for the critical wavelength, viz.,

$$\lambda_c' = 2\pi\sqrt{\frac{mc^2}{4\pi Ne^2 - m\gamma^2}} \qquad (5.156)$$

The correction by the term γ^2/m tends to reduce the gap between theory and measurement.

Finally, the real and imaginary components of $\hat{\epsilon}$ can be written in terms of $\omega_c = 2\pi c/\lambda_c$ as

$$\epsilon = n^2 - \kappa^2 = 1 - \frac{\omega_c^2 + \gamma^2}{\omega^2 + \gamma^2} \qquad (5.157)$$

$$\frac{\sigma}{\nu} = n \cdot \kappa = \frac{\gamma(\omega_c^2 + \gamma^2)}{2\omega(\omega^2 + \gamma^2)} \qquad (5.158)$$

Figure 5.24 shows the frequency behavior of the real and imaginary part of the dielectric constant for a metal according to Eqs. (5.157) and (5.158). The curves correspond to the example of rubidium for which λ_c was measured at 3500 Å and calculated at 3200 Å with Eq. (5.154). The value of γ then was chosen so as to compensate for this difference.

5.2.4.1 *Optics of Superconductors*

A number of metals and alloys exhibit a complete disappearance of resistivity with respect to a direct current at sufficiently low temperatures. If this property of superconductivity would remain unchanged at high frequencies, incident radiation would be totally reflected from the metal, and such a behavior was indeed expected and searched for as the exploration of this field extended beyond the realm of the purely stationary phenomena. However, detailed investigations established that loss mechanisms for alternating fields exist even in superconductors at, and

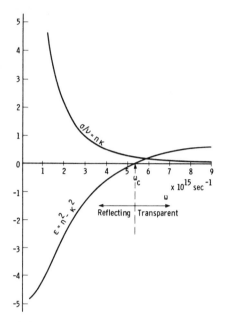

FIG. 5.24. Real and imaginary dielectric constants of alkali metals vs. angular frequency. The curves correspond to the data for rubidium. A damping term γ has been assumed of such value as to obtain agreement between the theoretical equation (5.156) and the observed value λ'_c for the onset of metallic transparency.

below, optical frequencies. In fact, the reflectivity of light remains unchanged in a metal which converts from the normal to the superconducting state (Wexler [27]). For wavelengths in the millimeter range, a type of superconducting behavior will assert itself for most materials. Thus, unloaded Q-values of cavities lined with superconducting metal such as lead may be as high as 10^8; yet even here energy is dissipated in the surface layers, and the alternating field conductivity is demonstrably not infinite.

In simplest terms, these various frequency dependent phenomena are explained with the so called *two-fluid* model of superconductivity (a term defined in far reaching analogy to the composition of superfluid helium). At $T = 0$, only superconducting electrons exist in the metals; at $T > T_c$ (the transition temperature to the normal state) only normal electrons are present; and in the superconducting region between these two points, the two components exist together in a temperature dependent ratio. A forbidden gap

$$E_{sg}(0) = 3.5kT_c \tag{5.159}$$

separates the higher energy band of the normal electrons from the "condensed" superconducting phase. The value of E_{sg} varies with temperature, Eq. (5.159) applying to $T = 0$ according to the theory of Bardeen, Cooper, and Schrieffer[28]. The existence of the gap explains at once the normal behavior of the optical properties. The gap energy is in the order of 10^{-3} V, corresponding to a wavelength of about 1 mm; hence photons of higher energy are absorbed by superconducting electrons which then become normal. In fact, the onset of radiative energy absorption at the frequency corresponding to Eq. (5.159) was one of the most direct proofs for the existence of a forbidden gap in superconductors. First, discontinuities of absorption could be demonstrated for microwaves in resonant cavities containing superconducting metals such as aluminum (Biondi *et al.*[29]). Second, optical methods in the far infrared showed similar changes for the reflectivity. These experiments consisted of measuring the intensity of radiation in the range between 0.1 and 2 mm with a bolometer after the light had been reflected repeated-

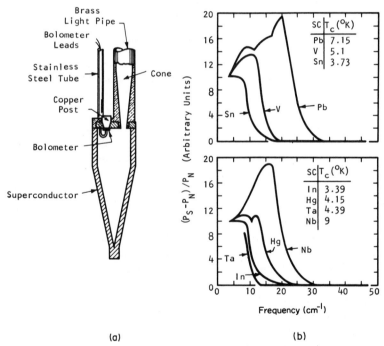

(a)   (b)

FIG. 5.25.   Optical determination of energy gap in superconductors. (a) Nonresonant conical cavity made of superconductor provides for many reflections of the infrared radiation (incident from the light pipe) before it is measured by the bolometer; (b) power measured by bolometer for seven pure superconductors approaches zero as incident frequency is raised to value E_{sg}/h [P. L. Richards and M. Tinkham, *Phys. Rev.* **119**, 575 (1960)].

ly from the superconducting walls of a conically shaped enclosure (Richards and Tinkham[30]). When the frequency selected exceeded a value of the order 3–4 kT_c/h, the power reflected by the walls would undergo the expected discontinuity and the power registered by the bolometer would diminish sharply (see Fig. 5.25). Of interest is also the transmission[31] through thin superconducting films which shows the behavior indicated in Fig. 5.26. All these and other experiments indicated

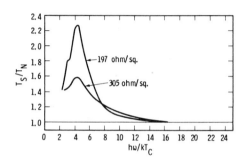

FIG. 5.26. Infrared transmission through superconducting lead films. The ratio of transmittances T_s/T_n for the superconducting and normal states is shown here for films of 305 ohms/square (lower peak) and 197 ohms/square (upper peak) of lateral resistance [D. M. Ginsberg and M. Tinkham, *Phys. Rev.* **118**, 990 (1960)].

that the optical behavior of superconductors was substantially that of the normal state for frequencies above E_{sg}/h, although the latter value could be shown to depend on temperature in a predictable manner[29].

That there exists for finite temperatures also an absorption mechanism for photon energies lower than E_{sg}, can be interpreted again with the two-fluid model. An alternating field produces a potential $L \cdot di/dt$ across a length of metal with inductance L. At $T = 0$, only supercon-ducting electrons exist in their ground state. The energy expended on their acceleration is reversibly returned to the electromagnetic field, i.e., reflection is total. For $T > 0$, however, a fraction of the conduction electrons is in the normal state and thus experiences irreversible (ohmic) losses in the presence of an alternating field. The ratio of superconducting to normal electrons can indeed be determined from the impedance behavior of the superconducting metal.

In all, the behavior of superconductors with respect to incident radiation can be summarized as follows according to frequency region:

(1) $0 < \nu < 10^6$–10^7 sec^{-1}. The complex impedance is zero in ideal superconductors up to radio frequencies. However, practical super-conductors of all types exhibit penetration and trapping of magnetic

flux. As a result, they also show hysteresis phenomena with finite, although often nondetectable loss.

(2) 10^6–$10^7 < \nu < E_{sg}/h$. The dominant feature at radio frequencies up to the superconducting gap threshold E_{sg} is a small, but detectable absorption due to the electrical energy expended on the acceleration of the normal electrons. Thus microwave cavities enclosed by super-conducting walls may have Q-values of about 10^8, whereas normal metals yield typically 10^3–10^4.

(3) $E_{sg}/h < \nu < E_{IB}/h$. In the optical region between the super-conducting gap threshold and the onset of interband transitions of energy E_{IB}, reflectivity and absorptivity remain the same for the normal and the superconducting state of a particular metal.

(4) $h\nu \approx E_{IB}$. Small differences (1–2%) of the optical indices n and κ in the normal and superconducting state may be expected for the onset of interband transitions $h\nu = E_{IB}$ (Dresselhaus *et al.*[32]). Specifically, these transitions occur from the first filled band to the energy region near the Fermi level of the conduction band. Therefore the optical parameters in the corresponding frequency domain should be sensitive to the existence of a superconducting gap.

5.2.5 Optics of Liquids and Liquid Crystals

Structually, the liquid state assumes an intermediate position, although in an exceedingly complex manner, between solids and gases. It shares with the solids the relatively high cohesion and high density, and it shares with the gases the ability to flow and the randomness of molecular orientation. A workable model of a liquid—substantiated to a certain extent by neutron scattering experiments—is that of molecules vibrating around positions which are fixed for a certain mean time and then abruptly assume a new configuration. This time of quasi-rigidity is large compared with the period of the molecular oscillations, and in fact large compared with the reciprocal of the optical frequencies.

Optically, therefore, the liquid state behaves like that of amorphous solids, and for this reason, it is not treated separately in the theory of propagation phenomena. Liquid solutions are, of course, convenient media for various kinds of optical measurements. Beer's law, Eq. (5.108), holds for liquids to the same extent to which it is valid for gases and solids of comparable optical path lengths. Analytical chemistry utilizes this fact qualitatively as well as quantitatively. Of particular importance, in this connection, is the measurement of *optical activity*, namely, the rotation of the electric wave vector as plane polarized light progresses through certain media. In simple cases, the rotation θ is found to be nearly proportional to the concentration n and path length L of the active material even in optically dense media, i.e., it obeys a law of the form

$$\theta \cong \rho_\lambda \cdot n \cdot L \qquad (5.160)$$

where ρ_λ is a measure for the specific *rotary power* of the active substance (ρ_λ is usually referred to 10-cm length of a liquid or, when applicable, to a solution containing 1 gram per cubic centimeter of the solute). The wavelength dependence of the rotary power follows approximately an empirical law of the form

$$\rho_\lambda = \frac{A}{\lambda^2} + \frac{B}{\lambda^4} + \frac{C}{\lambda^6} + \dots \qquad (5.161)$$

where usually only the first term on the right has to be considered. The effect must be ascribed to the electrical multipole structure of the active molecule. We may visualize an optical axis along which dipoles are twisted in helical progression. The incident electrical vector then will experience a rotation either clockwise or anticlockwise as it travels along the axis, regardless of the direction. Thus, despite the randomness of their molecular orientations, pure liquids as well as solutions may show a net rotary power which will increase, in general, with the complexity of the active structure. In contrast to certain crystalline solids such as quartz, which has in one of its modifications a rotary power of 15° to 50°/mm in the visible region, liquids have relatively low activity, e.g., about 0.4°/mm in the case of turpentine. Nevertheless, the optical activity of solutions is significant for chemical analysis which can determine concentrations, e.g., of sugar, by means of Eq. (5.160).

There exists a class of structures, represented by certain organic compounds, which is intermediate to liquids and crystalline solids. Such *mesomorphic* or *liquid crystal* states have the ability to flow, but their molecules are oriented along certain preferred patterns and directions (see, e.g., Brown et al.[33]). Among these mesomorphic substances, one differentiates among (1) the *nematic* (threadlike) phase with its molecules aligned in one preferred direction, (2) the *smectic* (soap bubble type) phase in which the molecules are not only aligned parallel, but moreover grouped in discrete layers, and (3) the *cholesteric* phase with the screw-like arrangement shown in Fig. 5.27. The latter structure is characteristic, as implied by its name, of cholesterol esters, and it is important for its optical properties.

Cholesteric liquid crystals, in fact, have remarkable optical properties in three respects. First, they exhibit a very large optical activity—in the order of 40,000°/mm. Secondly, they are notable for their iridescent colors, i.e., diffusely reflected spectral regions which depend on the angle of view and of incidence. Both of these properties can be understood on the basis of the helical progression of molecular alignment as shown in Fig. 5.27. As in the nematic structure, the molecules are arranged parallel in certain directions; however, progressing along a preferred axis (determined by the substrate of the liquid crystal film), these directions undergo a continuous and monotonic change as a result of the interparticle forces. The "pitch" d between isogonal planes is considerably smaller than, for instance, in quartz. This accounts for the large optical activity and also for a periodic change of the index of refraction such that the liquid crystal acts like a spatial reflection grating in the visible region. The dependence of wavelength on angle of incidence follows closely Bragg's law, although only the first order is observed,

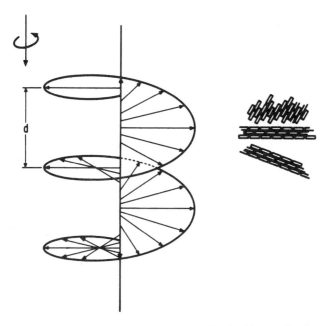

FIG. 5.27. Arrangement of molecules in cholesteric liquid crystals. On the right, layers of molecules are shown which are unidirectionally oriented for a given position along the optical axis. As shown on the left, this molecular orientation progresses helically with respect to the axis. A small strain along the axis produces relatively large rotations of the dipole directions as evidenced by the high optical sensitivity of cholesteric materials to changes of temperature, shear, and other factors.

presumably because of the harmonic variation of the index of refraction. A third characteristic of cholesteric substances is the high sensitivity with which the optical parameters of reflectance and polarization respond to disturbances of their molecular arrangement (Fergason[34]). Such disturbances can be produced chemically by exposure to certain vapors, mechanically by shear, electrically with a static voltage, and thermally by a change of temperature.[35] According to the helix model of liquid crystals, relatively large rotations of the molecular orientations can occur as the result of small strains along the optical axis. This furnishes an explanation for the singular property of the mesomorphic substances, namely, of responding strongly with a change of reflected color or angle of polarization to a variation of temperature, stress, voltage, or chemical environment.

REFERENCES

1. R. Becker and F. Sauter, "Theorie der Elektrizität," 16th ed., vols. I and II, Teubner, Stuttgart, 1957-1959.
2. M. Born and E. Wolf, "Principles of Optics," Pergamon Press, New York, 1959.
3. J. Strong, *J. Franklin Inst.* **232**, 1 (1941).
4. W. M. Elsasser, *Phys. Rev.* **54**, 126 (1938).
5. H. A. Lorentz, *Ann. Physik* [3] **9**, 641 (1880).
6. L. Lorenz, *Ann. Physik* [3] **11**, 70 (1881).
7. M. Born and K. Huang, "Dynamical Theory of Crystal Lattices." Oxford Univ. Press (Clarendon), London and New York, 1954.
8. B. Szigeti, *Trans. Faraday Soc.* **45**, 155 (1949).
9. G. B. B. M. Sutherland, *J. Opt. Soc. Am.* **50**, 1201 (1960).
10. A. Gilles, R. Bauplé, J. Romand, and B. Vordar, *Compt. Rend.* **229**, 876 (1949).
11. E. W. J. Mitchell and E. G. S. Paige, *Phil. Mag.* [8] **1**, 1085 (1956).
12. W. G. Spitzer and D. A. Kleinman, *Phys. Rev.* **121**, 1324 (1961).
13. R. B. Barnes, *Phys. Rev.* **39**, 562 (1932).
14. B. D. Saksena, *Proc. Indian Acad. Sci.* **A12**, 93 (1940).
15. E. Decamps and A. Hadni, *Compt. Rend.* **250**, 1827 (1960).
16. Y. Yamada, A. Mitsuishi, and H. Yoshinaga, *J. Opt. Soc. Am.* **52**, 17 (1962).
17. M. Gottlieb, *J. Opt. Soc. Am.* **50**, 350 (1960).
18. H. Y. Fan, in "Methods of Experimental Physics," (L. Marton, ed.), Vol. 6, Part B, p. 252. Academic Press, New York, 1959.
19. W. G. Spitzer, D. A. Kleinman, and D. Walsh, *Phys. Rev.* **113**, 127 (1959).
20. W. G. Spitzer, D. A. Kleinman, and C. J. Frosch, *Phys. Rev.* **113**, 133 (1959).
21. M. Gottlieb, *J. Opt. Soc. Am.* **50**, 343 (1960).
22. H. R. Philipp and H. Ehrenreich, *Phys. Rev.* **131**, 2016 (1963).
23. R. J. Zollweg, *Phys. Rev.* **97**, 288 (1955).
24. F. C. Jahoda, *Phys. Rev.* **107**, 1261 (1957).
25. R. W. Wood, *Phys. Rev.* **44**, 353 (1933).
26. C. Zener, *Nature* **132**, 968 (1933).
27. A. Wexler, *Phys. Rev.* **70**, 219 (1946).

28. J. BARDEEN, L. N. COOPER, and J. R. SCHRIEFFER, *Phys. Rev.* **108**, 1175 (1957).
29. M. A. BIONDI, M. P. GARFUNKEL, and A. O. McCOUBREY, *Phys. Rev.* **108**, 495 (1957).
30. P. L. RICHARDS and M. TINKHAM, *Phys. Rev.* **119**, 575 (1960).
31. D. M. GINSBERG and M. TINKHAM, *Phys. Rev.* **118**, 990 (1960).
32. G. DRESSELHAUS and M. S. DRESSELHAUS, *Phys. Rev.* **125**, 1212 (1962).
33. G. H. BROWN and W. G. SHAW, *Chem. Rev.* **57**, 1049 (1957).
34. J. L. FERGASON, *Scientific American* **211**, No. 2, 77 (1964).
35. J. L. FERGASON, T. P. VOGL, and M. GARBUNY, U. S. Patent No. 3114836, Dec. 17 (1963).

6

INTERACTIONS OF COHERENT RADIATION
WITH MATTER

6.1 Introduction and Fundamental Concepts

In the preceding chapter it has been of little consequence for the phenomena of radiative interaction under discussion whether the wavetrains constituting a beam of light had any relationship to each other or not—either in terms of phase or of polarization or of yet other characteristics. In fact, so far we have considered only elementary processes: the individual events of photons and atoms interacting or, in different language, the elementary process of an electromagnetic wave propagating through a medium or impinging on the interface between two materials. We must now turn to phenomena of cooperation between light waves, in other words, to situations in which it indeed does matter if a group of photons encountering a group of atoms are correlated to each other in some way or not.

The significance of the conditions under which the wavetrains of light are *coherent* (verbally, "hanging together") was first realized in classical optics in connection with the phenomena of interference. It is possible to define coherence by several alternative criteria. In classical presentation, the electromagnetic field generated by a vibrating dipole at any point $P(\mathbf{r})$ in the vacuum may be written as

$$\mathbf{E} = \mathbf{E}_0 e^{i\omega(t-\delta)} = (E_{x0}\mathbf{i} + E_{y0}\mathbf{j} + E_{z0}\mathbf{k})e^{i\omega(t-\delta)} \qquad (6.1)$$

$$\mathbf{H} = \mathbf{H}_0 e^{i\omega(t-\delta)} = (H_{x0}\mathbf{i} + H_{y0}\mathbf{j} + H_{z0}\mathbf{k})e^{i\omega(t-\delta)} \qquad (6.2)$$

where the phase δ is referred to some time point at the origin, viz.,

$$\delta = \mathbf{r} \cdot \mathbf{s} + \delta_0 \qquad (6.3)$$

We say then that two wavetrains at point P_1 and P_2 are completely coherent, if the vectors $\mathbf{E}$ and $\mathbf{H}$ at the two locations are the same at all times and their frequency ω has an infinitely narrow bandwidth. However, although all these conditions are sufficient, they may not be necessary. We still consider two wavetrains as coherent, if their phase relationship at the two points remains constant. Later on we shall define the *degree* of coherence by observable parameters.

Since **E** and **H** are perpendicular to each other and have equal magnitude in vacuum, it follows that Eqs. (6.1) and (6.2) contain six independent variables which define amplitude, polarization, direction of propagation, frequency, and phase δ at time $t = 0$. The radiation emitted from a *single* oscillator consists of coherent waves in which the surfaces of equal phase form concentric spheres, as shown in the Hertzian pattern of Fig. 3.3. At sufficiently large distances r, we may consider the light beam as formed of coherent plane waves over limited areas ($\ll r^2$). This situation can be realized almost ideally for microwave radiation, generated, for instance, at constant frequency by klystrons and transmitted by dipole antennas.

The sources of the optical spectrum, however, consist by definition of a multitude of oscillators—molecules, electrons, and atoms— and ordinarily the classical spherical waves emitted by each are more or less independent. In other words, there is little or no correlation between the sets of six variables from one wavetrain to the next. Thus the fields produced by the oscillators of an *extended* source oscillate at any point or points *incoherently*. Such incoherent light waves when superimposed on a screen will show no systematic enhancement or cancellation; rather will their intensities add without mutual "interference." In fact, in order to show interference effects with ordinary light, it is necessary to split the beam in some manner and then recombine the two components. What is done here in effect is the superposition of identical wavetrains, two from each point of the source, such that the phase difference between the two beam portions varies systematically across the screen and the familiar patterns of dark and bright fringes result.

This type of coherence has been known in classical optics for a long time; yet in degree and extent it is quite limited. When one speaks of coherence in connection with classical interference experiments, one refers to the correlation of only one parameter: the phase of corresponding pairs of wavetrains in the two beams, each pair being *self-coherent* since it is generated at the same point or from the same oscillator, although different pairs are more or less unrelated to each other. Moreover, because the lifetime of atomic vibrations is finite, even the self-coherence is limited. For the case of natural broadening, it was shown in Section 1.2.2 that the oscillator energy has dropped to $1/e$ of the initial value after $Q/2\pi$ periods of oscillation (or Q radians). Thus the wavetrain emitted during a time interval τ and propagating with the speed of light c has a *coherence length* L equal to $Q/2\pi$ wavelengths, to wit:

$$L = c\tau = \frac{c}{2\pi \, \Delta\nu_h} = \frac{\lambda^2}{2\pi \, \Delta\lambda_h} = \frac{Q}{2\pi} \cdot \lambda \tag{6.4}$$

As will be discussed further in the next section, the finite coherence length implies a limit of the optical path difference over which self-coherence can be maintained.

In contrast to the classical concepts and phenomena of coherence, limited as they are in scope and to interactions solely between wavetrains, there has emerged in more recent times a different field of correlation effects. These processes involve not only a far greater degree and variety of coherence than characteristic for the classical interference experiments, but they represent interactions of such radiation with matter, and thereby they inject an entirely new element into this field of optics. Classical interference deals mainly with the superposition of wavetrains, and such material components as mirrors and apertures merely constitute accessories to establish the boundary conditions, and it is characteristic that their chemical composition is not of primary concern. The new coherence phenomena, however, to be discussed in succeeding sections of this chapter are predominantly part of atomic transition processes, and the frequencies of interest are usually those of atomic resonance. Stated from a somewhat different viewpoint, the synchronism which is a condition of coherence also extends to the timing of the electronic interactions so that, instead of the randomized energy and momentum exchanges typical for the excited gas in an ordinary radiation field, a highly ordered mechanism results. This, in broad outline, is the nature of coherent interactions. It is clear that this field may be expected to exhibit a generality and a wealth of phenomena far greater than classical coherence phenomena could provide.

Historically, the first concept which implied a participation of matter in the generation and propagation of coherent light is that of negative absorption (cf. Section 3.1.4), which was postulated in classical electrodynamics at the turn of the century. It remained, however, for Einstein[1] to show the salient feature of this process, namely, that if a photon $h\nu$ encounters an atom in a state raised by an energy $E = h\nu$ above another level, it may stimulate the emission of a photon identical to itself in energy and momentum (cf. Section 3.3.1.1). Of course, this is an elementary process of almost complete coherence which must occur in the radiative interactions of excited atoms at a rate which is proportional to the density of upper states. In the normal Boltzmann distribution, given by the "barometric height formula" of energy states [cf. Eq. (3.59)], the number of excited levels is always smaller than the density of atoms in the ground state, and thus the coherent events of stimulated emission are completely obscured by the much more common inverse processes, viz., those of absorption.

However, in 1950 Kastler[2] showed that it is possible to upset the normal thermal distribution in such a manner as to actually maintain more atoms in certain higher energy states than in the lower lying levels. This *population inversion* corresponds to a negative temperature in the Boltzmann factor, and it can be attained by a variety of "pumping" methods which we shall discuss in detail later in this chapter. It was in inverted systems of this kind, consisting of nuclear spin systems, that Purcell and Pound[2a] first observed induced net emission. On the basis of these results, Townes[3] and others in 1951 suggested a principle of enormous potential, namely, the production of coherent radiation by encouraging modes of stimulated emission from inversely populated levels of atoms in suitable cavities. To be sure, the device first proposed and later built on the basis of this principle, the so-called *maser*, was only an amplifier in the microwave region,* and its main advantages were relatively noisefree operation and a narrowing of the resonant line by orders of magnitude below the width determined by the cavity Q. At that time, the restriction to microwave frequencies appeared natural because of the ν^3-relationship between spontaneous and induced transition probability [cf. Eq. (3.65)]. As a consequence, there exists a strong contribution of spontaneous emission at the frequencies of the infrared and visible region so that the advantages of noisefree amplification and perhaps even of narrowing of the spectral width could not be expected. However, in 1958, Schawlow and Townes[4] derived the analytical conditions under which a type of maser operation was possible in the near infrared and visible region. The arrangement proposed was that of an "optical resonant cavity", viz., a suitable maser material located between highly reflecting, plane-parallel mirrors, in other words, a Fabry-Perot interferometer construction enclosing the resonant atoms. In operation, pumping light was to invert the atom population over suitable levels. Then under certain minimum conditions of efficiency, coherent light in stimulated oscillatory modes would be emitted with positive gain so that coherent light would be generated rather than only amplified. Maiman[5,6] in 1960 demonstrated the first such *laser** with a strikingly simple arrangement containing a ruby rod pumped by a xenon discharge—incidentally an arrangement which for years held its place as the most important laser embodiment despite the growing number of competing devices subsequently developed and explored in several hundred laboratories. In fact, as will be described in detail later, optical

---

* The initials of "Microwave Amplification by Stimulated Emission of Radiation" yielded the acronym "maser," albeit with considerable limitation of its claims. The coherent optical oscillator is often called "optical maser," a name which provides continuity, although the more consistent term "laser" has found wide acceptance.

masers have been found to function with inverted populations in gases, liquids, amorphous solids and crystals, and even in semiconductors; in continuous, pulsed, or (with special storage mechanisms) "giant-spiked" operation; with a single optical mode or, simultaneously, with several of them; and with pumping schemes based on the transfer of an astounding variety of energy forms.

It is noteworthy that almost completely coherent light can be generated with laser systems, identical phase being maintained over areas large compared to λ^2 for relatively long times. This property is linked with narrowing of the emitted lines below the natural width and high directionality of the beam. Stimulated emission is capable of producing a beam which is parallel within the diffraction limit, contains high power density, and therefore can be focused with simple lenses so as to yield very large energy concentrations.

The availability of high flux densities makes accessible to experiment a class of propagation phenomena which are summarized under the name of *nonlinear optics*. We have seen in Section 4.5.3 that oscillators are capable of free harmonic oscillations only, if there exists a linear relationship between displacement and restoring force. This condition, which also includes the linearity between polarization and the incident oscillating field, usually is fulfilled for the phenomena of propagation discussed in the preceding chapter as a first order approximation, i.e., for sufficiently small fields, and it is important to realize that the validity of the laws governing reflection and refraction, etc., extends only to that limit. However, as we have seen in the case of diatomic molecules in Section 4.5.3, higher order approximations, e.g., between displacement and force, have to be considered for large fields and their effects. In this case, such phenomena as the generation of higher harmonics will occur. Indeed, frequency doubling and related effects have been observed under suitable conditions with the strong light intensities available from lasers. Finally, the phase coherence itself in combination with high field strengths can be expected to produce interactions with matter which cannot occur with incoherent light. All this is the subject of a new and rapidly expanding field as will be shown in the latter part of this chapter.

6.2 Criteria and Propagation Phenomena of Coherence

It will be obvious that the simplified viewpoint adopted in the preceding section for the coherence between two wavetrains has to be supplemented by more general criteria in the case of light beams containing a multitude of photons. Coherence between rays from ordinary light sources typified by thermal radiators can be only partially, never completely, attained since (1) the line width is not infinitely narrow; and (2) the source is of finite extent so that not all the wavetrains can be brought into superposition. Thus it is necessary to define a *degree of coherence*. Experimentally, such a criterion can be established by the visibility of interference fringes, viz., the relative brightness difference of the dark and bright bands in the interference pattern. The degree of coherence has a meaning with respect to time or to space, or to both, as we shall proceed to discuss in the following. It must be kept in mind, in all of this, that these traditional concepts and standards were formulated for light sources which over their extent were themselves incoherent and that a revision of such definitions is necessary for the radiation fields generated by optical masers.

6.2.1 COHERENCE IN TIME

We have seen in Section 1.2.2 that a finite lifetime τ of an oscillation is mathematically equivalent to a broadening of a purely harmonic vibration, by which we mean that a relation exists between the lifetime and the spectral half-width, to wit

$$\Delta \nu_h \cong \frac{1}{2\pi\tau} \tag{6.5}$$

Equation (6.5) follows, of course, classically from a Fourier analysis of the finite oscillation process. If the limitation of the lifetime is due to the process of communicating oscillator energy to the radiation field, $\Delta \nu_h$ is the broadening owing to radiation damping. Reduction of the lifetime by collisions leads to further broadening (cf. Section 3.4.3.1). Conversely, we may view the process of stimulated emission—since it exhibits the same frequency as the incident radiation— as a delay in the termination of the original oscillation. Therefore, in masers a sharpening of the line profile occurs below the limit of the natural line width. One describes this phenomenon as an increase in the coherence in time or "self-coherence," a concept which is directly accessible to measurement (Michelson [7,8]).

It is appropriate here to use the classical picture of propagating wavetrains. If an oscillator emits an electromagnetic wave during a vibration time τ and if this wave propagates with speed c, then we have to visualize a wavetrain of length

$$L = c\tau \cong \frac{\lambda^2}{2\pi \, \Delta\lambda_h} = \frac{Q}{2\pi}\lambda \tag{6.4a}$$

as already mentioned in the preceding section.

That such a wavetrain moving through space with a finite length has actually a physical significance and that therefore L represents a coherence length and τ a corresponding *coherence time*, can be demonstrated by the Michelson experiment (see Fig. 6.1). Assume we split the

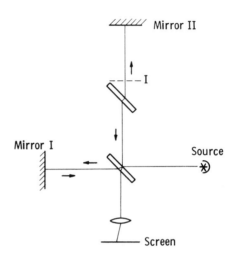

FIG. 6.1.   Michelson interferometer for the demonstration of coherence length. Semitransparent mirror splits beam from source into two parts which are reflected by mirrors I and II, respectively, and then brought to interference on screen. By increasing the path length to and from mirror II, the fringe pattern contrast is reduced until it disappears altogether. Plate in path to mirror II compensates for transmission through semitransparent mirror.

light from a nearly monochromatic source into two beams of variable optical path lengths such as shown for the Michelson interferometer arrangement. If we bring the two beams back into a common focus the various points of the image field will not experience the same difference in phase of the arriving wavetrains contained in the two beams. The intensity is given by the square of the linearly added fields [or, more generally, by the vector product of linearly added electric and magnetic fields, see Eq. (3.12)]. Now since this sum of linearly added fields will be a periodic function in the plane of the image, we will observe under certain circumstances an interference pattern such as in the form of dark and bright concentric rings. The interference pattern is sharpest when the path lengths along the axis of the two beams are equal. As we introduce and increase a difference in the lengths through which the two beams have to travel, the interference pattern is observed to lose its sharpness and disappear finally altogether. The explanation of this phenomenon is, of course, that the phase difference of the linearly added fields is at any point constant in time only if the beams are coherent. If the coherence length L by order of magnitude as given by Eq. (6.4) is exceeded by the optical path difference in the interferometer, the phase correlation between incident wavetrains is random in time. Therefore, the intensity given by the sum of the amplitude squares will show no systematic variation across the image field, and the brightness pattern will be uniform (except for vignetting effects), as indeed observed. Between total overlap of the wavetrains and total lack of such correlation, in other words, between complete coherence and incoherence, is the region of partial coherence which shows itself by a varying degree of pattern visibility. It will be seen that neither coherence nor incoherence can be complete under actual conditions, and that these concepts represent merely the extreme limits.

The coherence length of spontaneously emitted visible light is in the order of a meter corresponding to oscillation durations of about 10^{-8} sec. Classically, the natural line width is independent of wavelength, namely about 10^{-4} Å according to Eq. (3.22). It follows that the coherence length increases with λ^2. One of the features of stimulated emission is a potentially enormous increase in the coherence length. For instance, an increase by a factor of 1000 in the effective lifetime owing to a chain of induced emissions would produce coherence lengths of 1000 km for radiation at 10 μ wavelength.

In the definition and measurement of such values as coherence time and length, the duration of measurement may play an important role (Neugebauer[9]). Ordinary light presents no such problem. The lifetime of excited atoms—typically in the order of 10^{-8} sec—is much shorter than the time during which an interference pattern is observed either by eye or with the photographic plate. Even photoelectric recording with an amplifier bandwidth of 10^6 cps operates with an integration time which is still 10–100 times longer than the life of excited atomic states. Under these circumstances the coherence criteria such as the visibility of interference fringes can be applied without reservations as described, viz., τ and L are determined by the limit at which the fringes disappear. In the case of coherent sources, however, oscillation times may be longer than observation times, and then a more complex situation results. For example, a microwave oscillation may be generated by an electron tube for an unlimited period. Of course, during this operation the frequency will show a spread $\Delta\nu_h$, and as a result the phase is predictable —or coherent—only over a time τ in the order of $1/\Delta\nu_h$ in conformity with Eq. (6.5). Nevertheless, if two different microwave sources of the same mean frequency superpose their radiation fields, they will generate an interference pattern at all times. This conclusion, which has been verified by observation, implies as a direct consequence that if the Michelson experiment for time coherence is applied to maser sources, an interference pattern may be obtained even though the optical path difference introduces a delay far larger than the coherence time τ. Thus the existence of interference fringes is only a necessary, not a sufficient, condition for the existence of time coherence. However, in either case— i.e., that of the two coherent sources or that of the single source with split, time-delayed beams—the pattern will drift at a rate corresponding to the frequency half-width $\Delta\nu_h$. If no time coherence exists, the pattern will "wash out", if integrated over a sufficiently long time. It follows from this that the observation of fringes must be carried on over periods long compared with the coherence time, and it is only then that the visibility of the interference pattern represents a reliable criterion of coherence.

6.2.2 Coherence in Space

The coherence test performed according to Fig. 6.1 determines the correlation between light waves at a fixed point at succeeding times. An analogous inquiry can be made with respect to the coherence between two beams at two different points, but at the same time. The basic experiment for this purpose is that of Thomas Young (see Fig. 6.2) which differs from the time coherence experiment actually only in form rather than in principle. The spatial correlation of a wavefield at two points is established by placing in the path of the radiation source a screen S_c with two pinholes at points P_1 and P_2 and observing the resulting interference pattern in the image plane S_i behind the screen. It is known from the theory of double-slit diffraction that if two mono-chromatic point sources are viewed from S_c under a subtended angle θ, their patterns will mask each other when the slit distance d is increased to

$$d = \frac{\lambda}{2\theta} \qquad (6.6)$$

or odd multiples thereof.* By order of magnitude, Eq. (6.6) also holds

---

* The vanishing of the patterns in this case is the result, not of interference between the beams of the two sources (which are assumed to be incoherent), but of the shift of the two intensity patterns which each source independently produces by diffraction through the two slits. Maxima of one pattern will coincide with minima of the other, if the difference of path lengths from one of the two point sources is by $\lambda/2$ (or an odd multiple thereof) larger than for the other point source. It is easily seen that this condition amounts to Eq. (6.6).

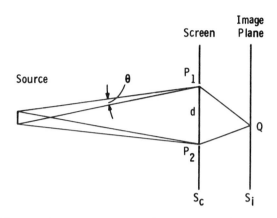

for an extended source ($\theta = 1.22\lambda/d$ for disk), and the arrangement shown in Fig. 6.2 is, in fact, basic to the Fizeau[10] and Michelson[11] stellar interferometers with which the distance between double stars and the diameter of single stars can be determined. In this type of measurement one varies d until the contrast between the intensity I_{max} of the bright fringes and the brightness I_{min} of the dark fringes goes through the first minimum. Michelson introduced as a measure of this contrast the visibility $\mathscr{V}$:

$$\mathscr{V} = \frac{I_{max} - I_{min}}{I_{max} + I_{min}} \qquad (6.7)$$

The question regarding spatial correlation is to all appearances different; namely, it asks for that distance d of the slits for which the *coherence* of radiation emitted from an extended source has vanished. In this case we expect a value d beyond which the visibility is, and remains, zero. A related experiment is this: for a given pinhole distance d, a nearly monochromatic source can always be made small enough so that a sharp interference pattern results on the image plane behind the screen. However, as the source dimensions and angle θ increase, the pattern becomes blurred and finally disappears altogether. We will find that for ordinary thermal sources this coherence limit is indeed also given by Eq. (6.6), at least by order of magnitude.

6.2.3 MUTUAL COHERENCE FUNCTION AND FRINGE VISIBILITY

The preceding discussion has shown a most important aspect of radiation coherence, namely, that it can be achieved only partially.

Therefore it is necessary to establish definitions and scales of reference for the concept of partial coherence such that a link between theory and experiment is created, making it possible to predict the correlation between light beams under various circumstances. It is also clear from the foregoing that under practical conditions, coherence in time and space cannot be strictly separated, and that its general function must contain both variables.

This need was met by the concept of *degree of coherence* and its application to visibility of fringes as first proposed by Zernike[12] (see also Born and Wolf[13], p. 490). The concept suggests itself quite naturally from a calculation of beam intensities under various conditions of interference. Suppose V_1 and V_2 represent the instantaneous electric fields at the points P_1 and P_2 in Fig. 6.2. For simplicity, we assume vacuum as the medium and nearly monochromatic, linearly polarized radiation, although V_1 and V_2 are complex[13a]. If we could measure the instantaneous field vibration at point P_1, we would find in the general case of thermal sources that the oscillations V_1 are not purely harmonic, but that there occur more or less sudden changes of amplitude and phase. The reason for this is, of course, that the light originates in independent atomic sources each emitting a wavetrain of finite length and duration. Thus we write

$$V_1 = a_1 e^{i(\omega t_1 - \phi_1)} \qquad (6.8)$$

where amplitude a_1 and phase ϕ_1 are functions of time which vary only slowly when measured on the scale of a period of oscillation. The quantity accessible to observation is the intensity which for a given oscillation is proportional to the mean square of the field. Thus, if constants are ignored, there results for the intensities at P_1 and P_2, respectively

$$I_1 = \overline{2(V_1{}^R)^2} = \overline{V_1 V_1{}^*} \qquad (6.9)$$

$$I_2 = \overline{2(V_2{}^R)^2} = \overline{V_2 V_2{}^*} \qquad (6.10)$$

where the mean taken during the observation time is presented in terms of the real part $V_1{}^R$ of the field as well as in terms of its complex conjugate. Now in a purely formal way we can introduce, in analogy to Eq. (6.9), a so-called mutual *coherence function* for the correlation between the vibrations at P_1 and P_2:

$$\Gamma_{12} = \overline{V_1 V_2{}^*} \qquad (6.11)$$

The statistical nature of this expression must be emphasized. It is typical for the partial coherence of the wavetrains which arrive at the two different points in space at different times that the phases and amplitudes fluctuate to a certain degree independently of each other (see Section 6.2.5). In particular, V_1 and V_2^* can assume positive and negative values, and if they fluctuate completely independently of each other, the correlation Γ_{12} will be zero. The degree of coherence is therefore given by the correlation of the fluctuations. The significance of the term Γ_{12} is at once apparent when we consider the interference of the waves passing through P_1 and P_2 and compute the intensities of the maxima and minima of the resulting pattern. For this purpose we first linearly add the fields resulting on the image plane S_i—allowing for a varying phase difference β along the pattern—then form the intensity product and determine its average over the time of observation. Omitting first such factors as the inverse distance square, we have

$$I = \overline{(V_1 e^{i\beta} + V_2)(V_1^* e^{-i\beta} + V_2^*)} = I_1 + I_2 + \Gamma_{12} e^{i\beta} + \Gamma_{12}^* e^{-i\beta} \qquad (6.12)$$

The last two terms of Eq. (6.12) add up to the real part of $2\Gamma_{12} e^{i\beta}$, and this is the intensity contribution which periodically varies across the image plane according to the optical path difference between the two beams so that it gives rise to the fringe pattern. Maxima occur for $\beta = 2n\pi$ and minima for $\beta = (2n + 1)\pi$, n being an integer number. Maxima and minima of the fringe intensities are thus

$$I_{\max} = I_1 + I_2 + 2\Gamma_{12} \qquad (6.13)$$

$$I_{\min} = I_1 + I_2 - 2\Gamma_{12} \qquad (6.14)$$

It is now possible to relate the pattern visibility, which is defined by Eq. (6.7) and which constitutes the experimental measure of coherence, to the theoretical correlation factor, viz.,

$$\mathscr{V} = \frac{I_{\max} - I_{\min}}{I_{\max} + I_{\min}} = \frac{2\Gamma_{12}}{I_1 + I_2} \qquad (6.15)$$

The values used for fields and intensities in Eqs. (6.12)–(6.15) correspond to those at P_1 and P_2 at the screen except for the interference effects, i.e. they correspond to integrals over zones of a given phase delay β. Actually the wave which produces a field V_1 at P_1 will generate in the image plane at a point Q a field $V_1(Q)$. A second field $V_2(Q)$ is superposed on the first at Q by the wave passing through P_2. We thus set

$$V_1(Q) = K_1 V_1; \qquad I_1(Q) = |K_1|^2 I_1 \qquad (6.16)$$

$$V_2(Q) = K_2 V_2; \qquad I_2(Q) = |K_2|^2 I_2 \qquad (6.17)$$

where the K-values represent geometric factors which account for the reduction of the field in proportion to the distance, the effect of hole size, and the effect of diffraction. If we take these K-values into consideration, we obtain for the visibility at Q in the image plane instead of Eq. (6.15)

$$\mathscr{V} = \frac{2 \sqrt{I_1(Q)/I_1} \cdot \sqrt{I_2(Q)/I_2}}{I_1(Q) + I_2(Q)} \cdot \Gamma_{12} \tag{6.18}$$

This expression can be simplified by introducing a normalization of the mutual coherence function, viz.,

$$\gamma_{12} = \frac{\Gamma_{12}}{\sqrt{I_1}\sqrt{I_2}} \tag{6.19}$$

The value γ_{12} is called the *complex degree of coherence*; it represents the mutual coherence function reduced to the geometric mean of the intensities. Introduced into Eq. (6.18), it yields

$$\mathscr{V} = 2 \frac{\sqrt{I_1(Q) \cdot I_2(Q)}}{I_1(Q) + I_2(Q)} |\gamma_{12}| \tag{6.20}$$

Of particular importance is the condition of intensity symmetry in which the average contribution of the two pinholes at P_1 and P_2 to the fields at Q is equal, so that $I_1(Q) = I_2(Q)$. Then we have

$$\mathscr{V} = |\gamma_{12}| \tag{6.21}$$

i.e., the visibility is equal to the degree of coherence.

The coherence criteria Γ and γ refer in general to two different points in time and space. In particular, if we let P_1 coincide with P_2, Γ_{11} represents the *self-coherence* function,

$$\Gamma_{11} = \overline{V_1(t + \tau)V_1{}^*(t)} \tag{6.22}$$

where τ is the phase delay between wavetrains produced, for instance, by the arrangement described in Section 6.2.1. If in particular $\tau = 0$ and $P_2 = P_1$, there result simply Eqs. (6.9) and (6.10). Thus the degree of coherence can assume values between zero for complete incoherence and unity for complete coherence.

6.2.4 COHERENCE PROPERTIES OF THERMAL LIGHT SOURCES

By virtue of the fact that the concept of partial coherence has been defined in mathematical terms, it is now natural to ask about the mutual coherence of light coming from a thermal source of known dimensions. The problem is to compute the correlation of wave-trains which emerge from elements or atoms of an extended radiator and which arrive on an image plane and to take the sum of the correlations over the entire source. Now it is important to distinguish between the case that the waves are produced by the various atoms randomly and the case that there exists some degree of mutual dependence of the individual events. The latter situation is, of course, that which prevails in reality; even thermal sources emit some stimulated light, although at a rate which is for most purposes negligible compared with that of spontaneous emission. The idealization that the elements of a thermal source are completely independent of each other can be used for the derivation of a theorem which was stated first by Van Zittert[14] and later shown to follow from the concept of coherence functions by Zernike.[12]

Assume that the mth element of a light source emits a wavetrain

$$V_{m1} = \frac{a_m}{R_{m1}} e^{i(\omega t - k R_{m1})} \tag{6.23}$$

where V_{m1} represents the field at a point P_1 which has a distance R_{m1} from the source, a_m represents the complex amplitude at $R = 0$, and $k = 2\pi/\lambda$. The total amplitude V_1 at point P_1 then results from a summation over all elements of the source. Similarly, V_2 at P_2 is obtained by summing over all V_{n2}. Of interest is, of course, the mutual coherence function $\overline{V_1 V_2^*}$ which is formed by adding all the $m \cdot n$ cross products of terms such as given by Eq. (6.23). However, the restriction to idealized thermal sources leads to a simplification since the conjugate complex product of the amplitudes $a_m a_n^*$ must be zero in the average. Thus there remains

$$I_{1,2} = \Gamma_{12} = \sum_m \frac{\overline{a_m a_m^*}}{R_{m1} R_{m2}} e^{-ik(R_{m2} - R_{m1})} \tag{6.24}$$

The product $\overline{a_m a_m^*}$ is a measure of the brightness $I(S)\,dS$ of an element dS on the surface of the source. Hence we can write for the mutual coherence function

$$\Gamma_{12} = \int I(S) \frac{e^{ik(R_1 - R_2)}}{R_1 R_2} \, dS \tag{6.25}$$

or for the complex degree of coherence

$$\gamma_{12} = \frac{1}{\sqrt{I_1 I_2}} \int I(S) \frac{e^{ik(R_1 - R_2)}}{R_1 R_2} \, dS \tag{6.26}$$

here

$$I_1 = \int \frac{I(S)}{R_1{}^2}\, dS \quad \text{and} \quad I_2 = \int \frac{I(S)}{R_2{}^2}\, dS \tag{6.27}$$

Now the significance of these expressions for Γ and γ is that they consist of integrals which are known from the theory of physical optics to also describe the amplitude of diffraction patterns. The *Van Zittert-Zernike* theorem, therefore, states that, given a thermal source $I(S)$, it will produce within a narrow wavelength interval a degree of coherence between a variable point P_2 and a fixed point P_1 equal to the normalized complex amplitude of the corresponding diffraction pattern around P_2; viz., that pattern which would be generated by a diffracting aperture of the same size, shape, and an amplitude distribution as the intensity distribution of the source. As a consequence, it is possible to apply the large number of results known for various boundary conditions in the theory of diffraction directly to problems of coherence.

As an example, we may ask what area in the image plane can be nearly coherently illuminated at a certain wavelength λ by a disk-shaped source of a given radius ρ and distance R. Demanding that $\gamma > 0.88$, one finds that the distance d between two points P_1 and P_2 must fulfill the condition (see Born and Wolf,[13] p. 509)

$$d < 0.16\,\frac{R}{\rho}\,\lambda = 0.32\,\frac{\lambda}{\theta} \tag{6.28}$$

where θ is the angle subtended by the source. Thus if λ is 6000 Å, and $\theta = 10^{-3}$, d must be smaller than $320\,\lambda$ or 1.92×10^{-2} cm so that the area of coherence is 2.88×10^{-4} cm^2.

6.2.5 PHOTON CORRELATION EXPERIMENTS

We shall describe in this section phenomena of correlation which differ from those shown in the classical interference experiments in various respects and thereby reveal additional, not generally expected interaction properties of light. The deviations from the classical conditions of interference can take various forms, but typically the observations deal with second order effects.

6.2.5.1 *Hanbury Brown-Twiss Experiment*

One of the most striking among these effects is the experiment by Hanbury Brown and Twiss[15–17] (see Fig. 6.3). The particular feature of the arrangement to be described is that correlation is established, not by the superposition of light beams, but by comparison of the signals from the two beams which they produce each independently in a detector. A practical motivation for this experiment came from an astronomical problem, namely, the diameter determination of distant stars by the Michelson method (cf. Section 6.2.2) either in the visible region or by radio waves. According to Eq. (6.6) the spacing d between mirrors may become very large and therefore excessively difficult to maintain in optical alignment, if the observed angle θ (given by the ratio of star diameter and distance) is too small. This led directly to the exploration of *intensity interferometry* in which the determination of phase relationship is abandoned and in which, instead, the correlation in photon arrival rate fluctuations is measured. Then an obvious practical advantage accrues from the fact that the two branches of the interferometer are connected to each other by a signal mixer rather than an optical path so that large separations can be resorted to without difficulty.

The apparatus shown in Fig. 6.3 tests the validity of this principle for thermal sources. Light from a mercury arc is focused on a small, well defined aperture which serves as the source in combination with filters. The light emerging from the aperture is split by a half-silvered mirror into two beams, each falling on the cathode of a photomultiplier through well defined, identical apertures. One of the photomultipliers is moved normal to the incident light, thereby varying the degree of coherence, just as it is done in the double slit experiment of Fig. 6.2 by changing the distance between P_1 and P_2. However, instead of adding the fields linearly, the beam intensities (i.e. the squares of the fields) are compared. Of course, no information would be supplied by this procedure, if the intensities were absolutely constant. The experiment relies on the fact that, because of the finite coherence lengths of the wavetrains or the finite number of photons in a light beam, the intensity fluctuates. As a result the two photomultipliers yield instantaneous currents $\bar{I}_1 + \Delta I_1(t)$ and $\bar{I}_2 + \Delta I_2(t)$, respectively. After the amplification of these signals within a selected bandwidth (3–27 Mc), they are multiplied in a mixer and integrated in a recorder over observation times in the order of an hour (see Fig. 6.3). This then yields the cross correlation function

$$G(d) = \overline{\Delta I_1 \, \Delta I_2} \tag{6.29}$$

If the signal fluctuations are independent of each other, the correlation product will be zero since ΔI_1 and also ΔI_2 by definition assume as much positive as negative values. $G(\infty)$ will therefore certainly vanish as will the correlation of any two independent thermal light

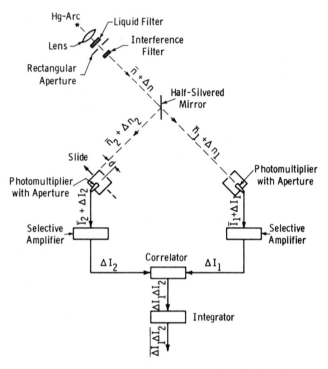

FIG. 6.3. Photon arrival correlation according to R. Hanbury Brown and R. Q. Twiss [*Nature* **177**, 27 (1956)]. Arrows indicate propagation of light beams with their fluctuations and their conversion into signal fluctuations. The latter, after being mixed and integrated, yield positive correlations depending on displacement d of photomultiplier from position of maximum coherence.

sources measured at two distant points. Now the significance of the Hanbury Brown–Twiss experiment is that it tests whether intensity correlations occur under conditions in which wavetrains are coherent. That a connection between these two phenomena exists, is by no means obvious, and was in fact at first strongly contested. For instance, the emergence of a photoelectron is contingent on the absorption of a single photon either reflected or transmitted by the mirror, but not both. Hence, if the photons are completely randomized, then the current fluctuations will be independent of each other and the correlation (6.29) will be zero.

The result of the Hanbury Brown and Twiss experiment was, however, positive; a correlation for the intensity fluctuations was established commensurate with the wave coherence. The result, furthermore, can be quantitatively accounted for by a fluctuation theory which can be constructed on the basis of either a wave or a corpuscular picture. The classical theory, as we have seen, assigns to finite times of oscillation (and observation) a Fourier distribution of frequencies spread over a certain line width. The various frequencies are producing beats with respect to each other so that a detector will register a statistical distribution of amplitude variations or fluctuations. It is plausible that the beats are preserved over a wavefront to the extent to which it is coherent, and thus the correlation as measured by two different phototubes can be assessed quantitatively[16]. A semiclassical treatment on the basis of photon statistics will be presented in the next section.

It is, of course, the practical objective of intensity interferometry to measure just those subtended angles θ of stars, according to Eq. (6.6), which are too small to be determined by the phases coherence interferometer of Michelson. For example, an Australian installation uses two low-quality mirrors of about 6 m diameter which can be brought to correlation at about 200 m distance; yet this system has a theoretical resolution limit of as low as 0.0005 sec of arc (3×10^{-9} rad).

6.2.5.2 *Quantum Statistics of Photon Correlations*

The positive results of intensity interferometry experiments can be explained also with the Bose-Einstein statistics of the photon gas[16,18-22]. The statistics has already been discussed in Sections 2.3.1 and 2.3.2. It was shown, in particular, that there exist in a volume V

$$dZ = \frac{8\pi\nu^2}{c^3} V \, d\nu \tag{6.30}$$

different cells of phase space or quantum states for the frequency interval between ν and $\nu + d\nu$. Each cell is occupied in the average by δ photons, δ being called the *degeneracy parameter*. It was also shown that in a thermal equilibrium this probability of finding a photon in the cell equals

$$\delta_T = \frac{1}{e^{(h\nu/kT)}-1} \tag{6.31}$$

The product

$$u(\nu) \, d\nu = h\nu \cdot \delta_T \cdot dZ \tag{6.32}$$

yields Planck's law of heat radiation. The value of δ_T is rather small for thermal emitters at temperatures up to $10^4 \,^\circ K$; in the case of the Hanbury Brown-Twiss experiment, for example, δ equals about 3×10^{-3}. For nonthermal sources such as masers, the degeneracy parameter can be very large as we shall see later.

Even in radiation from thermal emitters, however, there is a measurable "clumping" of photons. Stated somewhat differently, the degeneracy which distinguishes energy distributions of photons from those of a Boltzmann gas shows itself also in differences of the fluctuation phenomena (see Chapter 7). For example, if we count in the average n atoms occupying a given subdivision of a gas volume or passing per second through an orifice, we will find deviations $\Delta n = n - \bar{n}$ in the individual counts which for the case of complete randomness obey the simple law of *mean square deviation*:

$$\overline{(\Delta n)^2} = \bar{n} \qquad (6.33)$$

In the case of a Bose-Einstein gas the fluctuations are somewhat larger. If we count $\bar{n}$ unpolarized photons which arrive in the average during a time t_0 at a photosurface, the mean square deviation will be, as shown in Section 7.2.6

$$\overline{(\Delta n^2)} = \bar{n}\left(1 + \frac{\bar{n}}{N_B}\right) = \bar{n}(1 + \delta_B) \qquad (6.34)$$

where N_B is the available number of cells in phase space, i.e., the number of quantum states, so that $\delta_B = n/N_B$. Thus photon mean square fluctuations exceed those of a Boltzmann gas by a fraction equal to the degeneracy parameter, a situation which will arise simply through stimulated emission. In a beam of polarized photons, N_B is given by the ratio t_0/τ of the distance ct_0 traveled by the light during the observation time to the coherence length $c\tau$ ($t_0 \gg \tau$). Furthermore, a factor of $\frac{1}{2}$ has to be attached to δ_B, if unpolarized light is involved. Thus we have for the mean square deviation in Eq. (6.34)

$$\overline{(\Delta n)^2} = \bar{n}\left(1 + \tfrac{1}{2}\bar{n}\,\frac{\tau}{t_0}\right) \qquad (6.35)$$

The excess fluctuations then are the result of deviations from complete randomness which make it more likely that a photon assumes a certain set of phase space coordinates, if another photon is already in that state. This "bunching" of photons, which is typified by the process of stimulated emission, corresponds to the beats in the classical wave picture[23]. It was Purcell[18] who first pointed out that the excess fluctuations are capable of accounting quantitatively for the observed effects of intensity interference. Since Purcell's interpretation of the Hanbury Brown-Twiss experiment is particularly concise, we shall outline it here.

Let n photons be partitioned by the beam splitter in Fig. 6.3 into two groups of n_1 and n_2 quanta which reach the two photomultipliers, respectively, so that $n = n_1 + n_2$. The fluctuation of the initial beam is related to that of the two branches as follows:

$$\overline{(\Delta n)^2} = \overline{(\Delta n_1 + \Delta n_2)^2} = \overline{(\Delta n_1)^2} + \overline{(\Delta n_2)^2} + 2\overline{\Delta n_1 \Delta n_2} \qquad (6.36)$$

For each group of photons the same fluctuation law as that given for the total number in Eq. (6.35) applies so that

$$\overline{(\Delta n_1)^2} = \bar{n}_1 \left(1 + \frac{1}{2} \cdot \bar{n}_1 \cdot \frac{\tau}{t_0} \right) \qquad (6.37)$$

$$\overline{(\Delta n_2)^2} = \bar{n}_2 \left(1 + \frac{1}{2} \cdot \bar{n}_2 \cdot \frac{\tau}{t_0} \right) \qquad (6.38)$$

Now if the fluctuation terms in Eq. (6.36) are substituted by the expressions in Eqs. (6.35), (6.37), and (6.38), there results

$$\overline{\Delta n_1 \Delta n_2} = \tfrac{1}{2} \overline{\bar{n}_1 \bar{n}_2} \frac{\tau}{t_0} \qquad (6.39)$$

This equation proves the existence of a correlation function in terms of photon counts. It is seen that the correlation is positive, i.e., the fluctuations measured by the photocells within time intervals t_0 tend to assume equal sign. Obviously there would be no correlation and therefore no "intensity interference" for particles obeying Boltzmann statistics since then Eq. (6.33) rather than Eq. (6.34) applies to the mean square deviation and the cross terms $\overline{n_1 n_2}$ disappear. Thus in the particle model the correlation of photon arrival counts is explained by the degeneracy feature which permits photons in the same state to enter simultaneously the two branches of the interferometer. The property of "clumping" in a Bose-Einstein gas furthermore tends to increase the coherence of a light beam which coming, for instance, from a distant star was produced by entirely unrelated processes of generation. For this reason, as pointed out before, light cannot be completely incoherent.

The nonvanishing cross correlation $\overline{\Delta n_1 \Delta n_2}$ is a maximum at given n for $n_1 = n_2$, corresponding to maximum coherence under the conditions. Even then the correlation is small for thermal sources so that long integration times become necessary. Equation (6.39) is, of course, also valid for nonthermal sources, which may have a large δ, provided that they exhibit a Gaussian probability distribution of the photon yield.[23a] Practically, however, the observed value $\overline{\Delta n_1 \Delta n_2}$ will be smaller than that given by Eq. (6.39) for two reasons. First the process employed for counting photons (e.g., photoelectric detection) will have a quantum efficiency less than unity with a resulting decrease of correlation.

Secondly, we have yet to allow for the general case of partial coherence of degree $\gamma_{12}(0)$. We cite here without proof the result of a detailed calculation[23b] :

$$\overline{\Delta n_1 \Delta n_2} = \tfrac{1}{2} \overline{n_1 n_2} \frac{\tau}{t_0} \mid \gamma_{12}(0) \mid^2 \tag{6.40}$$

6.2.5.3 Photoelectric Mixing and Related Processes

It will be brought out in Chapter 7 that all detectors of optical radiation operating either by thermal or quantum effects measure energy directly, viz., as the square of the total incident electromagnetic field. This premise is basic to an experiment by Forrester, Gudmundsen, and Johnson[24] in which beats between incoherent light sources are reproduced in a stream of photoelectrons. The origin of such beats has already been discussed in the preceding sections. Forrester and his co-workers demonstrated that interference between Fourier components of unrelated wavetrains could indeed be demonstrated directly by means of a nonlinear detector.

In the experiment, the interaction between two lines of a Zeeman pattern were observed, using the transition at 5461 Å of the isotope Hg^{202}. The choice of this spectrum was motivated in part by the requirement that the coherence time $(\Delta\nu_h)^{-1}$ be long compared to the beat period $(\nu_2 - \nu_1)^{-1}$ or

$$\nu_i \equiv \nu_2 - \nu_1 \gg \Delta\nu_h \tag{6.41}$$

The line width under these (for incoherence) nearly optimal conditions was about 10^9 sec^{-1} so that the beat frequency ν_i was chosen to be 10^{10} sec^{-1}. This region for ν_i is still very favorable; first, because it corresponds to the Zeeman splitting and thus permits magnetic tuning; secondly, because it can be processed with typical microwave equipment. The actual apparatus is shown in Fig. 6.4. Essential features of the arrangement include a photomixer tube in which the light generates a photocurrent from a semitransparent cesium-antimony cathode and a grid structure (not shown) focuses the electrons into a cavity resonant at the beat frequency. The detection of beats in the current is improved by a suitable modulation technique, namely, by introducing into the optical path a polarizing plate rotating at a low frequency. The polarization of the Zeeman pattern components parallel and normal to the magnetic field is utilized here so that background intensity variations are minimized, while the beats experience the modulation. With this arrangement then the cavity output contained a signal of ν_i twice that of the noise. If photon mixing in nonlinear detectors is possible with beams

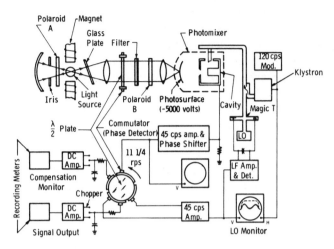

FIG. 6.4.   Arrangement for photoelectric mixing experiment [A. T. Forrester, R. A. Gudmundsen, and P. O. Johnson, *Phys. Rev.* 99, 1691 (1955)].

from independent quasi-thermal sources, considerably higher signal to noise ratios should be obtainable from coherent sources, such as radio transmitters or masers.[25,26] This has proved to be the case, and it is even possible to photograph interference patterns resulting from the superposition of two independent laser beams. Magyar and Mandel[27] performed this experiment by means of a "gated" image tube which passed the beams only for an interval small compared with the coherence time. Therefore an enormous number of photons coincide in phase space. The degeneracy parameter can then easily be made as high as 10^7–10^{12} or more.

6.2.6 COHERENCE CRITERIA FOR NONTHERMAL SOURCES

The fields produced by thermal radiators exhibit the largest mutual independence possible in a photon gas. This behavior is shared by all other sources in which the generation of photons is determined by Gaussian statistics, such as in glow discharges, although they are not really "thermal" in the sense applied in Chapter 2. The definitions and criteria for coherence were derived originally for the thermal type of randomness, and it remains to be seen to what degree these concepts are still adequate for the description of much more coherent sources such as lasers.

Obviously, there exist qualitative as well as quantitative differences of coherence between the two classes which are typified, respectively, by thermal radiators and lasers. Because the atoms in the first class emit photons largely independently of each other, every point in the radiation path receives a fluctuating spectrum of various frequencies; and it is only because such a spectral distribution may vary only slowly from point to point, that a measure of coherence is established between the various locations. The atoms of an ideal laser, however, radiate in unison, therefore the points in the radiation path experience a sharply monochromatic field variation (in other words, a single wavetrain), and this field is identical over large areas. Now the point is that the coherence criterion as stated in Eqs. (6.11) and (6.19) does not differentiate between these two classes, and that it of itself does not take account of the higher order coherence offered in the second case. A further index not considered by the original theory is the degeneracy parameter which assumes vastly different values among such nonthermal radiators as masers, microwave transmitters, or sources operating with special mechanisms as, e.g., the Purcell effect.* The need for a single, all-embracing coherence criterion is, of course, not easily satisfied. However, it is possible to list the various coherence parameters which together form a sort of merit factor for a light source.

The criteria used for highly coherent sources are extensions of those defined originally for thermal radiators. The concept of time coherence needs the stipulation that the observation time is large compared to the coherence time (cf. Section 6.2.1). Mutual coherence is defined in terms of various higher orders n, viz., for n pairs of points $x = \{\mathbf{r}, t\}$ in time and space (Glauber[29], Wolf[30]) as

$$g^n(x_1 x_2 \cdots x_{2n}) = \frac{G^{(n)}(x_1, x_2, ..., x_{2n})}{\prod_{j=1}^{2n}\{G^{(1)}(x_j, x_j)\}^{\frac{1}{2}}} \tag{6.42}$$

where $G^{(n)}$ is the n-th order correlation function normalized to $g^{(n)}$ such that $|g^{(n)}| \leq 1$. This equation† can be understood by the analogy of the first-order function

$$g^1(x_1 x_2) = \frac{G^1(x_1, x_2)}{\{G^1(x_1 x_1)G^1(x_2 x_2)\}^{\frac{1}{2}}} \tag{6.43}$$

---

* In the original experiment (Smith and Purcell[28]) a 300 kV electron beam is directed at grazing incidence over an optical metal grating normal to the grooves. Since the electrons induce image charges periodically in the metal crests, visible dipole radiation is emitted. The light [which has an angle θ with the beam for a ruling distance d and beam velocity v the fundamental wavelength $\lambda = d(c/v - \cos \theta)$] is strongly polarized and has coherence properties representing a special class.

† In this expression, the n-th order correlation function $G^{(n)}$ has been introduced by

to the normalized coherence function (6.19). Similarly, for complete coherence, the normalized function (6.42) must be unity; but for ideally coherent sources, this must be the case, in addition, for an infinite order n. Finally for an evaluation of the source, the degeneracy parameter has also to be considered.

It is in the context of these generalized criteria that the characteristics of laser radiation can be fully evaluated.

6.3 Optical Masers

The generation of coherent light by stimulated emission is a subject involving aspects of both pure physics and of radio frequency engineering, and for this reason the general field is also called *quantum electronics*. We shall begin the discussion of *optical masers* or *lasers* with a review of their operating principle.

It has been shown that the classical theory of absorption (Section 3.1.4), as well as the thermodynamics of quantized radiation processes (Section 3.3.1), lead to the postulate of negative absorption or stimulated emission. In particular, if we exclude for the moment the case of degeneracy in two atomic levels ϵ_2 and ϵ_1 ($\epsilon_2 > \epsilon_1$) so that there exists only one electronic configuration per energy level, then we can state that a photon of energy $h\nu_{12} = \epsilon_2 - \epsilon_1$ is as likely to stimulate emission in an encounter with an upper state ϵ_2 as it is to be absorbed in an encounter with a lower state ϵ_1 (see Fig. 3.8). Thus if there are more upper states ϵ_2 than lower states ϵ_1, an incident stream of photons will be augmented in number as long as this condition persists, provided that there is no prevalence of competitive processes which absorb or scatter the frequency ν_{12}. In other words, a negative coefficient of absorption results; an electromagnetic wave propagating in such inversion medium grows, rather than diminishes, in intensity; and we have instead of Eq. (5.52) a process of amplification, which with positive feedback leads to oscillation:

$$I(z) = I(0)e^{kz}, \quad k > 0 \tag{6.44}$$

Glauber[29] in the quantum theoretical terms of the field operator $E^{(+)}(\mathbf{r}, t)$ for positive frequencies (photon absorption) and $E^{(-)}(\mathbf{r}, t)$ for negative frequencies (photon emission):

$$G^{(n)}(x_1 \ldots x_n, x_{n+1} \ldots x_{2n}) = \text{tr}\{\rho E^{(-)}(x_1) \ldots E^{(-)}(x_n) E^{(+)}(x_{n+1}) \ldots E^{(+)}(x_{2n})\} \tag{6.42a}$$

where the trace is given by the sum of the diagonal matrix elements and ρ the density operator. $G^{(n)}$ is not restricted to time averages but represents general ensemble averages. As pointed out by Wolf[30], Eq. (6.42a) can be formally derived also from classical theory by analogy with the first order function.

The provision that there are more atoms in an upper state ϵ_2 than in a lower state ϵ_1 represents, of course, a perturbation of the thermal equilibrium. For under normal conditions the population N_2 in the upper state is related to the number N_1 in the lower level by the Boltzmann factor [cf. Eq. (3.59)]

$$\frac{N_2}{N_1} = \frac{g_2}{g_1} \exp\left[-\frac{\epsilon_2 - \epsilon_1}{kT}\right] \tag{6.46}$$

so that in thermal equilibrium $N_2 < N_1$. In fact, a temperature T_{21} can be measured by the population ratio according to Eq. (6.46), viz.

$$T_{21} = \frac{\epsilon_2 - \epsilon_1}{k \log_e(g_2 N_1/g_1 N_2)} \tag{6.47}$$

Here, the value T_{21} *defines* a specific temperature, namely, that corresponding to the excitation of the level ϵ_2 from ϵ_1 by interactions of the atom with photons, electrons, ions, and other atoms. Only in thermal equilibrium will T_{21} agree with the temperature T given by the kinetic energy of the atoms or with that defined by population densities of other levels. In this sense, T_{21} defines by Eq. (6.47) a negative temperature for inverted populations $N_2 > N_1$. The density ratio N_2/N_1 of upper and lower states is plotted schematically in Fig. 6.5 against temperatures in the positive and negative range. It will be seen that the condition of total inversion is approached as T_{21} tends to $(-)0$ from the negative side, whereas total depopulation of the upper level results, if T_{21} is reduced to $(+)0$ from the positive range.

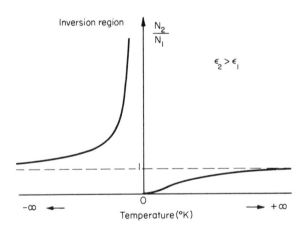

FIG. 6.5. Population ratio of two levels at positive and negative temperatures.

The disturbance of a thermal equilibrium to the extent of attaining negative temperatures appeared to be such an unlikely event that nearly a century passed between the initial concepts of statistical mechanics and the first active attempts[2] of inverting atom populations. However, once the maser principle and its dependence on negative temperatures was understood,[3, 31-33] a number of highly effective inversion techniques were found within a short span of time. Indeed, the development of maser technology is to a large extent paced by the progress of *pumping* methods as the inversion mechanisms are alternatively called in a mechanical analogy. We shall show later how these various techniques made a large number of materials, in various states of bonding, adaptable to maser operation.

A second condition for successful laser operation is the suppression of competing processes which detract from the buildup of coherent radiation by absorption or scattering so that the excited states will decay spontaneously before they chance to encounter a stimulating photon. We shall describe in the following sections how this difficulty is solved by means of optical cavities and how this measure affects the requirements for inversion efficiencies and pumping powers.

6.3.1 THEORY OF LASER OPERATION

Various quantitative considerations concern the production of coherent light: the conditions for laser oscillation, i.e., the needed concentration of excited states and the "pumping" power to maintain it; the requirements for the type of atomic transitions and for the optical design chosen from the theory; and the line narrowing, beam directionality, mode selection, and other phenomena connected with laser operation. The basic theory of laser operation was constructed by Schawlow and Townes[4] in advance of the first successful experiment.

6.3.1.1 *The Oscillation Conditions*

It was stated before that there are two major provisions required for laser operation. First, there must exist a population inversion; secondly, the stimulated radiation power P_M must be larger than the losses P_L resulting from absorption and scattering at the walls and other mechanism:

$$P_M \geqslant P_L \tag{6.48}$$

This inequality actually contains both requirements since it demands that the difference between the power generated by stimulated emission from n_2 upper levels per unit volume and that absorbed by n_1 lower levels be positive and larger than the losses which are dissipated outside the atomic systems. These losses are characterized by a time constant τ_c with which a radiation energy density $u(\nu)$ decays exponentially when left to itself. In other words, τ_c is an average lifetime of the photon (free photon time) as limited by cavity walls, mirrors and other lossy components. Thus, defining B-values as probabilities for negative and positive absorption as in Section 3.3.1, we have for the inequality (6.48)

$$(n_2 B_{21} - n_1 B_{12})u(\nu)h\nu = \left(n_2 - \frac{g_2}{g_1}n_1\right)B \cdot u(\nu) \cdot h\nu \geqslant \frac{u(\nu)}{\tau_c} \qquad (6.49)$$

On the left-hand side we have used the relation (3.63) between B_{12} and B_{21} for the general case that the statistical weights, g_1 and g_2, in the upper and lower level are not equal. In this event, the critical condition is that the population per quantum state rather than per energy level is inverted. We have thus for the required excess

$$\left(n_2 - \frac{g_2}{g_1}n_1\right) \geqslant \frac{1}{\tau_c B h\nu} \qquad (6.50)$$

This latter relationship becomes more informative, if we substitute the absorption probability, viz., replace B by A for spontaneous emission through Eq. (3.65). However, the energy density $u(\nu)$ is not constant throughout the frequency interval in which the atomic transition is responsive. In fact, we may assume that under typical operating conditions, $u(\nu)$ is concentrated in a much narrower interval than the half-width $\Delta\nu_r$ of the atomic transition ν_{12}, where $\Delta\nu_r$ is given by random broadening causes such as Doppler effect or statistical field distributions. As a result, the numbers of available modes have to be weighted with the probability distributions given by the spectral line profiles (see e.g. Yariv and Gordon[34]). In the case of a Lorentz distribution, we take this into account by writing for the transition probability

$$A_{21} = \frac{1}{\tau_s} = Bh\nu \int \frac{8\pi\nu^2}{c^3} \cdot \frac{(\Delta\nu_r/2)^2}{(\nu - \nu_0)^2 + (\Delta\nu_r/2)^2} \, d\nu \qquad (6.51)$$

Here τ_s is defined as the reciprocal of the spontaneous decay rate A_{21}, i.e., as the time that passes in the average before the transition ν_{21} occurs. The factor representing the line profile under the integral is adjusted so as to yield unity for the center frequency $\nu = \nu_0$ and c is the velocity of light in the medium. We thus obtain after evaluation of the integral

$$\frac{1}{B} = \frac{8\pi^2 h\nu^3 \tau_s}{c^3} \frac{\Delta\nu_r}{2} \qquad (6.52)$$

Introducing this value into Eq. (6.50) yields

$$\left(n_2 - \frac{g_2}{g_1} n_1\right) \geqslant \frac{4\pi^2}{c^3} \nu^2 \Delta\nu_r \frac{\tau_s}{\tau_c} \tag{6.53}$$

If we refer to the total number of states rather than to their densities in a volume V and assume that $g_1 = g_2$, we have for the excess number $N = N_2 - N_1$ that is needed for oscillation

$$N \geqslant \frac{4\pi^2\nu^2 V}{c^3} \Delta\nu_r \frac{\tau_s}{\tau_c} \tag{6.54}$$

For a Gaussian profile of the spectral line, this formula is changed to

$$N \geqslant \frac{4\pi^2\nu^2}{(\pi \ln 2)^{1/2}c^3} V \Delta\nu_r \frac{\tau_s}{\tau_c} \tag{6.55}$$

or if in general p designates the effective number of modes in a volume V,

$$N \geqslant p \frac{\tau_s}{\tau_c} \tag{6.56}$$

Thus to achieve oscillation, the excess number of excited states must be equal to the product of the effective number of modes and the ratio of the lifetime to the free photon time (so that there is always at least one photon per mode underway in the cavity). τ_s is given by Eq. (3.84), viz.,

$$\tau_s = \frac{3hc^3}{64\pi^4\nu^3e^2 \, | \, x_{21} \, |^2} \tag{6.57}$$

There remain then two variables to be adjusted for minimum requirements on n: (1) the number of eligible modes p which can be reduced by the choice of sharp line transitions and (2) the decay time τ_c which can be increased, for instance, by confining the interaction space between two parallel mirror arrangements (Fabry-Perot interferometer). If L represents the distance between such mirrors, c the speed of light in the medium, and α the fractional loss per traversal including that suffered in reflection, then the free photon time can be expressed as

$$\tau_c = \frac{L}{c} \cdot \sum_{n=0}^{\infty} (1 - \alpha)^n = \frac{L}{\alpha c}, \qquad \alpha \ll 1 \tag{6.58}$$

Typical values for some of the first used lasers are $L = 10$ cm, $c = 10^{10}$ cm/sec, and $\alpha = 10^{-2}$ so that $\tau_c = 10^{-7}$ sec. Hence with $\tau_s = 10^{-8}$ sec, the condition for oscillating is $n \geqslant 10^{-1}p$ per unit volume.

At a frequency of $\nu = 3 \times 10^{14} \sec^{-1}$ and $\Delta\nu_r = 10^9 \sec^{-1}$, $n \geqslant 4 \times 10^8$ cm^{-3}. This represents a very small excess population compared with the density of normal atoms even of a gas in the pressure range of the order of millimeters of mercury, in which such density amounts to 10^{17} cm^{-3}. It follows that transitions to the ground state are relatively uneconomical for laser operation since the number of excited levels is large compared to the net number used for stimulated emission. Of course, the decisive criterion for the question, whether a particular laser system is operable, concerns not directly the number of excited levels required, but the power needed to maintain it and the means of supplying it by a suitable pumping mechanism.

A lower limit of the power necessary for a stationary inverted population, in which each atom has the natural lifetime τ_s, is given by

$$P \geqslant \left(n_2 - n_1 \frac{g_2}{g_1}\right) V \frac{h\nu}{\tau_s} \tag{6.59}$$

or, if $g_1 = g_2$, $(n_2 - n_1)V = N$,

$$P \geqslant \frac{Nh\nu}{\tau_s} \geqslant \frac{ph\nu}{\tau_c} \tag{6.60}$$

With the values cited in the preceding paragraph, the minimum per cm^3 is 4×10^{16} eV/sec, or about 6 mW/cm^3. The power actually needed will always exceed, often by many orders of magnitudes, the limit given by Eq. (6.60), and this for several reasons. First, the efficiency of injecting or frequency-matching the required pumping power will be normally quite low. Appreciable losses will also occur, if radiationless energy transfer from the upper level is dominant or if spontaneous radiation escapes to a large extent through the walls. At any rate, the excitation frequency will have to be higher than that emitted. Therefore, although the inherently required lower limit of laser input power is quite modest, a number of loss mechanisms may make operation either impractical or very inefficient. It is possible to set forth some general rules and alternatives to reduce these losses, and as a result quality and variety of laser devices have steadily increased.

6.3.1.2 Alternatives of Operation

The inequality (6.59) or the simpler Eq. (6.60) represent a necessary and, in theory, sufficient condition of laser oscillation, demanding that the population of states be inverted (N positive) and that the stimulated emission power be larger than the losses. Despite the simplicity of the expressions, a large choice exists among the possible parameters. Some affect the free photon time τ_c which should be as large as possible. Here belong for instance the design of the mirrors and the length of the cavity. Other parameters essentially determine the number of competing modes which should be as small as possible, as for instance the width of the spectral line used for the stimulated transition. Largely, then, p is determined by the material, τ_c by the system. We shall discuss later some of the possible alternatives for small mode number and long photon lifetime. The methods of providing the needed excitation power, too, are multifold. Finally, important differences result from the term configuration of the resonant atom and, in particular, the number of levels sharing in the sequence of transitions. This question is most important, and the following alternatives exist (see Fig. 6.6):

(1) *The two-level maser*, which involves only the ground state and an excited level with inverted population (Fig. 6.6a). As described in Section 6.3.2.1, such a scheme was indeed the first embodiment in the maser field. An obvious difficulty of two-level *laser* schemes is the larger energy transition which requires excitation methods of a different sort such as optical pumping or collisions. Many of these will have equal

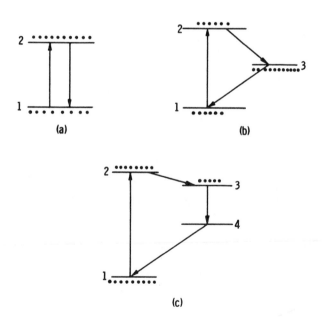

Fig. 6.6. Multilevel systems for maser techniques. (a) 2-Level system, (b) 3-level system, (c) 4-level system.

probabilities of excitation as of de-excitation and this at best will approach, but not reach, population inversion.

(2) Inclusion of an *intermediate level*, such that inversion by optical excitation is feasible either by creating, in the highest level, a larger population than in the intermediate state or, as shown in Fig. 6.6b, exhausting the ground state density in favor of the intermediate level. Typically in such operation, pumping light is used to raise populations from the ground state to level 3 whence spontaneous transitions occur, not—or mostly not—to the ground state, but to level 2 from which the system returns to the ground state by emission of the fluorescent line ν_{21} or by stimulated emission at that frequency. The transition from level 3 to level 2 is often radiationless, if the energy can be converted into lattice vibrations. An obvious advantage of the *three-level maser* is that the processes governing stimulated emission and fluorescence can be made more or less independent of those which determine the emptying of the ground state. This is the reason why it is feasible to store more atoms in level 2 than in the ground state.

(3) *The four-level maser* (see Fig. 6.6c) which appreciably reduces the pumping power needed in the preceding scheme. This advantage results from the fact that the lower level of the induced transition ν_{32} can be depopulated with respect to the ground state in a ratio which is approximately the Boltzmann factor $\exp(-E_2/kT)$. Therefore, if the energy E_2 of level 2 is sufficiently large compared to kT, the total number of inverted levels 3 needed for oscillation is given directly by N in Eq. (6.60).

In general, the equilibrium density n_j which a level j assumes in relation to that of lower levels i and higher levels k is given by the positive and negative absorption rates $B_{ij}u_{ij}$ and $B_{jk}u_{jk}$, the spontaneous emission probabilities from and to level j, and the "thermalizing" rates S_{jk} $(= S_{kj}\exp[-h\nu_{jk}/kT])$ per atom and unit time. The S_{jk}-values represent the probabilities for nonradiative transitions caused, for instance, by collisions between gas atoms or by lattice vibrations in crystals.* We can thus write a rate equation for each level j in the stationary case:

$$0 = \frac{dn_j}{dt} = \sum_i [(n_i - n_j)B_{ij}u_{ij} - A_{ji}n_j + S_{ij}(n_i - n_j e^{h\nu_{ij}/kT})]$$

$$+ \sum_k [(n_k - n_j)B_{jk}u_{jk} + A_{kj}n_k - S_{jk}(n_j - n_k e^{h\nu_{jk}/kT})] \qquad (6.61)$$

---

* According to the principle of detailed balancing, the thermalizing rates of transitions between two states for the general case of different statistical weights are related by $S_{jk} = (g_k/g_j)\exp(-h\nu_{jk}/kT) \cdot S_{kj}$.

subject to the condition

$$\left(\sum n_i + n_j + \sum n_k\right) = n_0 \tag{6.62}$$

where n_0 is the number of atoms involved in the active component. For a sequence of m levels, there are m simultaneous equations to be solved. Usually, several simplifications are applicable (see e.g. Heavens[35] or p. 34 in Lengyel[36]). For example, among the processes determining the term populations in the three-level ruby laser (see Section 6.3.2.2), it is possible to ignore all downward transitions from the highest level except the radiationless energy transfer to the intermediate state since that process is by far the fastest among those competing. For similar reasons, the rates determined by B_{23}, S_{13}, S_{23}, and S_{12} do not have to be considered. The rate equations (6.61) for the second and third level then become

$$0 = \frac{dn_3}{dt} = B_{13}u_{13}n_1 - (B_{13}u_{13} + S_{32})n_3 \tag{6.63}$$

$$0 = \frac{dn_2}{dt} = B_{12}u_{12}n_1 - (B_{12}u_{12} + A_{21})n_2 + S_{32}n_3 \tag{6.64}$$

and according to Eq. (6.62)

$$n_1 + n_2 + n_3 = n_0 \tag{6.65}$$

Now the relaxation rate S_{32} in ruby is ordinarily very large compared with the pumping speed $B_{13}u_{13}$. This leads to the conclusion that the stationary population n_3 is much smaller than either n_1 or n_2. Equations (6.63)–(6.65) yield then for the excess density

$$n_2 - n_1 = n_0 \cdot \frac{B_{13}u_{13} - A_{21}}{B_{13}u_{13} + A_{21} + 2B_{12}u_{12}} \tag{6.66}$$

The first condition for laser oscillation, viz., that of inversion between levels 1 and 2, therefore, is fulfilled, if $B_{13}u_{13} > A_{21}$. The second condition demands that, in addition, $(n_2 - n_1) > (\rho/V)(\tau_s/\tau_c)$, according to Eq. (6.56).

6.3.1.3 *Spectral Profile of Maser Oscillation*

The narrowing of the radiation spectrum produced by stimulated emission is intimately connected with the coherent power produced. This follows from the fact that the atoms of the medium, which in thermal equilibrium emit randomly, are in maser operation stimulated to radiate as a single oscillator. Furthermore, the coherence time increases with the number of individual systems participating, hence as the power involved. Instead of noise power P_r randomly emitted over a half-width $\Delta\nu_r$ (defined as before as the full width of the spectral profile at the half-power points), we now have a coherent maser power P_M contained in a much smaller interval $\Delta\nu_M$. It is plausible that these widths stand in inverse ratio to the powers they contain. For if the noise in a thermal source is dominated by the random distribution of spontaneous transitions, we should expect the approximate relationship $P_M \Delta\nu_M \approx P_r \Delta\nu_r$ or, with $P_r = h\nu/\tau_r \approx 2\pi h\nu\,\Delta\nu_r$.

$$\frac{\Delta\nu_M}{\Delta\nu_r} \approx \frac{P_r}{P_M} \approx \frac{2\pi h\nu\,\Delta\nu_r}{P_M} \tag{6.67}$$

An exact calculation was carried out for the case of the ammonia beam maser (see Section 6.3.2) which operates in the microwave region. Spontaneous emission in this case is negligible, and the noise behavior is dictated by thermal radiation power which, according to the Rayleigh-Jeans law, equals kT per unit frequency and per mode. The half-width $\Delta\nu_r$ will be much smaller than the cavity bandwidth $\Delta\nu_c$ in the microwave region (since, typically $\nu/\Delta\nu_r \approx 10^6$, whereas $Q_c \approx 10^4$). The spectral profile of the maser power $P_{M\nu}$ per unit frequency is then (according to Gordon et al.[37]):

$$P_{M\nu} = \frac{4kT\,\Delta\nu_r^2}{(\nu - \nu_0)^2 + \left(\dfrac{4\pi kT}{P_M}\,\Delta\nu_r^2\right)^2} \tag{6.68}$$

where ν_0 is the oscillation frequency at the maximum and P_M is the total maser power, i.e., $P_{M\nu}$ integrated over all frequencies. By comparison with Eq. (1.10) we find for the half-width of the maser line

$$\Delta\nu_M = \frac{8\pi kT}{P_M}(\Delta\nu_r)^2 \qquad \text{(for microwaves)} \tag{6.69}$$

For laser operation, at optical frequencies, spontaneous transitions rather than thermal radiation contribute to the noise spectrum, and $h\nu$ replaces kT as random power per mode per unit frequency. Assuming again that $\Delta\nu_r \ll \Delta\nu_c$, the laser oscillation half-width equals

$$\Delta\nu_M = \frac{8\pi h\nu}{P_M}(\Delta\nu_r)^2 \qquad \text{(for optical frequencies)} \tag{6.70}$$

For example, if $P_M = 1$ mW, $h\nu = 1$ eV $= 1.6 \cdot 10^{-19}$ watt-sec, and $\Delta\nu_r = 10^9$ sec^{-1}, the half-width for this type of laser operation amounts to 10^3 sec^{-1}, i.e., about a millionth of the atomic bandwidth.

In the optical region, the bandwidth of the cavity may be much smaller than that of the atomic transition, $\Delta\nu_r$. According to Eq. (6.58), the cavity Q is given by

$$Q_c = 2\pi\tau_c\nu = \frac{2\pi}{\alpha}\frac{L}{\lambda} \qquad (6.71)$$

which is essentially equal to Eq. (4.58), if the losses are confined to those of the reflection process, i.e., $\alpha = 1 - R$, and are small, i.e., $\alpha \ll 1$. Under the typical conditions of gas maser operation (see Section 6.3.2.3), $L \approx 10^2$ cm, $\alpha \approx 10^{-2}$, and $\lambda = 10^{-4}$ cm; hence, $Q_c \approx 10^9$, and $\Delta\nu_c \approx 10^6$. This means that $\Delta\nu_r$ in Eq. (6.70) has to be replaced by $\Delta\nu_c$, and that more generally (see e.g. Bennett[38]) an effective bandwidth has to be substituted in (6.70) instead of $\Delta\nu_r$:

$$\Delta\nu_M = \frac{8\pi h\nu}{P_M} \cdot \left(\frac{\Delta\nu_c\,\Delta\nu_r}{\Delta\nu_c + \Delta\nu_r}\right)^2 \qquad (6.72)$$

For the gas laser conditions, then, with $h\nu \approx 1$ eV, $P_M \approx 10^{-3}$ W, the maser frequency width $\Delta\nu_M$ will be as low as $\approx 10^{-3}$ sec^{-1}. Experimental determination by measuring the beat notes between two independently operating gas masers yielded a half-width of less than 2 sec^{-1}, amounting to a frequency spread in the maser transition of less than one part in 10^{14} (Javan *et al.*[39]).

6.3.1.4 *Beam Directionality*

An aspect which belongs to the general coherence performance of a laser is that of beam directionality. We have seen (Section 3.3.1.1) that the elementary process of stimulated emission generates a photon which has the same momentum as that which is incident. More precisely, the two photons occupy the same cell in phase space, and thus their position in phase space is identical to the extent of the volume $\hbar^3$. Actually, however, a much more severe limitation to directionality exists in the Rayleigh diffraction limit

$$\theta = \frac{1.22\lambda}{d} \qquad (6.73)$$

where d is the diameter of the aperture, assumed to be circular, which defines the beam as it leaves the cavity and θ is the angle by which the beam spreads from the axial direction. Thus for $d = 1$ cm and $\lambda = 1\mu$, θ is in the order of 10^{-4} rad. The observed angles of diffraction are considerably larger as a rule, partly because the beam may occupy only a fraction of the available cross section, and also because of multimode operation and the existence of additional scattering centers.

6.3.2 EXCITATION METHODS AND PRACTICAL MASERS

It is possible to classify the various maser schemes either according to the materials or according to the inversion processes used. To a certain extent, these two viewpoints amount to the same, that is, excitation methods must be specifically adapted to the class of substance in which the population has to be inverted. We shall discuss the subject as far as possible from this unified aspect.

6.3.2.1 *Molecular Beam Methods*

The first successful maser operation was achieved with beams of ammonia molecules on the basis of a two-level system (Gordon, Zeiger, and Townes[33,37]). Beam methods operate on the principle of the selective deflection which dipoles experience in nonuniform fields. Ammonia (NH_3) is a suitable molecule in this respect since it is capable of a transition between levels for which the force exerted by an electric field differs markedly. This transition corresponds to the "inversion" line in the NH_3 microwave spectrum and results from a flip of the nitrogen atom through the triangle formed by the three hydrogen atoms. The observed line, at 23,870 Mc/sec, has considerable structure and hyperfine splitting owing to rotations, spins, etc., of the atoms involved.

The scheme of the ammonia maser is shown in Fig. 6.7. The molecules first enter into a cylindrical quadrupolar field where the upper states

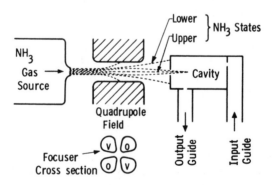

FIG. 6.7. Level separation of ammonia molecular beam in electric quadrupolar field [according to J. P. Gordon, H. J. Zeiger, and C. H. Townes, *Phys. Rev.* **95**, 282 (1954)].

of the inversion levels experience a radial force toward the axis, while molecules in the lower states are defocused away from the axis. Thus upper and lower states separate radially.

As a result, there are mainly only the upper states left in that axial portion of the beam which enters a cavity tuned to the frequency band of the transition. If now input power from a klystron is fed into the cavity and swept through the resonant frequency range, narrow lines (6 kcps) are observed in the output. One of the original motivations for this experiment was the aspect of low-noise amplification. This was, in fact, demonstrated by this method in that the noise from highly ordered, low temperature molecular states compares favorably with that produced by thermionic currents. A contributing factor is the low operational frequency due to which spontaneous emissions are virtually nonexistent, whereas they represent the dominant noise component in the optical region because of the ν^3-dependence of the spontaneous transition probability [cf. Eqs. (6.69) and (6.70)].

The ammonia maser represents a hybrid between microwave and optical methods insofar as it operates in the wavelength region and with the techniques of microwaves, but generates its coherent radiation in large numbers of atomic systems. However, beam masers are not easily suited for the optical region because they function by the separation of hyperfine structure components of the ground state. In other words, the excitation energies involved are in the order of kT. It is difficult to conceive of methods by which gas atoms are first excited to a more energetic higher state which is later isolated in a beam separation process. Hence, other pumping schemes had to be found for the optical region.

6.3.2.2 *Optical Excitation in Dielectric Crystals*

The concept of inverting populations by means of optical energy is due to Kastler[2] who applied this scheme to distributions over Zeeman fine structure levels. Either three-level or four-level lasers can be operated by optical excitation (see Fig. 6.6b and c). A number of general rules govern *a priori* the choice of suitable materials[40]: (1) the pump transition should accomodate a wide range of frequencies, i.e., the highest level should be quite broad, to assure a good match to the optical excitation source; (2) the fluorescent transition, however, should have a $\Delta\nu_r$ which is as narrow as possible since the necessary number of excited states, the needed power, and the maser line width decrease with $\Delta\nu_r$ according to Eqs. (6.56), (6.59), and (6.70), respectively; (3) the probability should be high that the energy of the highest level is released through the fluorescent transition rather than in bypassing it through other channels to the ground state; (4) the upper laser level, while radiative, should have a relatively long lifetime to facilitate inversion; and (5) the material should be free of scattering or absorbing centers which compete for or with the excitation or fluorescence energies. *Cet. par.*, four-level masers are, of course, preferable over the three-level schemes; but to be effective, the lower laser level must be located sufficiently high in energy above the ground state compared with kT, and its lifetime should be much smaller than that of the upper level.

It is of interest to explore which classes of materials match all these conditions. The first of these, demanding a strong broad absorption band for excitation, is not easily fulfilled for gases, at least those in the monatomic state. Liquids present a large problem because of condition (5): the thermal coefficient of the index of refraction and perhaps the mobility of microdomains in the liquid state set up scattering centers.[40a] There remain thus the solids. Most of the laser conditions for optical excitation favor the solid state since it provides strong broad absorption bands, large density of centers and mechanisms of lattice interactions which may be useful in the rapid depopulation of a state. Then again high purities are possible by means of established crystallization processes. Among the solids, however, only a small class is eligible under condition (2), insofar as narrow line spectra are limited to a few specialized situations. As will be seen from Section 3.6, the conditions favoring sharp fluorescent transitions under optical excitation are met in the spectra of the transition metals and the rare earths embedded in host lattices of such materials as Al_2O_3 (sapphire), SiO_2 (quartz), C (diamond), but also the alkaline earth fluorides and tungstates. Sharp lines are available for transitions between d- or f-electron configurations of these materials. As described in Section 4.6, such transitions are more or less shielded, and although they are forbidden in free ions, the selection rules are relaxed in the crystal fields of the host lattice. Materials then such as pink ruby (about 0.05% Cr_2O_3 in Al_2O_3) are well suited for optically pumped laser operation. These considerations, of course, do not exclude other substances. However, the demands on optical pumping intensity and the likelihood of successful operation become increasingly unfavorable, the less conditions (1)–(5) are fulfilled.

a. Ruby Laser. The first laser operation was achieved by Maiman[5,6] with optical excitation of ruby in the arrangement shown in Fig. 6.8. The heart of the apparatus consists of the ruby crystal in form of a small rod with optically flat, parallel, and silvered end faces—one side almost completely reflecting, the other partially transmitting, the laser beam.

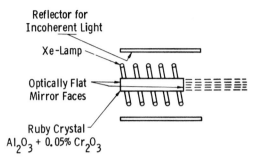

Reflector for
Incoherent Light

Xe-Lamp

Optically Flat
Mirror Faces

Ruby Crystal
Al_2O_3 + 0.05% Cr_2O_3

FIG. 6.8. Ruby Laser of T. H. Maiman [*Nature* 187, 493 (1960); *Brit. Commun. Electron.* 7, 674 (1960)]. Crystal may contain up to 0.5% Cr_2O_3, vary in length from 2 to 20 cm, and in diameter from millimeters to centimeters. Xenon flashlamp may receive pulses of as high as 10^5 J from various combinations of capacitance and voltage. Left side of ruby as shown here has maximum reflectivity and is opaque, right side is partially transmitting.

The crystal is surrounded by a helical flash tube or other source and its reflector. The lamp is designed to produce intensive pumping light in pulses, the electrical energy being obtained from a condenser of several hundred microfarads at 3–4 kV.

The operation of the ruby laser can be understood on the basis of the three-level scheme shown in Fig. 6.9. Three transitions are involved.

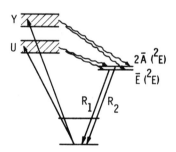

Y

U

$2\bar{A}$ (^2E)

$\bar{E}$ (^2E)

R_1 R_2

FIG. 6.9. Term scheme of ruby. Designations refer to point symmetry groups belonging to Cr^{3+} levels in Al_2O_3 lattice.

(1) Absorption centered at 5500 Å brings the Cr^{3+} ions to the broad level designated as U. Alternatively, an absorption band centered at 4000 Å allows excitation to Y.

(2) Lattice interactions rapidly depopulate U and Y by radiationless transfer to the narrow intermediate levels $\bar{E}$ and $2\bar{A}$.

(3)   Normally, the centers return from here to the ground state with
the fluorescent emission of two lines in the red region, viz., R_1 and R_2
(6943 and 6929 Å, respectively, at 20 °C). However, the relaxation time
between the two levels is very small. Thus when the excitation power
reaches the critical value for the start of laser oscillations in the transition
with the larger oscillator strength—which is R_1—all stimulated emission
occurs at the frequency of that line (although laser operation at R_2 is
possible[41] by using interference filters selectively reflecting at 6929 Å).
    The onset of laser activity is characterized by a number of abrupt
changes[42] in the light output of the crystal: (1) most noticeable, a sharp
narrowing of the originally diffuse fluorescent glow to an angle of a
fraction of a degree (more or less, depending, e.g. on crystal perfection);
(2) narrowing of the spectral line width (e.g. for the R_1 line from 6 cm^{-1}
to 0.2 cm^{-1} or less); (3) a fast emptying of the intermediate levels[43]
(which have ordinarily a lifetime of several milliseconds), the decay
occurring in a type of relaxation oscillation; (4) mutual coherence of
radiation coming from various points of the front face (as demonstrated
by interference of beams masked out at variable distances from each
other); and (5) a sharp increase in the ratio of axially to radially emitted
light.
    These observations are in excellent agreement with the expected
behavior of maser operation. In particular, the relatively high threshold
power and low yield—several hundred joules of input deliver only
10^{-1} J of coherent light output—can be understood as a consequence
of the three-level operation and other sources of inefficiency in the ruby
system. Of course, noninherent losses may be reduced drastically.

b. Four-Level Systems.   For the reasons given before, the narrow band
transitions between the 4f- and 5f-electron configurations of the lanthanides and actinides, respectively, are especially suitable for laser
operation. In addition, these materials lend themselves to four-level
systems in which the lower laser level may be a component of the ground
term splitting, yet high enough above the ground state so that its
population may be quickly emptied. Hence, the density of the upper
level in the laser transition is directly the excess population for which
Eqs. (6.56) and (6.60) have been solved. As a result, pulse input energies
in the order of joules or less have proved sufficient to start laser operation in these materials, and efficiencies in excess of 2% have been
reported.*

---

* Laser efficiencies are defined here as the ratio of energy in the emergent laser beam
to the electrical energy necessary to produce it. Over-all efficiencies of 2 to 5% may
imply unit quantum efficiency per photon absorbed by the ion.

The first-four-level laser operation was reported, a few months after Maiman's discovery, by Sorokin and Stevenson,[44] who observed stimulated emission at about 2.5 μ in a CaF_2 lattice with trivalent uranium ions substituted for calcium. The term scheme and pertinent transitions of $CaF_2:U^{3+}$ are shown in Fig. 6.10. The gap between the lower laser level and the ground state amounts to 609 cm^{-1} and thus is $\approx 3kT$ at room temperature. For this reason, and also because the metastable state at 4436 cm^{-1} decays by competitive passage at higher temperature, cooling with liquid nitrogen or helium is necessary. This measure

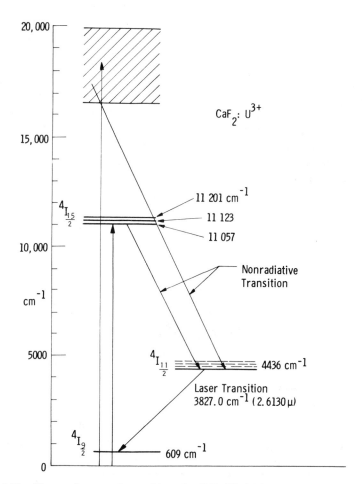

FIG. 6.10. Term scheme and transitions in $CaF_2:U^{3+}$. Maser transition occurs at 2.613 μ [G. D. Boyd, R. J. Collins, S. P. S. Porto, A. Yariv, and W. A. Hargreaves, *Phys. Rev. Letters* **8**, 269 (1962)].

reduces the oscillation threshold by orders of magnitude. Boyd and his co-workers[45] find thresholds of 2.0 J at 20 °K, 3.78 J at 77 °K, 4.35 J at 90 °K, and 1200 J at 300 °K, the energies referring to the input values of a xenon flash lamp. It was possible in this investigation to achieve also continuous operation of the uranium laser with an arrangement of which the important components are sketched in Fig. 6.11. The necessary excitation flux was provided by a mercury arc lamp in the focal axis of a reflecting elliptical cylinder with the laser crystal in the complementary focal position.

Other rare earth ions have been used successfully in various lattices, notably Nd^{3+}, Ho^{3+}, Sm^{2+}, Tm^{3+}, and Dy^{2+} in suitable crystals such as $CaWO_4$, BaF_2, CaF_2, SrF_2, and others. Among these, the neodymium trivalent ion is of special importance because it has a ground term splitting in the order of several thousand cm^{-1} (i.e. $E_2 \gg kT$), hence tolerates operation at room temperature. In combination with calcium tungstate as the host lattice, it offers a particularly low threshold[46-48], namely typically 1–2 J at room temperature. Of further interest is the fact that Nd^{3+} also functions as a laser material when it is embedded in various glasses.[49] Stimulated emission of the important Nd^{3+} transition occurs at 1.06 μ and can be maintained continuously at room temperature in glass as well as in calcium tungstate and, with high efficiency, also in certain garnets,[49a] e.g., $Y_3Al_5O_{12}$. Other rare earth ions have sub-

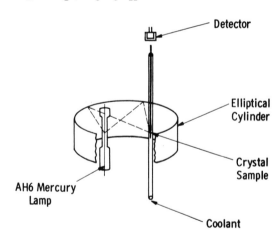

FIG. 6.11. Experimental arrangement for continuous operation of $CaF_2:U^{3+}$ laser of Boyd et al. [G. D. Boyd, R. J. Collins, S. P. S. Porto, A. Yariv, and W. A. Hargreaves, *Phys. Rev. Letters* **8**, 269 (1962)]. Pumping source and crystal are aligned with the two focal axes of water cooled elliptic mirror. Crystal is kept at 77°K by liquid oxygen cooled to nitrogen temperatures so as to avoid gas bubbles. Laser beam from sample is transmitted axially towards PbS detector.

sequently been used in combination with various glasses, although pulsed operation and cooling have been found necessary in general. A summary of dielectric crystal laser materials is found in Table 6.2 of Section 6.3.3.3.

6.3.2.3 *Collision Excitations and Gas Lasers*

Gases offer a number of important advantages as laser materials. They exhibit spectral lines of narrow width so that the demands on inversion and excitation power are low; they fill homogeneously a prescribed cavity volume which preferably has the form of a long cylinder, and thus high cavity Q-values can be combined with single-mode operation; as a result, a spectral purity and beam directionality are obtained unmatched by any other optical system; and they offer a large number of elementary interaction processes and of possible compositions so that a correspondingly large number of laser transitions are available in the infrared, visible, and ultraviolet spectrum.

Gas lasers, however, are inherently low-power devices, and the reason lies partly in the low density of the radiating centers, partly in the nature ot ᴧe excitation mechanisms. For although there exists a considerable variety of possible interactions, really effective inversion methods have been limited to processes of collision. Unlike solid state lasers, gases are not too easily adapted to optical pumping, the fact notwithstanding that the first proposals for practical lasers envisioned alkali vapors excited by light. One of the early operating gaseous lasers was indeed optically pumped cesium.[50,51] In this case, a strong spectral line—the helium transition 3^3P-2^3S at 3888 Å—was found to qualify for excitation since it nearly coincided with a suitable resonance line of cesium ($6S_{1/2}-8P_{1/2}$). The resulting oscillation at 7.18 μ yielded about 25 μW. Even this performance, however, must be considered exceptional since the coincidence of a strong emission line with a *suitable* resonance line for absorption is quite rare in view of the narrow band-widths of gaseous spectra. To be suitable, a sequence of transitions must obey selection rules, which—contrasting with the case of solid state spectra—cannot be relaxed. For these and other reasons, optical excitation has seen only limited use for gas lasers.

An entirely different situation exists for inversion processes by collision which match the requirements of the gaseous state, whereas they, in turn, are in general not suitable for solids. In gases, various particles have the mobility to excite higher levels by collisions of the first and second kind. Furthermore, as already known from critical potential measurements of electron bombardment (Franck-Hertz experiment), the threshold for excitation of an atomic state is sharp, but the response to excess energy is broad. Thus collisions can be effective over a wide range of kinetic and potential energies. Various kinds of energy exchange processes between particles exist. Examples for three important classes are shown schematically in Fig. 6.12a, b, and c. We shall discuss these mechanisms in the following paragraphs at some detail following the chronological order in which they were developed.

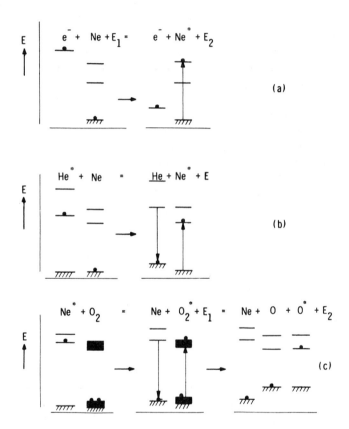

FIG. 6.12. Gas laser excitation by collisions. Representative examples for three important excitation mechanisms are shown: (a) electron impact on rare gas atoms (collisions of the first kind); (b) inelastic collision between metastable and ground state atoms of different species (collisions of the second kind), exemplified by the helium-neon laser; and (c) inelastic collision between metastable atom and molecule with subsequent dissociation into excited and normal atomic states. In all three cases energy is transferred in broad resonance processes, the excess being distributed among the participating particles as kinetic energy E according to their mass. The excited states (indicated by star) may build up to inverted populations.

a. The Helium-Neon Laser. Fluorescence by resonance transfer has long been an active technique for a combination process by which (1) energy is first excited and stored in metastables such as can exist in the rare gases; (2) transferred to other atoms (in a mechanism by which the metastable atom returns to the ground state after exciting the atom of the other species in a collision of the second kind to a level within a few kT); and (3) emitted as the fluorescence line of the second gas. This process has the virtue of efficiency since mainly the desired levels are excited with the conversion of only a small excess into kinetic energy (Fig. 6.12b). Resonance transfer processes thus present an attractive possibility as a population inversion technique.

On this principle is based the helium-neon laser of Javan, Bennett, and Herriott[52] which was the first to operate continuously and the first to utilize a gaseous medium. An electrodeless glow discharge is maintained by means of a radio frequency generator (≈ 30 Mc/sec) in a mixture of about 1 mm Hg helium and 0.1 mm Hg neon. The electrons produced and accelerated in the electric field will excite primarily the metastable 2^3S level of helium under certain conditions (see Fig. 4.18). The energy thus stored can be released by inelastic collisions. and in the presence of neon this process occurs to a major extent by exciting atoms of that gas to the configuration $1s^22s^22p^54s$ which is in resonance with the metastable helium level within ≈ 0.1 eV (see Fig. 6.13). The configuration can give rise to 4 P-levels, viz. $^1P_1$ and $^3P_{0,1,2}$. Because the electronic orbits in neon do not too well interact by L-S coupling, but rather by a type of j-l coupling (Racah[53]), the levels are still described by the Paschen terminology (somewhat ambiguously) as $2s_2$, $2s_3$, $2s_4$, and $2s_5$, respectively.* All 4 levels are eligible to be energized by resonance transfer from the helium 2 ^{3}S term. Here it is noteworthy that the $s_2(^1P_1)$ level is particularly likely to be excited even though it involves a change of spin by unity.[38] However, the s_2 level is closest to the helium state (within kT), and the rules of LS coupling are not strictly maintained in neon. By contrast, the next lower term in neon, consisting of 10 levels of the configuration $1s^22s^22p^53p$ (designated as $2p_1$–$2p_{10}$), is too distant from the helium level to be excited by it. Thus the result of the resonance transfer is an inversion of the 2s and 2p densities unless the latter are

---

* In the Paschen notation, levels are indicated by the orbital state of the excited electron and numbered by subscript according to energy position within the term (e.g. p_1, ..., p_{10}). The first excited states in the noble gases are called 1s levels, although they have been termed also 2s elsewhere in the literature. By contrast, the Racah notation[53] uses the designation $nl[K]_J$, where nl is the orbital state of the excited electron, K the resultant of l and the total core momentum, and J the total momentum obtained from coupling K and the spin of the excited electron.

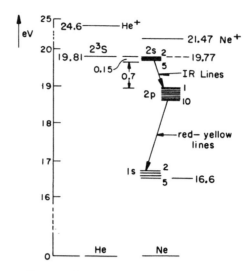

FIG. 6.13.   Energy diagram of the He-Ne mixture. Resonance transfer by collisions of the second kind between the 2 ¹S He and the 3s Ne and between the 2 ³S He and 2s Ne levels, respectively, are indicated together with the subsequent laser transitions [from A. Javan, W. R. Bennett, Jr., and D. R. Herriott, *Phys. Rev. Letters* **6**, 106 (1961) and W. R. Bennett, Jr., *Appl. Optics* Suppl., p. 24 (1962)].

enriched by some detrimental mechanism. Yet, such interference with the pumping process is avoidable for the following reasons: (1) the lifetime of the 2s levels is in the order of 10^{-7} sec (determined almost exclusively by radiative decay to the 2p term), and consequently about 10 times as large as that of the 2p levels which live about 10^{-8} sec (limited by radiation into the 1s term); (2) although electron bombardment will excite also lower neon levels and thereby enrich the 2p term population, the latter will remain smaller than that of the 2s term for a certain practical range of electron densities. Under such conditions, then, continuous laser operation was observed for the various 2s–2p transitions, which lie in the near infrared, the strongest effect occurring for the $2s_2$–$2p_4$ line at 1.1523 μ.

Enrichment of the 2p term population by electron collisions, which tends to upset inversion, can be directly demonstrated by the laser oscillation itself. When the radio-frequency field is interrupted, a sequence of *afterglow* phenomena begins its regime. First the steady-state interactions cease after a time of 10 to 20 μsec during which the electrons abandon their role as the primary agent of the discharge processes and through many inelastic collisions assume thermal velocities. As a result, the energy stored in the He 2 ³S level begins to be depleted by transfer to the Ne 2s term and from there by radiation to the 2p levels, and since the lifetimes of the three levels decrease in that order, the three terms can be observed to decay in unison. Nevertheless, during this process the maser oscillation actually *increases* in output power since 2p levels are no longer generated by electrons and population inversion is heightened. Conversely, beyond certain limits of input power, laser oscillation will cease altogether because inversion is foiled by the density increase of the 2p level. Such effects must be expected for all lasers of this kind, causing one of the power limitations of the gaseous medium.

Several other factors determine the balance of level densities and their rate of change. Among these belong such effects as resonance imprisonment (see Section 3.3.4.1) which increases the density of levels commuting directly with the ground state, diffusion of metastables to the walls, and collisions between metastables. Of course, before details of the helium-neon and other gas lasers were proposed, a thorough analysis of these factors had to precede, based on line strength measurements as those mentioned in Section 3.3.4, on the behavior of the metastables (Phelps[54,55]), on the processes of resonance transfer, and on the measurement of other parameters pertaining to glow discharges.

The helium-neon laser can be used for transitions other than the 2s–2p lines, and in the past various frequencies were generated using substantially the original experimental arrangement of Javan and co-workers[52] (see Fig. 6.14). This first embodiment consisted of an interferometer with 100 cm spacing of the mirrors between which the helium-neon discharge tube was positioned coaxially. Great care in mode restriction was exercised by ensuring a high degree of flatness for the end plates and a reflectance of 98.9% with 0.3% transmittance of the coating which consisted of 13 evaporated dielectric layers. As a result, already the first such device exhibited a high degree of self-coherence, a line width of 10^4 cps (measured by photoelectric mixing and described

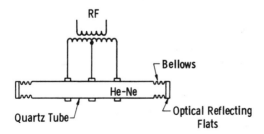

Fig. 6.14. Gaseous laser arrangement of A. Javan, W. R. Bennett, Jr., and D. R. Herriott, *Phys. Rev. Letters* **6**, 106 (1961).

in Section 6.2.5.3), and a beam divergence of less than half a minute of an arc. Since then, of course, the performance of gaseous lasers has improved considerably as indicated by a measured line width of less than 2 cps (cf. Section 6.3.1.3).

Among the other important transitions which could be brought to stimulated emission in the helium-neon system, we mention the visible line $3s_2 - 2p_4$ at 6328 Å. This type of oscillation was obtained by White and Rigden[56] with a 10:1 He-Ne mixture energized by direct current from a thermionic cathode (1700 V, 25-50 mA). The peak reflectance of the mirrors was adjusted to 6350 Å. Here the upper laser level receives its excitation from the metastable $2\ ^1S$ level of helium which is located somewhat above the triplet state $2\ ^3S$ in a position suitable for resonance with the 3s level of neon (see Fig. 6.13). However, competitive transitions are also possible from the 3s to the 3p states, and laser operation was observed for the $3s_2$-$3p_4$ line at 3.39 μ. Thus there exists a considerable number of combinations qualifying for stimulated emission, provided the optical cavity and the dielectric film reflectance is tuned to them. Such optimization is possible simultaneously for more than one frequency, and Ridgen and White[57] were able to produce simultaneous oscillation with the $2s_2$-$2p_4$ and the $3s_2$-$2p_4$ lines. Certain variations and refinements of the original helium-neon laser arrangement had to be introduced to achieve these alternative processes, such as confocal, instead of plane, mirrors and windows inclined at the Brewster angle characteristic for the index of refraction (see Section 6.3.3.2).

b. Dissociative Excitation Transfer. A method which appears to offer a large variety of level combinations involves the dissociation of molecules. Certain excited states of diatomic molecules are unstable in that a repulsive force is set up between the two atoms so that they fly apart with one or both forming an excited state. Obviously population inversions can be achieved in such states with respect to other levels of the same atom, especially since most atoms in the ground state will return to the molecular bond.

Such a scheme was successfully developed by Bennet et al.[58] in a neon (or argon) discharge containing oxygen as an impurity. The basic pumping process is here, as in the case discussed in the preceding section, impact excitation to a metastable state of the inert gas and subsequent transfer of this energy by collisions of the second kind. However, the essential feature is the energy transfer to a molecule such that the latter dissociates into atoms in various states of excitation (see Fig. 6.15). One advantage over the resonance transfer between states that must coincide within a few kT (as the $2\ ^3S$ He and 2s Ne states) accrues to the

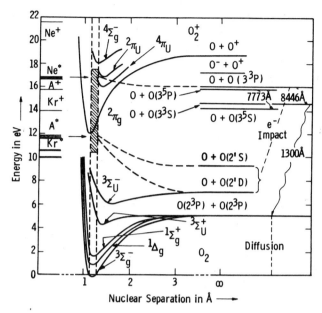

Nuclear Separation in Å ⟶

Fig. 6.15. Energy levels involved in transfer of excitation from rare-gas atoms to oxygen molecules with subsequent dissociation of the latter [W. R. Bennett, Jr., W. L. Faust, R. A. McFarlane, and C. K. N. Patel, *Phys. Rev. Letters* **8**, 470 (1962)].

dissociation process in that it is much less sensitive to a discrepancy of the energy levels which may be 1–2 V apart. Although the energy response of dissociation into repulsive states is broad, the excited atoms funnel into narrow energy levels after separation.

Laser operation was first observed for the transition between the 3 ^{3}P and 3 ^{3}S state of atomic oxygen at 8446 Å as shown in Fig. 6.15. The 3 ^{3}P state was produced primarily by the reaction

$$\text{Ne}(^3P_1,\,^3P_0) + O_2 \rightarrow \text{O}(3\,^3P) + O + \text{Ne}$$

as was established by decay time measurement of the 8446 Å line.

c. Excitation by Electrons in Pure Gases. The inherently simplest method of population inversion in gases is that of excitation by electron collisions (cf. Fig. 6.12a). The first observations of laser action in pure gases based on this process were made by Bennett[59] and co-workers (see also review article by Bennett[38]). The probability of exciting an atom by electrons from a level 1 to a level 2 is for higher energies proportional to the transition matrix between the two states; in other words, it behaves as the probabilities of the optical transitions, although there are important exceptions.* Nevertheless, resonance is much broader: an excess of many volts over the excitation threshold does not interfere with the energy transfer process unless other levels begin to compete. Because of the analogy to optical transitions however, the following reactions are most suitable in the noble gases[38]:

$$e + (np)^6 \rightarrow (np^5)ms + e$$
$$e + (np)^6 \rightarrow (np^5)md + e$$

where n corresponds to the quantum number of the ground state configuration and m can assume the values

$$m = n + 2, \quad n + 3, \tag{6.74}$$

Laser transitions then are possible from the ms or md levels to the lower lying $(np)^5 m'p$ configurations ($m' < m$). Since optical transitions to the ground state can reverse the pumping process, it is necessary to use a pressure which is high enough ($>10^{-2}$ mm Hg) for resonance imprisonment and which thus prolongs, in effect, the radiative lifetime. However, a compromise must be reached with a pressure dependent destructive effect, viz., that of undesired collisions between excited atoms.

Because of the simplicity of the electron impact process and the broad resonance with which it functions, it has been applied to all noble gases. Most of the laser lines in question lie in the region between 1 and 10 μ for reasons connected in general with the position of the long-lived terms and the frequency dependence of the transition probabilities. However, green and blue line oscillations could be obtained[63] with singly ionized argon which was excited by high power pulses in the voltage range up to 20 kV. Laser action from ionized atoms was also observed[64] by using 10 kV pulses in mercury vapor (mixed with helium as a buffer). Even ultraviolet laser activity has been reported[65] for 20 strong simultaneous lines between 3000 and 4000 Å in nitrogen. In this case, pulses of 100 to 150 kV of less than a microsecond duration were imposed on the nitrogen discharge tube, and light beams confined to 1 millirad and 10 W peak power were obtained as a result.

6.3.2.4 Inversion of Carrier Populations in Semiconductor Lasers

It has been shown, e.g., by Eq. (6.53), that the threshold requirements for laser oscillations increase in proportion to the bandwidth of the

---

* The probability of exciting a state by electron collision is given by the Born approximation (see e.g. Massey and Burhop[60]) which at high energies in first order is proportional to the transition probability.[38] However, the Born approximation was found not to apply to certain levels such as the triplet series of helium (Schulz,[61] St. John et al.[62]).

normal fluorescent process. The resulting need for narrow linewidths is fulfilled relatively easily in the spectra of gases and in the transitions between d- and f-configurations of the transition metal or rare-earth ions, respectively, embedded in dielectric solids. Gases and impurity dielectrics were therefore the first substances in which laser operation could be demonstrated. For practical reasons, it is of course desirable to widen the list of eligible materials, and in particular, to semiconductors since these have functioned in the past as electronic circuit elements with high economy in terms of space and power demands and in terms of their operational simplicity. Semiconductors are characterized by energy bands, i.e. by regions of continuous density of energy states. However, as we have seen in Section 4.6.5, there exist narrow levels and narrow bandwidth transitions under certain circumstances even in these materials, although the spectra are considerably coarser than those of the isolated atom or ion configurations. Consequently, the possibility of maser action in semiconductors was explored rather early, and at the end of 1962, several groups succeeded almost simultaneously in producing an infrared laser beam from gallium-arsenide p-n junctions.[66,67]

a. Excitation by Injection. The physical situation existing in the process of injection luminescence of forward-biased p-n junctions (cf. Section 4.6.5.1) fulfills, in fact, *a priori* the major part of the conditions necessary for maser oscillations. First of all, injected minority carriers represent an inverted population with respect to certain energy states. Thus electrons in the conduction band which have drifted across the junction into a p-type region face either empty acceptor levels or holes at the top of the valence band so that they are in a state of population inversion. Furthermore, the efficiency of radiative recombination can be between 10 and 100% corresponding to a favorable line strength. Finally, although bands rather than sharp energy levels are involved, the fluorescent lines in injection luminescence may be tolerably narrow. The reason for this is the fact that the carriers involved are all near the band edge within kT and that transitions may occur without change of the momentum vector. At 77 °K, $kT \approx 0.01$ eV so that for a gap E_g of 1 eV (1.24 μ), $E_g/\Delta E \approx 100$. Actually, transitions take place mainly between the bottom of the conduction band and acceptor levels or between the top of the valence band and the donor levels or between donor and acceptor levels.[68] As a consequence, the emitted frequency is somewhat smaller than that corresponding to the band gap, and that improves the prospects for laser operation because of the reduced probability of reabsorption of the emitted photon.

b. Diodes of GaAs and Other III-V *compounds.* All the conditions cited
in the preceding are particularly well fulfilled in the case of gallium
arsenide, hence its common choice by the various groups [66,67,69] for the
first semiconductor lasers. There is also little difference among these first
experiments in the structural details of the resonant cavity (see Fig.
6.16). A GaAs crystal, doped strongly n-type (e.g., by 10^{18} donors/cm³),

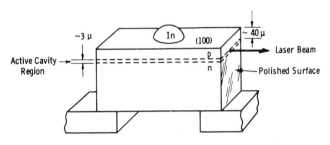

FIG. 6.16.   Junction region and configuration in typical gallium-arsenide diode laser.

receives in the order of 10^{19} acceptor atoms/cm³ (e.g., zinc) by diffusion
through a polished, well-defined plane, e.g., the (100) plane shown in
Fig. 6.16. This operation creates a junction region, a few microns thick,
about 50 μ below the surface. The crystal subsequently is shaped into
the form of a cube or parallelepiped so dimensioned that the junction area
amounts to about 10^{-3} to 10^{-4} cm². At least two opposing sides normal
to the junction area are polished plane-parallel to define the resonant
modes. The resulting structure is operated as a semiconductor diode,
a few volts being applied in pulses through ohmic contacts to the
p- and n-region.

The onset of laser oscillations is characterized by typical symptoms.
The light intensity emitted axially through the optical cavity increases
linearly with current density as long as the latter does not exceed the
order of 10^3 to 10^4 A/cm², depending on the temperature (see Fig. 6.17).
For higher current densities, however, the beam brightness increases
sharply by up to two orders of magnitude and may then continue again
linearly with current. This second linear domain must be interpreted as
a saturation effect, in that all eligible carriers created during a pulse
recombine by stimulated emission. Since the quantum efficiency of the
spontaneous luminescence is already very high, most of the brightness
gain in the axial beam is simply that geometric factor of directionality
by which the angle of random directions is reduced to that of the laser
beam. Most indicative of laser action, furthermore, is the narrowing of
the spectral width from about 10^2 Å to 1/10 or less of that value as

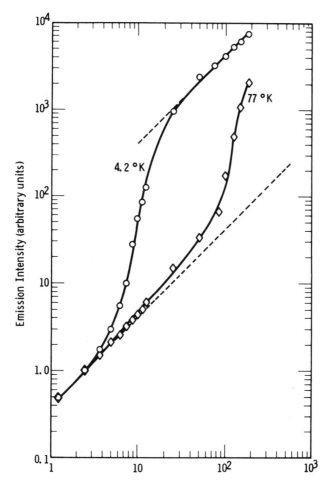

Fig. 6.17. Beam intensity vs. current in gallium arsenide laser. Transition from random emission to coherent radiation is shown for two temperatures [T. M. Quist, R. H. Rediker, R. J. Keyes, W. E. Krag, B. Lax, A. L. McWhorter, and H. J. Zeiger, *Appl. Phys. Letters* 1, 91 (1962)].

schematically indicated in Fig. 6.18. The GaAs oscillation line lies near 8400 Å, corresponding to 1.47 eV or 0.04 eV less than the energy gap of 1.51 eV from the top of the valence band to the bottom of the conduction band. This implies a transition from donor to acceptor levels,[68] and from other indications it follows that electrons are the carriers.

Direct transition diode lasers could be made to operate with other materials and in other wavelength regions. First, the addition of gallium

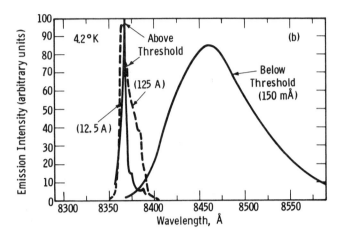

FIG. 6.18.   Narrowing of gallium arsenide emission spectrum after threshold current is reached. Grating spectrometer has resolution of 4 Å [T. M. Quist, R. H. Rediker, R. J. Keyes, W. E. Krag, B. Lax, A. L. McWhorter, and H. J. Zeiger, *Appl. Phys. Letters* 1, 91 (1962)].

phosphide to gallium arsenide in a structure written as $Ga(As_{1-x}P_x)$ increases the band gap, and consequently oscillations were obtainable in the visible region.[70] Operation at 0.91 μ was attained with InP[71]; at 1.77 and 2.07 μ with $As(Ga_{1-x}In_x)$[72]; at 3.1 μ with InAs[73]; and at 5.2 μ with InSb.[74] Then, too, materials for diode lasers are not limited to the III-V compounds but include other semiconductors as well (see Table 6.3 in Section 6.3.3.3).

c. Specific Properties.   Neither the narrowing of the beam nor that of the spectral width is as spectacular as that of gas lasers or even that of ion impurity type lasers, presumably, because of multimode operation which, in turn, is a result of the relatively large fluorescent line width $\Delta \nu_r$. The great advantages of semiconductor lasers lie, however, in the ease of population inversion—requiring as it were the application of a small pulsed voltage,—the high efficiency, and the remarkably small volume of the device itself.

Of at least equal importance is the high speed of response with which the diode laser light is capable of following the impressed voltage. The lifetime τ of spontaneous recombination is about $2 \cdot 10^{-9}$ sec in GaAs.[75,76,77] The onset of laser oscillations after the abrupt increase of a diode current from 0 to I is delayed by a time t_d which is needed to establish the inverted population in the presence of the decay time τ.

Konnerth and Lanza[75] find experimental confirmation for the relationship

$$t_d = \tau \ln \left(\frac{I}{I - I_{th}} \right)$$ (6.75)

where I_{th} is the current threshold for laser oscillation. Accordingly, very small delay times are possible, and values below the resolution limit of the measuring oscilloscope ($2 \cdot 10^{-10}$ sec) were observed. Thus by proper biasing of the diode, modulation frequencies in the order of 10^9 cps appear to be attainable. This fast response compares very favorably with that of the other laser types. The oscillation response time in gases, for instance, is determined by the slowest of a series of processes which may be the electron thermalization time of 10^{-5} sec. In the case of transition metal or rare earth ions in solids, the time constant of the pumping light or that of radiationless transitions is likely to be limited to 10^{-7}–10^{-6} sec.

Another feature of semiconductor lasers is that their resonant frequency can be affected to a certain extent by a variation of several physical parameters. For instance, an increase of diode current generally shifts the frequency of the laser oscillations toward higher values. The explanation of this effect lies in "band filling": as the excitation of the diode increases, the zone of occupied states near the band edges widens, and thus the transition energy becomes larger. A slight effect on frequency is also noticed as a result of pressure[78] and of magnetic fields,[74] the latter as a result of Landau-level excitation which will be discussed at the end of this section.

To summarize, the small size of the injection diode laser, its efficiency and operational simplicity, high speed of response, and the possibility of adjusting its resonant frequency by external means—these are properties, in combination with those of coherence, which can be expected to become of significant potential usefulness for the technology of communication.

d. Cyclotron Resonance Masers. A certain magneto-optical effect is of interest here, although its importance for maser operation has been so far less than for detector applications to be discussed later. In the presence of a magnetic field H, the continua of the valence and conduction bands are split into a sequence of discrete levels (see Fig. 6.19). These levels result from circular orbits which holes in the valence band and electrons in the conduction band describe in planes normal to the magnetic field direction. The orbital, so-called *cyclotron frequency* is determined classically by the equilibrium condition between

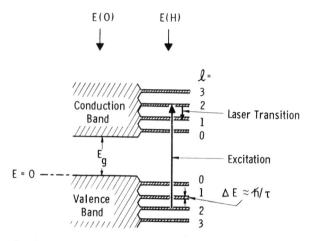

FIG. 6.19. Landau levels and cyclotron resonance maser scheme. Energy bands for holes and electrons shown on the left are split into discrete levels in the presence of a magnetic field H as shown on the right. Optical or injection excitation may produce population inversion with subsequent maser action, if level spacings are not degenerate.

the centrifugal force and the magnetic Lorentz force on the charge, viz.,

$$m^*\omega_c^2 r = \frac{e}{c} H\omega_c r \tag{6.76}$$

where m^* is the effective mass either of the hole or of the electron. It is seen that the orbital frequency ν_c is independent of the radius and thus of the total energy:

$$\nu_c = \omega_c/2\pi = \frac{eH}{2\pi m^*c} \tag{6.77}$$

The energy associated with this motion is quantized as $h\nu_c$, and the continuum is split into a series of levels separated in first approximation by $h\nu_c$ (*Landau*[79] *levels*). If the field is in the z-direction, the band energy given by Eq. (4.111) now is changed to

$$E_l = \frac{h^2}{8\pi^2 m^*} k_z^2 + (l + \tfrac{1}{2})\frac{eh}{2\pi m^*c} \cdot H \tag{6.78}$$

as shown in greater detail in energy band theory (see e.g. Callaway[80]), Setting the energy at the top of the valence band as zero, we have for the energy in the conduction and valence bands, respectively,

$$E_c = E_l + E_g \tag{6.79}$$

$$E_v = -E_l \tag{6.80}$$

The existence of these levels can be demonstrated by absorption of the energy hv_c. In fact, the mechanism of transitions between Landau levels has been used successfully for the detection of infrared radiation within narrow and tunable wavelength intervals. As will be described in Chapter 7, this was possible with high signal-to-noise ratios so that a large transition matrix element can be inferred. It is therefore theoretically feasible to obtain semiconductor maser operation by induced emission between Landau levels. Such a *cyclotron resonance maser* would be magnetically tunable between the microwave and infrared region. For example, indium antimonide ($m^* = 0.0145\, m$.) requires $1.57 \cdot 10^3$ G for resonance at 1 mm and $1.57 \cdot 10^5$ G for $10\,\mu$, in other words, a magnetic field range which is, by order of magnitude, accessible with the aid of superconducting magnets. A lower limit for the linewidth below oscillation threshold is given by the orbit life or relaxation time τ of the electrons, viz.,

$$\Delta E \approx \frac{\hbar}{\tau} \approx \hbar \frac{v}{\Lambda} \qquad (6.81)$$

where v is the velocity of the carriers and Λ their mean free path. The latter should be as large as possible, and since it is determined by collisions with impurities and phonons, the sample should be very pure and operated at low temperatures. Typical line widths are 0.001 eV, corresponding to a relaxation time of 10^{-12} sec.

In anticipation of a program for the development of a cyclotron resonance maser, Lax[81,82] analyzed its requirements and proposed details of its design. Optical pumping across the gap is feasible under the selection rules $\Delta k_z = 0$ and $\Delta l = 0$. In fact, excitation by laser light pulses such as from ruby is particularly suitable because of beam directionality, intensity, and spectral purity. Alternatively, pumping by injection is a possibility. However, a difficulty arises, if the Landau levels are exactly equidistant, as indeed indicated by Eq. (6.79), because although some levels l, $l - 1$, $l - 2$ may exhibit population inversion, others such as $l + 1$, $l + 2$ will be empty, and thus the net effect is absorption rather than maser oscillations. However, in many semiconductors, the energy surfaces are such as to prevent equidistant magnetic levels. Consequently, the problem is not insoluble.

In vacuo, a cyclotron maser has been demonstrated by Hirshfield and Wachtel[83] with an electron beam in a magnetic field. Relativistic effects at 5 kV change velocities and hence spacings of the levels sufficiently so as to remove the degeneracy mentioned before. Oscillation powers in excess of 10 mW at $v = 5.800 \times 10^9$ cps could be demonstrated in a cavity.

6.3.2.5 *Excitation Transfer in Metal-Organic Lasers*

To what extent the maser principle is applicable, is perhaps best illustrated for the case of organic compounds. We cite here as a successful example the rare-earth chelate laser which is of interest for the two reasons that it is based on a high-viscosity liquid and that the latter is a metal-organic compound (Lempicki and Samelson[84,85]). Chelates consist of organic molecules which are attached to metal ions by multiple bonds. The rare-earth chelate under discussion in europium tris-benzoylacetonate. The organic part of this molecule is known to absorb ultraviolet radiation with a broad resonance (600 Å) centered at 3200 Å, the excitation leading to a singlet which in turn decays into a triplet level. In benzoylacetonate, recombination of the triplet with the ground state occurs with slow phosphorescence. However, if the molecule is bonded to europium, the stored energy is transferred to the 4f-configuration of the europium ion. The resulting excitation of the Eu^{3+} ion reveals itself by strong fluorescence of visible lines between 6125 and 6140 Å. The role of the organic compound is therefore not unlike that of the rare gas metastables which excite another gas component by collisions of the second kind. To achieve laser operation it is convenient to dissolve the chelate in alcohol (such as a mixture of ethanol and methanol) and to cool the liquid to a glassy state below 140 °K in a suitable optical cavity with highly reflecting and movable pistons as boundaries (see Fig. 6.20). The purpose of the solvent is to provide a host environment similar to that of crystals or glasses in which the rare earth ions are embedded.

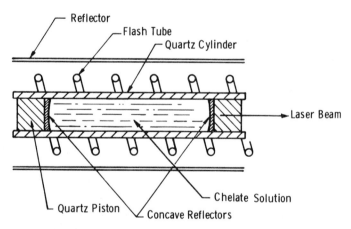

FIG. 6.20. Optical cavity arrangement for europium chelate laser [A. Lempicki and H. Samelson, *Phys. Letters* **4**, 133 (1963); *Proc. Symp. Optical Masers, New York, 1963* p. 347 (Polytechnic Press, New York, 1963)].

In particular, the absorption of the excitation light is sufficiently reduced so that it can penetrate the cavity more uniformly. Exposed to flashes from a helical tube, the chelate shows the typical laser behavior known from ruby, i.e., spikes of intensity which grow at higher excitations into relaxation oscillations. This is accompanied by a narrowing of the spectrum to a line 0.3 Å wide at 6129 Å. Later developments of chelate lasers have shown that operation at room temperature is possible and that thermal expansion of the liquid can be minimized so that pistons may not be necessary.

Far greater difficulties are encountered with rare earth chelates in a low-viscosity solution. The index of refraction of alcohols, for instance, varied too much with temperature to allow the optical stability needed for the attainment of the threshold condition.[40a] Furthermore, scattering in microdomains of the liquid state may represent an additional barrier.

Chelates, of course, are not the only organic compounds capable of laser action. Various other organic materials and the excitation mechanisms suitable for them have been suggested, including the energy of exothermic chemical reactions.[86]

6.3.3 TECHNOLOGY AND SYSTEMS FOR LASERS

There exists a number of methods which are of interest in their own right rather than only in connection with the specific laser for which they had been introduced originally. In the following we shall discuss these techniques with a view towards their greater generality along with phenomena which apply to the laser field as a whole.

6.3.3.1 *Techniques of Optical Excitation*

a. Sources. The incoherent light used for optical pumping is obtained normally from high-brightness plasmas because they represent the most effective radiators in terms of total yield, speed of buildup, and matching spectral selectivity. Most of these lamps operate with xenon discharges, although other gases have come into use for pumping purposes, sometimes with the addition of metal iodides to generate additional spectral lines. The requirements of optical excitation run of course parallel to those of a great many other applications, particularly in the field of pulsed flash lamps (see e.g., the detailed review of Marshak[87]). Typically, current densities in xenon flash tubes amount to 10^3 A/cm^2 for large helices and up to 10^5 A/cm^2 for short linear structures. The emphasis on power has led to experimentation with discharge energies up to 400,000 J per pulse[88] and to the application of the "magnetic pinch" principle in gas discharge tubes. For example, a discharge of 400 J in 1 to 2 μsec was obtained in a "theta pinch lamp" through induction by a coil at 600 kcps. Shaped as a toroid, this lamp surrounded a ruby rod.[89] For special purposes, it is feasible to use excitation sources other than plasma tubes. Here belong incandescent lamps which are suitable for continuously operated four-level lasers (since these have a low excitation threshold), electron-bombarded phosphors[90] and exploding wires.[91] For example, certain Nd^{3+} doped garnets, such as $Y_3Al_5O_{12}$, are capable of reaching threshold conditions at 300°K in short crystal lengths (1.5–3 cm) with continuous operation in response to tungsten lamp radiation (400 W).[49a]

b. Radiator Configurations. There are three basic methods by which it is possible to maximize the light transfer from an incoherent source to a cylindrically shaped laser cavity, and they are shown, respectively, in Fig. 6.21a,b, and c: (1) the helix surrounded by a diffusely reflecting

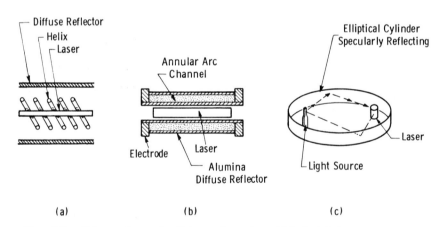

(a)   (b)   (c)

FIG. 6.21.   Scheme of pumping light configurations: (a) helical flash tube arrangement with diffuse reflector (alumina); (b) laser surrounded by annular arc discharge [C. H. Church, J. Ryan, and J. P. Lesnick, *Laser Flash Lamp Conf., Stanford Res. Inst., 1964*]; (c) arrangement in focal axes for source and laser in elliptical cyclinder mirror.

cylinder; (2) the laser imbedded in, or at least completely surrounded by, the luminous medium; and (3) sources and laser aligned with the focal axes of an elliptic, specularly reflecting cylinder or similar confocal arrangements. Pumping by laser light itself is a most efficient method in which the beam can be made to traverse directly the second laser cavity, e.g., GaAs laser radiation to excite $CaF_2 : U^{3+}$ (Keyes and Quist[91a]).

In addition, refractory elements for focusing may be involved. In particular, we mention here two methods which are related insofar as they both use a section of undoped host crystal to concentrate the incident pumping light into a section doped with the active ion. Figure 6.22a shows such a combination for sapphire (Al_2O_3) and ruby $(Al_2O_3 + Cr_2O_3)$. The clear crystal of conical shape serves to collect the pumping light with large aperture and to funnel it into the second section of ruby. The gain in intensity obtained by this structure and its attending feature

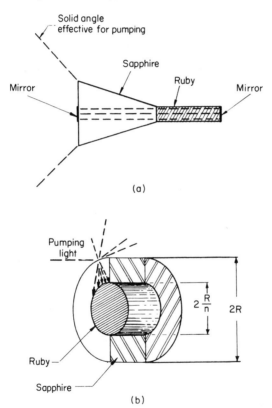

Fig. 6.22. Sapphire-ruby combinations for the focusing of pumping light. (a) Conical design of sapphire section funnels radiation within a large solid angle into Cr_2O_3-doped cylindrical extension [D. F. Nelson and W. S. Boyle, *Appl. Optics* 1, 181 (1962)]. Laser mode between Fabry-Perot mirrors is indicated by dashed lines. (b) Coaxial design of sapphire and ruby gathers excitation light uniformly into central Cr_2O_3-doped section [G. E. Devlin, J. McKenna, A. D. May, and A. L. Schawlow, *Appl. Optics* 1, 11 (1962)]. Dashed lines indicate focusing action of sapphire shell.

of multiple internal reflections is so large that Nelson and Boyle[92] could demonstrate the first continuous operation of a ruby laser shaped in trumpet form. Another combination[93] (see Fig. 6.22b) consists of a sapphire cylinder in which only the core is doped with Cr_2O_3. The design is predicated on the fact that light which is radially incident on a cylinder of index n and radius R will be refracted in such a manner that all of it passes through a smaller cylindrical region of radius R/n, whereas the outer shell is only partially illuminated. Extending the region of ruby (which has the same n as sapphire) only to a radius of R/n provides greater and more uniform intensity in the space where it is needed.

6.3.3.2 Phenomena and Methods of Mode Selection

Aside from the problems of excitation (i.e. the pumping methods, the supply of sufficient power, and the matching of energy levels), the problem of proper mode encouragement ranks next in importance for a given laser material. The significance of manipulating the field configuration lies in the fact that there exist phenomena of mode pulsation and mode competition which usually—but not always—are undesired. In remedying the cause, a number of interesting improvements have been made.

a. Relaxation Oscillations and Q-Switching. The existence of competing modes leads to more or less irregular shifting of the available laser power between them, although generally the successful configurations feed and grow at the expense of the less developed oscillations. The result is random pulsing of any individual mode. However, there is superimposed in addition a more regular rhythm of the total intensity which was observed in individual pulses of ruby laser emission in some of the early experiments.[42] This phenomenon results from the periodic build-up of the inverted population to—and even beyond—the threshold level and the subsequent more or less abrupt decay by stimulated emission (see Fig. 6.23a). Each burst of stimulated radiation power is followed by a period of recovery so that a series of spikes is observed in this relaxation oscillation (see Fig. 6.23b).

The spiking process can be very much enhanced by a method[94] called Q-*switching* (also Q-*spoiling*). It has been shown in Section 6.3.1.1 that the excess number of excited states and hence the power needed for the threshold condition of oscillation is inversely proportional to the free photon time τ_c and therefore to the cavity Q_c, according to Eqs. (1.14), (6.55), (6.58), and (6.59). McClung and Hellwarth[95] utilized this fact by first building up a large excess population under conditions of low

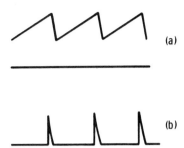

FIG. 6.23.   Schematic diagram (a) of relaxation oscillation in the density of the inverted population; and (b) of synchronous pulses of laser emission.

effective reflectivity, i.e. low Q_c, and then releasing the stored energy in a single "giant pulse" by a sudden increase of the reflectivity and, Q_c to high values. The original means of switching the Q-value consisted of a Kerr cell which was introduced into the optical path between the external mirrors M_1 and M_2 of a ruby laser (see Fig. 6.24). With voltage

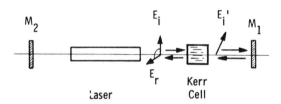

FIG. 6.24.   Q-switching of optical cavity by rotating plane of polarization to generate giant pulses. [F. J. McClung and R. H. Hellwarth, *J. Appl. Phys.* 33, 828 (1962)].

applied, the Kerr cell will turn the polarization of the radiation from the ruby crystal by 90° total for the two passes to and from mirror M_1. Polarization is obtained in the beam along the cavity axis, if the latter is normal to the *c*-axis of the crystal, although other means, such as a polarizer in front of mirror M_2, can be used, and in either case the acti-vated Kerr cell will foil reflection by mirror M_1. However, as soon as the cell voltage is turned off—which is possible within 10^{-8} sec—the cavity operates with a high Q_c and the accumulated inverted population is released under stimulated emission. The storage effect is large because with high regeneration the *excess* needed in the excited state for the threshold condition may be quite small and is always much smaller than the ground state density. By contrast, if the Q_c-value is small, only the spontaneous emission rate A (and other ordinary decay processes) serve to balance the build-up rate $B \cdot u$ (cf. Section 6.3.1.2) so that the saturation excess population N_s equals

$$N_s = \frac{Bu - A}{Bu + A} N_0 \qquad (6.82)$$

which implies total inversion for $Bu \gg A$, within the limits of its validity. Experimentally, the first reported peak intensity of giant pulses was 600 kW or about 100 times that normally obtained under otherwise equal conditions.

The sudden variation of the Q_c-value can be accomplished by a variety of other techniques, for instance, with rotating mirrors. Furthermore, methods equivalent to Q-switching are available such as the "detuning" of the laser material by a magnetic or electric field, which varies along the optical axis and is suddenly turned off. Another technique is that of *bleachable filters* which absorb the laser radiation up to a limit of intensity build-up. At that point, the population in the absorbing state of the filter has been sufficiently emptied so that laser light is passed to the reflectors and a "giant spike" results. A variation of this method is the "bleachable" internal filter action of $UO_2{}^+$ ions (10^{20} cm^{-3}) embedded in glass which is doped with Nd^{3+} ions (Melamed et al.[95a]).

c. Mirrors and Brewster Angle Windows. The Kerr cell method of Q-switching is made possible by placing the Fabry-Perot mirrors external to the space filled by the laser material. This feature was introduced by Rigrod and his co-workers[96] originally for gaseous systems where it has some additional advantages, notably those of flexibility, independence of bake-out cycles or chemical treatment from restrictions (imposed by considerations for the sensitive dielectric reflectors), and

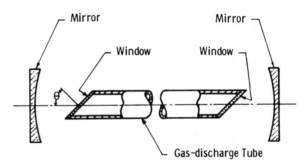

FIG. 6.25. Optical cavity arrangement with external concave mirrors and windows inclined at Brewster angle [W. W. Rigrod, H. Kogelnik, D. J. Brangaccio, and D. R. Herriott, *J. Appl. Phys.* **33**, 743 (1962)].

simplifications of construction. The windows of the gas discharge tube were inclined at the Brewster angle with respect to the optical axis (see Fig. 6.25). As described in Section 5.1.3.2, rays reflected and refracted for this angle are polarized: the electric vector component in the plane of incidence is missing in the beam reflected by the window. Consequently, after many passes through the optical cavity, modes polarized in the plane of incidence are selectively encouraged. Another technique, introduced[96] along with these advances, was the replacement of the plane mirrors by confocal, concave mirrors. The advantages of the latter had already been known through an analysis of field distributions and power losses for plane and confocal spherical mirrors by Fox and Li.[97,98] In particular, it can be shown[99] that the simplest mode between confocal concave mirrors of radius has a beam-width angle between the half-intensity points of

$$\theta = 0.939 \left(\frac{\lambda}{r}\right)^{\frac{1}{2}} \quad [\text{rad}] \qquad (6.83)$$

For $r = 10^2$ cm and $\lambda = 1\ \mu$, there results an angle $\theta = 0.939 \times 10^{-3}$ rad or approximately 3 min of arc which corresponds approximately to the values measured in the work cited[96]. A higher beam directionality is, of course, achievable with plane parallel mirrors, but in the latter arrangement the beam pattern changes for mirror displacements as small as 4 sec of arc.

d. Frequency Fluctuations and Micropulses. A phenomenon occurs in solid state lasers which is sometimes referred to as "hole-burning." The first observation of this effect was made by Hughes[100,101] who photographed the time-variant spectrum of a ruby laser by sweeping the output of a Fabry-Perot interferometer with a spinning mirror. With this arrangement he could resolve to some extent the spectral content of individual emission spikes and found that whereas each individual spike contained only one or at most a few very narrow wavelength components, the latter varied from one spike to the next in a rather systematic manner (cf. Fig. 6.26). Later experiments which followed up this work (DeMaria and Gagosz[102]) showed that the laser pulses in ruby, which typically follow each other in periods of several microseconds, themselves consist of "micropulses" with durations and repetition periods in tenths of microseconds.

These, as well as related phenomena can be explained by a model which postulates that oscillation at narrow frequency intervals within the broad spectral profile of the atomic transition rapidly depletes the inverted states capable of contributing to that particular mode so that

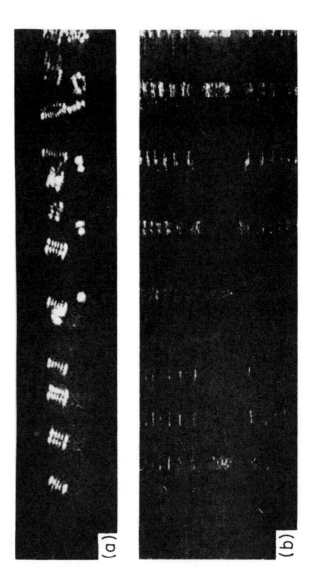

FIG. 6.26. Time-resolved beam pattern from ruby laser: (a) variation of intensity distribution across beam from spike to spike; (b) same as in (a) with addition of Fabry Perot, displaying changes of gross spectrum.

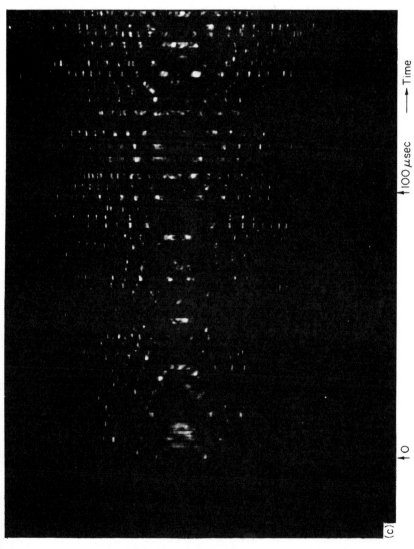

Fig. 6.26. (c) Highly resolved spectrum of laser spikes showing that each spike contains only one or at most a few highly monochromatic frequencies which vary systematically from one spike to the next. Photographs obtained by courtesy of T. P. Hughes [*Nature* **196**, 332 (1962): (a) and (b); *ibid.* **195**, 325 (1962): (c)].

the laser resonance shifts to other, still undisturbed parts of the line (see Fig. 6.27a). The resonances are of course determined by the discrete sets of mode configurations which are admitted in the cavity. The most important class of these modes is given by the condition that $n \cdot \lambda/2 = L$, in other words, by the stipulation that the standing wave configuration in the cylinder of length L comprise an integer number of half-wavelengths. Among these proper modes, those for which λ is closest to the maximum of the atomic line profile are most likely to begin laser oscillation. In solids, the inverted population fixed at the sites of the antinodes (see Fig. 6.27b) will be depleted relatively fast[103] since the rate of stimu-

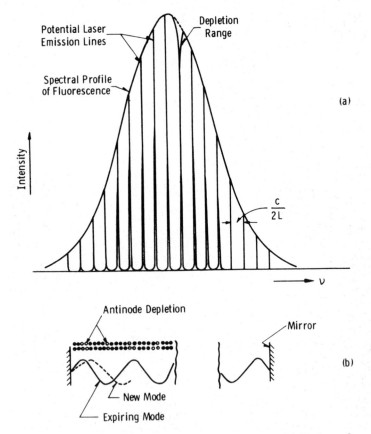

FIG. 6.27.   Hole-burning model in solid state laser spectrum. (a) Mode configurations define a set of sharp laser lines within relatively broad spectral profile of atomic transition (assumed Gaussian). Depletion pulls mode frequency from one line to the next. (b) Standing wave model interprets [P. F. Browne, to be published] mode pulling as the result of level depletion for excited atoms fixed in antinode positions.

lated emission is proportional to the square of the electric field. As a result, the mode subsides; another builds up in its stead, and its integer number is likely to be $n - 1$, $n - 2$, etc., or the neighbors on the positive side. The frequency may thus migrate from micropulse to micropulse in a single direction. Ultimately, however, the states at the original antinodes are being replenished by the pumping light, and the frequency may flick back to the starting value. The frequency change for $\Delta n = 1$ is $c/2L$ (where c is the velocity of light in the medium). Variations of this magnitude, or small multiples thereof, from spike to spike were indeed measured,[100] and beats between these modes could be established revealing their simultaneous existence.[102,104]

Mode pulling effects in gases are more difficult to interpret because the atoms, unlike in solids, have a Maxwell-Boltzmann velocity distribution and may move, for instance, a distance of 100 μ during 10^{-7} sec at room temperature. An investigation of hole-burning effects in the helium-neon laser was made by Bennett.[105]

6.3.3.3 *Summary of Laser Materials and Energy Level Transitions*

We conclude this section on optical masers by listing the more important materials and spectral lines for which oscillations have been observed (see Tables 6.1–6.3).

6.4 Nonlinear Optics

The classical phenomena of propagation—for instance, the laws of reflection and refraction or the superposition of wave-trains—follow from linear relationships between the electromagnetic field and certain observable, basically atomic, quantities such as the induced dipole moment and polarization. However, it is obvious that relationships of higher than linear order will become important for sufficiently large fields or forces, just as this is the case for the mechanical analog of Hooke's law. Thus, in a wide sense, nonlinearities are quite common. We have encountered them in diatomic molecules (see Section 4.5.3) where the force between two atoms must be expressed as a power series of the internuclear distance to explain either the strong repulsion when the distance is small or dissociation when it is large. It was shown that, as a result, harmonics of the fundamental frequency appear in such vibrators. Thus there exists a *nonlinear optics* which must become valid for sufficiently large light intensities, whether the light is coherent

TABLE 6.1

GAS LASERS

Material	Excitation	Transition	Wavelength in air (μ)	References[a]
Helium-neon	Resonance transfer from He 2 ^{1}S state	$3s_2-2p_4$ $3s_2-3p_4$	0.6328 3.3913	1 2
	Resonance transfer from He 2 ^{3}S	$2s-2p^b$, e.g. $2s_2-2p_4$ $2s_3-2p_5$ $2s_2-2p_1$	1.1523 1.1614 1.5231	3
	Resonance transfer from He 2 ^{1}P$_1$ and other levels	7p–6d [Racah]	35.602, 37.231 53.486, 54.019 54.117, 57.355	4
Helium-xenon	Resonance transfer, cw	$3d''-2p_7$ ^{b} $3d_4-2p_9$ $3d_6-2p_{10}$	3.3667 3.5070 3.9955	5
Helium-mercury	High voltage, pulsed	$7p^2P_{3/2}-7s\ ^2S_{1/2}$ (H$_g$ II) $5f\ ^2F_{7/2}-6d^2D_{5/2}$ (H$_g$ II)	0.61495 0.56772	6
Helium	e$^-$ impact, cw	$7^3D-4\ ^3P$	2.0603	7
Neon	e$^-$ impact, cw	$2s_2-2p_4$ ^{b} 6p–5d [Racah]	1.1523 31.928, 34.679, 41.741	5 4
Argon	e$^-$ impact, cw	$3d_3-2p_6$ ^{b} $3d_3-2p_3$	1.6941 2.0616	5
	High voltage e$^-$ impact-pulsed	4p–4s, e.g. $4p\ ^4D_{1/2}°-4s\ 2P_{3/2}$	0.487986	8
Krypton	e$^-$ impact, cw	$3d_3-2p_7$ ^{b} $3d_3-2p_6$	2.1165 2.1902	7
Xenon	e$^-$ impact	$3d_2-2p_7$ $3d_4^1-2p_8$	2.0261 5.5738	5
Mercury	e$^-$ impact	$6p^1\ ^3P_2°-7s\ ^3S_1$ ^{c}	1.52954	9, 10
Nitrogen	High voltage (150 kV) e$^-$ impact, pulsed	$C^3\pi_u-B^3\pi_g$ ^{c}	0.3-0.4 (30 lines) 0.3371^b	11
Neon-oxygen	Dissociative resonance transfer cw	$3\ ^3P_2-3\ ^3S_1$ (oxygen)	0.84462	12
Argon-oxygen	Dissociative resonance transfer, cw	$3\ ^3P_2-3\ ^3S_1$ (oxygen)	0.84462	12
Cesium	Optical pumping	$8\ ^2P_{1/2}-8\ ^2S_{1/2}$	7.1821	13

TABLE 6.2

Dielectric Crystal Laser Materials

Material[a]	Conditions of optical excitation	Wavelength in air (μ)		References[b]
Cr^{3+} in Al_2O_3	0.01–0.05% pulsed and cw at 4.2–300°K	0.6943		*1, 2*
	0.5%, 77°K pulsed	0.701		*3*
U^{3+} in CaF_2 (also BaF_2, SrF_2)	0.05%, pulsed and cw, 4.2–300°K	2.24 2.5	2.6	*4*
Nd^{3+} in $CaWO_4$ (also $SrWO_4$, $SrMO_4$, $Y_3Al_5O_{12}$, BaF_2, LaF_3)	0.1%, 77–300°K, pulsed and cw	1.06		*5*
Ho^{3+} in $CaWO_4$ (also CaF_2)	0.5%, 77°K pulsed	2.05		*6*
Sm^{2+} in CaF_2 Sm^{2+} in SrF_2	{0.05%, 4.2°K{ { pulsed }	0.7083 0.6967		*7*
Tm^{3+} in $CaWO_4$ in SrF_2	{0.5%, 77°K{ { pulsed }	1.91 1.97		*8*

[a] Properties and wavelengths listed are valid also for host lattices shown in brackets to the extent of the significant figures listed here.

[b] References: (*1*) T. H. Maiman, *Nature* **187**, 493 (1960); (*2*) R. J. Collins, D. F. Nelson, A. L. Schawlow, W. Bond, C. G. B. Garrett, and W. Kaiser, *Phys. Rev. Letters* **5**, 303 (1960); (*3*) I. Wieder and L. R. Sarles, *ibid.* **6**, 95 (1961); (*4*) P. P. Sorokin and M. J. Stevenson, *ibid.* **5**, 557 (1960); (*5*) L. F. Johnson and K. Nassau, *Proc. IRE* **49**, 1704 (1961); (*6*) L. F. Johnson and K. Nassau, *ibid.* **50**, 87 (1962); (*7*) P. P. Sorokin and M. J. Stevenson, *IBM Journal Res. Develop.* **5**, 56 (1961); (*8*) L. F. Johnson, G. D. Boyd, and K. Nassau, *Proc. IRE* **50**, 86 (1962).

[a] References: (*1*) A. D. White and J. D. Rigden, *Proc. IRE* **50**, 1697 (1962); (*2*) A. L. Bloom, W. E. Bell, and R. E. Rempel, *Appl. Optics* **2**, 317 (1962); (*3*) A. Javan, W. R. Bennett, Jr., and D. R. Herriott, *Phys. Rev. Letters* **6**, 106 (1961); (*4*) C. K. N. Patel, W. L. Faust, R. A. McFarlane, and C. G. B. Garrett, *Appl. Phys. Letters* **4**, 18 (1964); (*5*) W. L. Faust, R. A. McFarlane, C. K. N. Patel, and C. G. B. Garrett, *ibid.* **1**, 85 (1962); (*6*) W. E. Bell, *Appl. Phys. Letters* **4**, 34 (1964); (*7*) C. K. N. Patel, W. R. Bennett, Jr., W. L. Faust, and R. A. McFarlane, *Phys. Rev. Letters* **9**, 102 (1962); (*8*) W. B. Bridges, *Appl. Phys. Letters* **4**, 128 (1964); (*9*) J. D. Rigden and A. D. White, *Nature* **198**, 774 (1963); (*10*) R. A. Paananen, C. L. Tang, F. A. Horrigan, and H. Statz, *J. Appl. Phys.* **34**, 3148 (1963); (*11*) H. G. Heard, *Nature* **200**, 667 (1963); (*12*) W. R. Bennett, Jr., W. L. Faust, R. A. McFarlane, and C. K. N. Patel, *Phys. Rev. Letters* **8**, 470 (1962); (*13*) P. Rabinowitz, S. Jacobs, and G. Gould, *Appl. Optics* **1**, 513 (1962).

[b] Only the strongest lines among those already known are indicated here.

[c] Tentative assignment.

TABLE 6.3

SEMICONDUCTOR LASER MATERIALS

Materials	Conditions of excitation by injection	Wavelength in air (μ)	References[a]
GaAs	77°K, pulsed	0.840	*1, 2, 3*
Ga(As$_{1-x}$P$_x$)	77°K, pulsed	0.662–0.840	*4*
GaAs$_{0.63}$P$_{0.37}$	77°K, pulsed	0.662	*5*
GaAs$_{0.76}$P$_{0.24}$	77°K, pulsed	0.700	*5*
InP	4.2°K, pulsed 77°K, pulsed	0.905–0.907 (7 components) 0.907	*6*
InAs	4.2–77°K, pulsed	3.1	*7*
InSb	1.7–4.2°K, at magn. fields of 20 to 90 kG	5.2	*8*
PbTe	12°K, pulsed	6.5	*9*

[a] References: (*1*) R. N. Hall, G. E. Fenner, J. D. Kingsley, T. J. Soltys, and R. O. Carlson, *Phys. Rev. Letters* 9, 366 (1962); (*2*) M. I. Nathan, W. P. Dumke, G. Burns. F. H. Dill, and G. Lasher, *Appl. Phys. Letters* 1, 62 (1962); (*3*) T. M. Quist, R. H, Rediker, R. J. Keyes, W. E. Krag, B. Lax, A. L. McWhorter, and H. J. Zeiger, *ibid.* 1, 91 (1962); (*4*) N. Holonyak and S. F. Bevacqua, *ibid.* 1, 82 (1962); (*5*) N. Ainslie, M. Pilkuhn, and H. Rupprecht, *J. Appl. Phys.* 35, 105 (1964); (*6*) K. Weiser and R. S. Levitt, *Appl. Phys. Letters* 2, 178 (1963); (*7*) I. Melngailis, *ibid.* 2, 176 (1963); (*8*) R. J. Phelan, A. R. Calawa, R. H. Rediker, R. J. Keyes, and B. Lax, *ibid.* 3, 143 (1963); (*9*) J. F. Butler, A. R. Calawa, R. J. Phelan, Jr., T. C. Harman, A. J. Strauss, and R. H. Rediker, *Appl. Phys. Letters* 5, 75 (1964).

or not. Nevertheless, there are two reasons for treating this subject as such in connection with coherent radiation.

First, to be observable, nonlinear propagation phenomena require fields upward from 10^5 to 10^6 V/cm. Since power flux density S in the vacuum is related to the electric field E (in esu) by

$$S = \frac{c}{8\pi}(E^2 + H^2) = \frac{c}{4\pi}E^2 \tag{6.84}$$

it follows that from 10^7 to 10^9 W/cm^2 are needed to demonstrate such effects. Brightnesses of this magnitude can in general be achieved only with laser beams.

The second reason for the connection of this field with lasers lies in the other properties of coherent radiation: for strong nonlinear effects, high monochromaticity and directionality of the beam may be required or, again, it may be desirable to maintain a constant phase relationship in time and space over dimensions of many periods and wavelengths, respectively. This will become apparent from practical examples as we proceed to discuss some manifestations of nonlinear optics.

Any nonlinearity involving terms of the electric or magnetic field also changes the linear relationship existing in first order between such observables as fluorescent intensity, absorbed power, or photoelectric yield and the incident radiation power since the latter is proportional to E^2 and H^2. In other words, nonlinear optics includes the breakdown of proportionality for certain effects which ordinarily increase in magnitude directly as the number of incident photons. This, in turn, implies multiphoton processes. For example, an event depending on the cumulative action of two photons will have a probability given by the square of the incident intensity. Thus harmonic generation of the frequency 2ν from radiation at frequency ν may be interpreted as the interaction of two photons resulting in the creation of a single photon with twice the incident energy. Multiple photon processes also show themselves in a type of absorption which involves an intermediate "virtual" level and thus produces spectral lines not known at weak incident intensities. Similarly, certain secondary radiation effects, such as photoemission, show an anomalous behavior at high radiation levels which in some cases is explained with the cascaded excitation by several photons through virtual levels. These and other phenomena are described by nonlinear optics.

6.4.1. OPTICAL HARMONICS

The first experimental demonstration of harmonic light generation and its theoretical interpretation is due to Franken and his co-workers.[106] The experiment consisted of focusing the light from a pulsed ruby laser into a suitably oriented quartz crystal. The incident coherent radiation was thus a narrow red spectral line at 6943 Å, the crystal being shielded from the shorter wavelengths of the pumping source by a red filter. Under these conditions, the light reemerging from the quartz sample contained a component at 3472 Å and thus gave direct proof for frequency doubling.

The origin of harmonic generation can be understood in simple mathematical terms. Assume a one-dimensional nonlinear relationship between polarization P and the electric field E, as it occurs, for instance, in diatomic molecules:

$$P = a_1 E + a_2 E^2 + a_3 E^3 + \cdots \tag{6.85}$$

where $a_1 = N\alpha$ (in the terminology of Section 5.2.1). An oscillating field $E = E_0 \sin \omega t$ due to the incident light wave will then produce a polarization:

$$P = a_1 E_0 \sin \omega t + a_2 E_0^2 \sin^2 \omega t + a_3 E_0^3 \sin^3 \omega t$$

$$= a_1 E_0 \sin \omega t + \frac{a_2}{2} E_0^2 (1 - \cos 2\omega t) + \frac{a_3}{4} E_0^3 (3 \sin \omega t - \sin 3\omega t) + \cdots$$

(6.86)

It is seen that the presence of nonlinear terms in the relationship between P and E generates not only terms containing the second and yet higher harmonics, but in addition (for the even powers) a constant bias polarization. If P is an odd function of E, frequency doubling is not possible. This is the case for dielectrics which are either isotropic or which have in their structure a center of symmetry. Glasses and liquids in general are examples of the former class and calcite is an example for the latter.

The simple scalar relationship (6.85), is not, of course, typical for dielectric crystals. Instead, depending on the structure of the material, the polarization is a vector of which each component is a specific (nonlinear) function of the impressed electric field components. In other words, tensors replace the scalars $a_1, a_2, \cdots$, in close analogy to the relationships known from the theory of piezoelectricity (see Cady,[107] pp. 187–192 for the tensor relationships between polarization and the six stress components in the 32 crystal classes and Franken and Ward[108] for the application of piezoelectric coefficients to optical polarization). In particular, one can write for the second-order polarization which is responsible for frequency doubling:

$$P_i(2\omega) = A_{ijk} E_{jk}^{(2)}(\omega) \tag{6.87}$$

Here the subscripts stand for the coordinates, i.e., $P_1(2\omega)$ represents the components P_x, P_y, P_z; $E_{jk}^{(2)}(\omega)$, a matrix of 1 column and 6 rows, stands for the 6 second-order fields, viz., E_x^2, E_y^2, E_z^2, $2E_y E_z$, $2E_x E_z$, $2E_x E_y$ (analogous to the 3 compression and 3 shear stresses in the piezoelectric case); and A_{ijk} is a matrix of 6 columns and 3 rows containing 18 elements, viz., the coefficients of the three polarization components with respect to the 6 second-order fields. Depending on the degree of crystal symmetry, the tensors can assume various simplified forms. Thus for $BaTiO_3$, A_{ijk} has the form[109]

i $\backslash$ jk	xx	yy	zz	yz	xz	xy
x	0	0	0	0	d_{15}	0
$A_{ijk} = y$	0	0	0	d_{15}	0	0
z	d_{31}	d_{31}	d_{33}	0	0	0

(6.88)

where the subscripts of the nonlinear coefficients d_{ijk} are enumerated according to the system used in piezoelectricity. In addition, in the theory of piezoelectricity the symmetry relationships have already been worked out for all crystal classes.[101] Although this comparison yields information only on the form of the matrix and the interrelationship of its elements, but not on their amounts as optical coefficients, tensor systems such as Eq. (6.88) indicate the preferred directions of the primary beam with respect to a chosen axis for frequency doubling.

6.4.1.1 *Effects of Frequency Doubling and Dispersion*

The process of harmonic generation by an incident wave of frequency ν has to be viewed as consisting of two steps. First, to the extent of the nonlinearity of the medium, a second-order polarization wave is produced with a wavelength

$$\lambda_1 = \frac{c}{2\nu n_1} \tag{6.89}$$

Here n_1 is the index of refraction for the fundamental frequency, i.e., $n_1 = n(\nu)$; in other words, the polarization wave follows the fundamental wave simply with twice its wave number. The second step is the result of the harmonic polarization, viz., the generation of radiation at a frequency 2ν and a wavelength, in the medium, of

$$\lambda_2 = \frac{c}{2\nu n_2} \tag{6.90}$$

where n_2 is the index of refraction for the doubled frequency, $n_2 = n(2\nu)$, which will in general be different from n_1. Hence, λ_1 and λ_2 will also be different, the harmonic polarization wave will propagate out of phase with the electric field wave, and the intensity of the latter will go through a series of maxima and minima for points progressing along the direction of the secondary beam. The detailed calculation[108] yields for the intensity of harmonic generation in a slab of thickness L the expression

$$I(2\omega) \propto \frac{\sin^2[2\pi(L/\lambda)(n_1 - n_2)]}{(n_1 - n_2)^2} \tag{6.91}$$

where λ is the wavelength *in vacuo* corresponding to the fundamental frequency ν. Equation (6.91) states that, beginning with thickness zero, the second harmonic intensity initially increases with the square of L, but then quickly reaches a maximum for a thickness

$$L_{\max} = \frac{\lambda}{4(n_1 - n_2)} \tag{6.92}$$

The intensity vanishes for $2L_{max}$, and maxima and minima follow with the period $2L_{max}$, except for the effects of attenuation. With typical dispersion values in the visible region, thicknesses beyond about 10 microns do not contribute to the harmonic intensity, and for this reason, the effect is small. Only if n_1 can be made substantially equal to n_2, relatively high efficiencies of frequency doubling will obtain. The oscillatory character of the harmonic intensity with L was demonstrated by Maker and his co-workers[110] in an experiment in which they passed a ruby laser beam through a quartz platelet at varying inclination and, therefore, varying optical path length (see Fig. 6.28d).

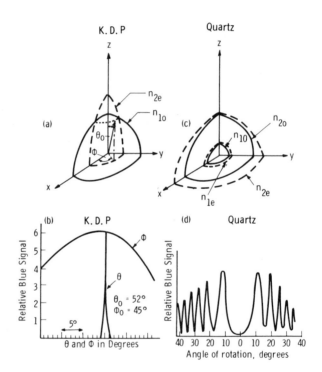

FIG. 6.28.   Index of refraction surfaces $n(x, y, z)$ for KDP and quartz and the resulting efficiency of harmonic generation by ruby laser beam as a function of angle of incidence. (a) Surface n_{2e} for extraordinary rays in the near ultraviolet (2ν) intersects those of the ordinary rays in the red (ν), n_{1o}.   (b) As a result, index matching is possible at specific angles θ_0 and ϕ_0.   (c) In quartz, the index surfaces do not intersect, and dispersion cannot be cancelled by birefringence.   (d) As a consequence, rotation with respect to the incident beam produces only weak maxima and minima corresponding to the phase differences between the harmonic polarization and the radiation wave [P. D. Maker, R. W. Terhune, M. Nisenhoff, and C. M. Savage, *Phys. Rev. Letters* **8**, 21 (1962); also J. A. Giordmaine *ibid.* **8**, 19 (1962).

The importance of nonlinear systems for frequency conversion schemes and the attending need for phase velocity equalization had, of course, been realized by a number of other workers.[111, 112]

6.4.1.2 *Index Matching*

An effective method of providing equal phase velocities for the fundamental and second harmonic wave in the nonlinear medium was developed simultaneously by Giordmaine[113] and by Maker and his coworkers.[110] The technique, in essence, utilizes the fact that in some anisotropic crystals normal dispersion is smaller than birefringence. Under such conditions, the (spherical) surface, $n_{1o}(x, y, z)$, representing the index of refraction for ordinary rays at one frequency, will intersect the surface $n_{2e}(x, y, z)$ of the extraordinary refraction index at the second frequency. This is the situation for potassium dihydrogen phosphate (KDP) as indicated in Fig. 6.28a where the optical axis is assumed to coincide with the z-direction. It will be seen that the condition $n_{2e} = n_{1o}$ is fulfilled for rays entering the crystal at an angle θ_0 with the optical axis. By extrapolation of published data, θ_0 was estimated[113] at $50.4°$, and the observed angle for peak emission of harmonic radiation agrees with this prediction in both observations within the expected errors. Actually, the measured response (see Fig. 6.28b) is quite sharp with respect to θ, and there exists also a broad maximum of intensity for changes in the azimuthal angle ϕ. By contrast, this method of *index matching* is not suitable for quartz. As shown in Fig. 6.28c, the n_2-surfaces do not intersect those of n_1. Hence, it is not possible to compensate dispersion by a choice of incidence angle, and in fact, rotation merely produces the oscillatory behavior shown in Fig. 6.28d.

In all, index matching increases the conversion efficiency by several orders of magnitude, and further gains are possible by repeated passage of the primary light through the crystal at the focus of a suitable optical cavity. Furthermore, as the pulsed power of the laser beam is increased to the order of megawatts, coherence properties are heightened considerably, and thus conditions for index matched harmonic generation are improved. By such means, Terhune *et al.*[114] have achieved frequency doubling efficiencies of 20% with ammonium dihydrogen phosphate.

6.4.2. SECONDARY NONLINEAR EFFECTS

A rapidly growing number of nonlinear optical phenomena other than harmonic generation have been discovered or at least predicted. We can enumerate here only a few of these developments.

6.4.2.1 *Maser Stimulated Raman Radiation*

Woodbury *et al.*[115] observed the emission of highly monochromatic, directional radiation in the infrared at 7660 Å (also at 8515 Å and 9610 Å) from nitrobenzene while it was exposed to a ruby laser beam at 6943 Å in a Kerr cell for the purpose of producing giant pulses. Subsequent studies identified this emission as stimulated Raman scattering.[116] The effect was also found in other organic liquids, initially at longer wavelengths (Stokes radiation), later in addition, however, with higher energy[117] than the primary light (anti-Stokes radiation). The various frequencies and angular dependence of the stimulated Raman radiation are explained[118, 119] in terms of a two-step process: (a) the production of diffuse first-order Stokes scattering which then (b) diminishes, or adds to, the primary laser photon energy so as to produce Stokes and anti-Stokes emission of higher order.

6.4.2.2 *Self Trapping of Optical Beams*[120]

When a sufficiently intense laser beam progresses through a nonlinear medium, higher orders of the index of refraction may become important:

$$n = n_0 + n_2 E^2 + \cdots \tag{6.93}$$

If the second term on the right-hand side of Eq. (6.94) is sufficiently large, the sheath of dielectric medium surrounding a circular laser beam may totally reflect any light diverging because of aperture diffraction. In this manner, a self-guiding action of strong laser beams may be predicted.

REFERENCES

1. A. EINSTEIN, *Physik Z.* **18**, 121 (1917).
2. A. KASTLER, J. *Phys. Radium* **11**, 255 (1950).
2a. E. M. Purcell and R. V. Pound, *Phys. Rev.* **81**, 279 (1953).
3. C. H. TOWNES, *Symposium on Sub-Millimeter Waves, Univ. of Illinois, 1951*, through A. H. NETHERCOT.
4. A. L. SCHAWLOW and C. H. TOWNES, *Phys. Rev.* **112**, 1940 (1958).
5. T. H. MAIMAN, *Nature* **187**, 493 (1960).
6. T. H. MAIMAN, *Brit. Commun. Electron.* **7**, 674 (1960).
7. A. A. MICHELSON, *Phil. Mag.* [5] **31**, 338 (1891).
8. A. A. MICHELSON, *Phil. Mag.* [5] **34**, 280 (1892).
9. H. E. J. NEUGEBAUER, *Appl. Optics* Suppl. **1**, 90 (1962).
10. H. FIZEAU, *Compt. Rend. Acad. Sci.* **66**, 934 (1868).
11. A. A. MICHELSON, *Phil. Mag.* [5] **30**, 1 (1890).
12. F. ZERNIKE, *Physica* **5**, 785 (1938).

13. M. Born and E. Wolf, "Principle of Optics." Macmillan (Pergamon), New York, 1959.

13a. D. Gabor, *J. Inst. Electr. Engrs.* **93**, Pt. III, 429 (1946).

14. P. H. Van Zittert, *Physica* **1**, 201 (1934).

15. R. Hanbury Brown and R. Q. Twiss, *Nature* **177**, 27 (1956).

16. R. Hanbury Brown and R. Q. Twiss, *Proc. Roy. Soc.* **A242**, 300 (1957).

17. R. Hanbury Brown and R. Q. Twiss, *Proc. Roy. Soc.* **A243**, 291 (1958).

18. E. M. Purcell, *Nature* **178**, 1449 (1956).

19. F. K. Kahn, *Optica Acta* **5**, 93 (1958).

20. L. Mandel, *Proc. Phys. Soc. (London)* **72**, 1037 (1959).

21. L. Mandel, *Proc. Phys. Soc. (London)* **74**, 233 (1959).

22. U. Fano, *Am. J. Phys.* **29**, 539 (1961).

23. A. Einstein, *Physik Z.* **10**, 185 (1909).

23a. E. Wolf *in* "Quantum Electronics III" (P. Grinet and N. Bloembergen, ed.), p. 13. Columbia Univ. Press, New York, 1964.

23b. M. Mandel *in* "Progress in Optics" (E. Wolf, ed.) Vol. 2, p. 181. North Holland Publishing Company, Amsterdam, 1963.

24. A. T. Forrester, R. A. Gudmundsen, and P. O. Johnson, *Phys. Rev.* **99**, 1691 (1955).

25. L. Mandel, *J. Opt. Soc. Am.* **51**, 797 (1961).

26. L. Mandel, *J. Opt. Soc. Am.* **52**, 1407 (1962).

27. G. Magyar and L. Mandel, *Nature* **198**, 255 (1963).

28. S. Smith and E. M. Purcell, *Phys. Rev.* **92**, 1069 (1953).

29. R. J. Glauber, *Phys. Rev.* **130**, 2529 (1963).

30. E. Wolf, *Proc. Symp. Optical Masers, New York, 1963,* p. 29. (Polytechnic Press, New York, 1963).

31. J. Weber, *IRE Trans. Electron Devices PGED* **3**, 1 (1953).

32. N. G. Basov and A. M. Prokhorov, *Zh. Eksperim. i Teoret. Fiz.* **27**, 431 (1954).

33. J. P. Gordon, H. J. Zeiger, and C. H. Townes, *Phys. Rev.* **95**, 282 (1954).

34. A. Yariv and J. P. Gordon, *Proc. IEEE* **51**, 4 (1963).

35. O. S. Heavens, *Appl. Optics* Suppl. 1, 1 (1962).

36. B. A. Lengyel, "Lasers." Wiley, New York, 1962.

37. J. P. Gordon, H. J. Zeiger, and C. H. Townes, *Phys. Rev.* **99**, 1264 (1955).

38. W. R. Bennett, Jr., *Appl. Optics* Suppl. 1, 24 (1962).

39. A. Javan, E. A. Ballik, and W. L. Bond, *J. Opt. Soc. Am.* **52**, 96 (1962).

40. S. Sugano, *Appl. Optics* Suppl. 1, 92 (1962).

40a. E. P. Riedel, *Appl. Phys. Letters* **5**, 162 (1964).

41. F. J. McClung, S. E. Schwarz, and F. J. Meyers, *J. Appl. Phys.* **33**, 3139 (1962).

42. R. J. Collins, D. F. Nelson, A. L. Schawlow, W. Bond, C. G. B. Garrett, and W. Kaiser, *Phys. Rev. Letters* **5**, 303 (1960).

43. I. Wieder and L. R. Sarles, *Phys. Rev. Letters* **6**, 95 (1961).

44. P. P. Sorokin and M. J. Stevenson, *Phys. Rev. Letters* **5**, 557 (1960).

45. G. D. Boyd, R. J. Collins, S. P. S. Porto, A. Yariv, and W. A. Hargreaves, *Phys. Rev. Letters* **8**, 269 (1962).

46. L. F. Johnson and K. Nassau, *Proc. IRE* **49**, 1704 (1961).

47. L. F. Johnson, G. D. Boyd, K. Nassau, and R. R. Soden, *Phys. Rev.* **126**, 1406 (1962).

48. K. Nassau, *Proc. Symp. Optical Masers, New York, 1963* p. 451 (Polytechnic Press, New York, 1963).

49. E. Snitzer, *Phys. Rev. Letters* **7**, 444 (1961).

49a. J. E. GEUSIC, H. M. MARCOS, and L. G. VAN UITERT, *Appl. Phys. Letters* 4, 182 (1964).
50. S. JACOBS, G. GOULD, and P. RABINOWITZ, *Phys. Rev. Letters* 7, 415 (1961).
51. P. RABINOWITZ, S. JACOBS, and G. GOULD, *Appl. Optics* 1, 513 (1962).
52. A. JAVAN, W. R. BENNETT, JR., and D. R. HERRIOTT, *Phys. Rev. Letters* 6, 106 (1961).
53. G. RACAH, *Phys. Rev.* 61, 537 (1942).
54. A. V. PHELPS, *Phys. Rev.* 99, 1307 (1955).
55. A. V. PHELPS, *Phys. Rev.* 110, 1362 (1958).
56. A. D. WHITE and J. D. RIGDEN, *Proc. IRE* 50, 1697 (1962).
57. J. D. RIGDEN and A. H. WHITE, *Proc. IRE* 50, 2366 (1962).
58. W. R. BENNETT, JR., W. L. FAUST, R. A. MCFARLANE, and C. K. N. PATEL, *Phys. Rev. Letters* 8, 470 (1962).
59. W. R. BENNETT, JR., *Bull. Am. Phys. Soc.* [2] 7, 15 (1962).
60. H. S. W. MASSEY and E. H. S. BURHOP, "Electronic and Ionic Impact Phenomena" Oxford Univ. Press (Clarendon), London and New York, 1952.
61. G. J. SCHULZ, *Phys. Rev.* 112, 150 (1958).
62. R. M. ST. JOHN, F. L. MILLER, and C. C. LIN, *Phys. Rev.* 134, A888 (1964).
63. W. B. BRIDGES, *Appl. Phys. Letters* 4, 128 (1964).
64. W. E. BELL, *Appl. Phys. Letters* 4, 34 (1964).
65. H. G. HEARD, *Nature* 200, 667 (1963).
66. R. N. HALL, G. E. FENNER, J. D. KINGSLEY, T. J. SOLTYS, and R. O. CARLSON, *Phys. Rev. Letters* 9, 366 (1962).
67. T. M. QUIST, R. H. REDIKER, R. J. KEYES, W. E. KRAG, B. LAX, A. L. MCWHORTER, and H. J. ZEIGER, *Appl. Phys. Letters* 1, 91 (1962).
68. B. LAX, *Proc. Symp. Optical Masers, New York, 1963* p. 119. (Polytechnic Press, New York, 1963).
69. M. I. NATHAN, W. P. DUMKE, G. BURNS, F. H. DILL, and G. LASHER, *Appl. Phys. Letters* 1, 62 (1962).
70. N. HOLONYAK and S. F. BEVACQUA, *Appl. Phys. Letters* 1, 82 (1962).
71. K. WEISER and R. S. LEVITT, *Appl. Phys. Letters* 2, 178 (1963).
72. I. MELNGAILIS, A. J. STRAUSS, and R. H. REDIKER, *Proc. IEEE* 51, 1154 (1963).
73. I. MELNGAILIS, *Appl. Phys. Letters* 2, 176 (1963).
74. R. J. PHELAN, A. R. CALAWA, R. H. REDIKER, R. J. KEYES, and B. LAX, *Appl. Phys. Letters* 3, 143 (1963).
75. K. KONNERTH and C. LANZA, *Appl. Phys. Letters* 4, 120 (1964).
76. W. P. DUMKE, *Phys. Rev.* 132, 1998 (1963).
77. G. LASHER and F. STERN, *Phys. Rev.* 133, 553 (1964).
78. T. A. FULTON, D. B. FITCHEN, and G. E. FENNER, *Appl. Phys. Letters* 4, 9 (1964).
79. L. LANDAU, *Z. Physik* 64, 629 (1930).
80. J. CALLAWAY, "Energy Band Theory." Academic Press, New York, 1964.
81. B. LAX, *in* "Quantum Electronics" (C. H. TOWNES, ed.), p. 428. Columbia Univ. Press, New York, 1960.
82. B. LAX, *in* "Quantum Electronics" (J. SINGER, ed.), p. 465. Columbia Univ. Press, New York, 1961.
83. J. L. HIRSCHFIELD and J. M. WACHTEL, *Phys. Rev. Letters* 12, 533 (1964).
84. A. LEMPICKI and H. SAMELSON, *Phys. Letters* 4, 133 (1963).
85. A. LEMPICKI and H. SAMELSON, *Symp. Optical Masers, New York, 1963*, p. 347 (Polytechnic Press, New York, 1963).
86. R. YOUNG, *J. Chem. Phys.* 40, 1848 (1964).
87. I. S. MARSHAK, *Pribory i Tekhn. Eksperim.* 3, 5 (1962).

88. S. CLAESSON, *Laser Flash Lamp Conf., Stanford Res. Inst., 1964.*
89. J. M. FELDMAN, *Laser Flash Lamp Conf., Stanford Res. Inst., 1964.*
90. J. W. OGLAND, C. H. BAUGH, and W. E. HORN, *Bull. Am. Phys. Soc.* [2] **9**, 270 (1964).
91. C. H. CHURCH, R. D. HAUN, JR., T. A. OSIAL, and E. V. SOMERS, *Appl. Optics* **2**, 451 (1963).
91a. R. J. KEYES and T. M. QUIST, *Appl. Phys. Letters* **4**, 50 (1964).
92. D. F. NELSON and W. S. BOYLE, *Appl. Optics* **1**, 181 (1962).
93. G. E. DEVLIN, J. McKENNA, A. D. MAY, and A. L. SCHAWLOW, *Appl. Optics* **1**, 11 (1962).
94. R. W. HELLWARTH, *in* "Quantum Electronics" (J. Singer, ed.), p. 334. Columbia Univ. Press, New York, 1961.
95. F. J. McCLUNG and R. H. HELLWARTH, *J. Appl. Phys.* **33**, 828 (1962).
95a. N. T. MELAMED, C. HIRAYAMA, and P. W. FRENCH, *Bull. Am. Phys. Soc.* [2] **9**, 729 (1964).
96. W. W. RIGROD, H. KOGELNIK, D. J. BRANGACCIO, and D. R. HERRIOTT, *J. Appl. Phys.* **33**, 743 (1962).
97. A. G. FOX and T. LI, *Proc. IRE* **48**, 1904 (1960).
98. A. G. FOX and T. LI, *Bell System Tech. J.* **40**, 453 (1961).
99. G. D. BOYD and J. P. GORDON, *Bell System Tech. J.* **40**, 489 (1961).
100. T. P. HUGHES, *Nature* **195**, 325 (1962).
101. T. P. HUGHES and K. M. YOUNG, *Nature* **196**, 332 (1962).
102. A. J. DEMARIA and R. GAGOSZ, *Appl. Optics* **2**, 807 (1963).
103. P. F. BROWNE, to be published.
104. B. J. McMURTY and A. E. SIEGMAN, *Appl. Optics* **1**, 51 (1962).
105. W. R. BENNETT, JR., *Phys. Rev.* **126**, 580 (1962).
106. P. A. FRANKEN, A. E. HILL, C. W. PETERS, and G. WEINREICH, *Phys. Rev. Letters* **7**, 118 (1961).
107. W. G. CADY, "Piezoelectricity." McGraw-Hill, New York, 1946.
108. P. A. FRANKEN and J. F. WARD, *Rev. Mod. Phys.* **35**, 23 (1963).
109. R. C. MILLER, *Phys. Rev.* **134**, A1313 (1964).
110. P. D. MAKER, R. W. TERHUNE, M. NISENOFF, and C. M. SAVAGE, *Phys. Rev. Letters* **8**, 21 (1962).
111. P. K. Tien, *Appl. Phys.* **29**, 1347 (1958).
112. R. D. HAUN, *Westinghouse Res. Rept.* 6-41003-R2 (1959).
113. J. A. GIORDMAINE, *Phys. Rev. Letters*, **8**, 19 (1962).
114. R. W. TERHUNE, P. D. MAKER, and C. M. SAVAGE, *Appl. Phys. Letters* to be published.
115. E. J. WOODBURY and W. K. NG, *Proc. IRE* **50**, 2367 (1962).
116. G. ECKHARDT, R. W. HELLWARTH, F. J. McCLUNG, S. E. SCHWARZ, D. WEINER, and E. J. WOODBURY, *Phys. Rev. Letters* **9**, 455 (1962).
117. R. W. TERHUNE, *Bull. Am. Phys. Soc.* **8**, 359 (1963).
118. E. GARMIRE, F. PANDARESF, and C. H. TOWNES, *Phys. Rev. Letters* **11**, 160 (1963).
119. N. BLOEMBERGEN and Y. R. SHEN, *Phys. Rev. Letters* **12**, 504 (1964).
120. R. Y. CHIAO, E. GARMIRE, and C. H. TOWNES, *Phys. Rev. Letters* **13**, 479 (1964).

7

SECONDARY EFFECTS OF LIGHT
AND PROCESSES OF DETECTION

We have seen that from a practical viewpoint the interactions of radiation with matter can be subdivided into three major processes: (1) those involved in the emission of light; (2) the phenomena of propagation, and (3) the effects of light on matter. In fact, in applications such as systems of communication, sensing, or measurement, one customarily differentiates the three corresponding subdivisions of (1) sources, (2) optical components, and (3) detectors.

We have been following this sequence of subjects only to a limited extent, and this for the simple reason that the physical aspects of the three areas are often inseparable. This is in particular true of the elementary processes of emission and absorption which are in most respects symmetrical events and must be described together. There are, of course, many other examples. The phenomena of propagation and especially dispersion cannot be treated without a discussion of absorption. Then again the subject of lasers, which we have dealt with in the chapters on propagation, contains many elements of emission theory. Thus when we finally come to the effects of optical radiation, this general subject has already been anticipated in many respects: in the discussion of F-centers, for instance, or in the treatment of elementary absorption processes.

Nevertheless, there exists a range of phenomena and their applications in this third area of optical physics which stands apart from the other two. We refer here to the class of secondary effects of light—so-called because they involve more than mere mirror images of the light emission process—and their use in detectors. Here belong the various phenomena of photoemission and photoconduction and the effects of light on the chemical bond. This subject will be treated in the first part of this chapter; the end of the chapter will deal with its salient application, viz., the processes of detection. These, for reasons of completeness will include other principles as well. In passing, the subject of fluctuations and noise will be taken up because from it follow the ultimate limitations to detection and also because it represents an interesting aspect of photon and electron behavior in its own right.

382

7.1 Photoeffects

We shall discuss in this section a very general class of quantum effects all of which have in common that a photon of energy $h\nu$ encounters a particle bonded to another particle, or a system of particles, with an energy $E_0 < h\nu$. In all these cases, there exists a probability—which often approaches unity in solids—that the energy of the quantum is converted into (a) disruption of the bond and (b) an excess which usually appears as kinetic energy E_k of some form. This energy balance is expressed by the Einstein equation[1]

$$h\nu = E_0 + E_k \qquad (7.1)$$

The most important among these phenomena is the external photoelectric effect which was discovered by Hertz[2] and which amounts to the liberation of photoelectrons from the surface of a solid by light into vacuum or an atmosphere with an excess energy $E_k = mv^2/2$. This excess may, of course, be dissipated by collision processes, and thus E_k represents the *maximum* kinetic energy observed in the emerging electron stream. The effect was first observed with metals. Later work showed that certain semiconductors offered much higher photoemissive yield and a response to considerably smaller photon energies than was attainable with metals. By its very nature, the photoelectric effect is in principle not limited to any state of matter and even appears in the interaction of photons with free atoms as *photoionization*. The minimum energy E_0 or *work function* which light requires to liberate an electron depends, of course, on the type of bonding encountered. E_0 is in the order of 1 eV for the solids with smallest binding energy and amounts to several electron volts for most metals. The ionization potential of the free atoms, ranging from 3.8 V for cesium to 24.5 V for helium, is in the order of twice the work function in the case of metals. Splitting off electrons from ions requires yet higher energies. Thus the external photoelectric effect is primarily a phenomenon of the ultraviolet and visible region, with only one type of compound offering a wavelength threshold as long as $1.25\ \mu$.

A somewhat different situation exists for the *internal photoelectric effect*, so called because here light does not entirely free the electron from the bonds of a lattice, but merely moves it across an energy gap to a new level. If in this new energy state, as is usually the case, the electron (or hole) has also a different mobility, then the internal photoelectric effect reveals itself as *photoconductivity*. It is at once obvious that photoconductivity, unlike the external effect, is limited to those substances—

notably among the semiconductors and insulators—in which the appearance of relatively few current carriers produces a measurable change of electrical resistivity. In further contrast to the external effect, photoconductivity can occur for the frequencies of the entire optical spectrum. Of course, these possibilities have been opened up only gradually over many decades since they required, in increasing measure, control over the composition of materials and their operation.

Certain other photoeffects are more or less related to the preceding two classes. Examples are the photographic process, the quenching or enhancing of luminescence by incident light, the blackening of crystals by photon induced free carriers, and magneto-optical processes such as the inverse of the transitions between Landau levels which were described in connection with the cyclotron maser (see Section 6.3.2.4). In the following, details of the various photoeffects will be discussed. Because of Eq. (7.1), the most important parameter in each case is the energy distribution of the eligible charged particles. The distributions follow from the structural models of the various kind of matter and have been described in the preceding pages, particularly in Chapter IV.

7.1.1 The Photoelectric Effect

7.1.1.1 *Electron Energy Distribution and Response in Metals*

As discussed in detail in Section 4.6.1.3, the distribution $dN(E)$ of electrons over various energies E in metals can be understood on the basis of the band structure of the latter and the Fermi-Dirac distribution law

$$f(E) = \frac{2}{e^{(E-E_f)/kT} + 1} \tag{7.2}$$

where we count two opposite spin states per cell and where the Fermi level E_f is defined by the condition of being half occupied. This function is shown for absolute zero and finite temperatures, respectively, in Fig. 7.1a. It is important to emphasize the implications of the probability function (7.2) for the electron theory of metals. First of all, the Fermi-Dirac distribution law was derived, in many respect parallel to the Bose-Einstein statistics (see Section 2.3.2), by computing the most probable distribution of N electrons over the various available cells in phase space each of volume h^3. However, in contrast to "bosons," only two "fermions" of opposite spin can occupy a cell in phase space. Suppose the electron ensemble is held at absolute zero, $T = 0$. The electrons will then assume the lowest possible positions in energy, two

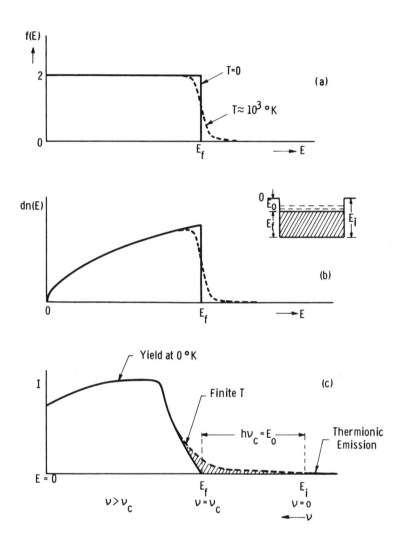

FIG. 7.1. Fermi-Dirac distribution of electron energies in metals and its consequences for the photoelectric effect. (a) Probability function of occupancy for cells in phase space with various energy E (including a factor 2 for electrons of opposite spin). (b) Number of electrons in various energy intervals between E and $E + dE$. Insert shows energy level diagram with respect to vacuum. (c) Photoemissive yield for absolute zero and finite temperatures versus frequency (from right to left). Scale of "forward" energies $m\dot{x}^2/2$ of metal electrons is superimposed (from left to right).

in each cell, so that starting with $E = 0$ all levels are filled until all N electrons are accommodated. This is exactly the meaning of Fig. 7.1a or Eq. (7.2) which was derived on this premise and which states that the probability of occupation for any energy level up to a limit E_f is unity and that for larger energies—discontinuously at $T=0$—the cells remain empty. Therefore, the Fermi level $E_f(0)$ at absolute zero follows from the number of available electrons N and the density of states at the various energies. The number of cells in momentum space between p and $p + dp$ is, in analogy to Eq. (1.32),

$$dZ = 4\pi V \frac{p^2\,dp}{h^3} \tag{7.3}$$

Since energy and momentum are related by

$$E = \frac{p^2}{2m} \quad \text{and} \quad dE = \frac{p\,dp}{m} \tag{7.4}$$

we can compute the distribution $dN(E)$ of electrons over the various energy intervals between E and $E + dE$ by means of Eqs. (7.2) and (7.3) and the density of states function, $g(E)\,dE$, Eq. (4.98):

$$dN(E) = f(E)g(E)\,dE = 2\,\frac{4\pi V}{h^3} \cdot \frac{\sqrt{2m^3E}}{e^{(E-E_f)/kT} + 1}\,dE \tag{7.5}$$

This function is shown for absolute zero and finite temperatures by the solid and dashed curves, respectively, of Fig. 7.1b. The boundary condition

$$\int dN = N = nV \tag{7.6}$$

determines the Fermi level $E_f(T)$ in terms of the electron density n. In general, $E_f(T)$ can be evaluated only by approximation methods. For $T = 0$, however, the calculation is simple, since it is known that all those cells in phase space are filled for which the momentum $p < p_f$, the limit corresponding to the Fermi level. Therefore, p_f is the radius of that sphere in phase space which contains $N/2$ volume elements of size h^3:

$$\frac{N}{2} = \frac{V}{h^3}\frac{4\pi}{3}p_f^3 = \frac{4\pi}{3}\frac{V}{h^3}[2mE_f(0)]^{\frac{3}{2}} \tag{7.7}$$

or

$$E_f(0) = \frac{h^2}{8m}\left(\frac{3n}{\pi}\right)^{\frac{2}{3}} \tag{7.8}$$

For relatively low temperatures, $kT \ll E_f(0)$, $E_f(T)$ can be approximated by a series.[3] We cite here the result:

$$E_f(T) \cong E_f(0) \left[1 - \frac{\pi^2}{12} \left(\frac{kT}{E_f(0)} \right)^2 \right] \tag{7.9}$$

For the electron densities typical of metals, the Fermi level $E_f(0)$ is in the order of 10 eV. Thus the second term in the bracket of Eq. (7.9) remains small even at the highest temperatures which a metal can tolerate, and for practical purposes the Fermi level can be considered a constant of the material, $E_f \cong E_f(0)$.

The principal effect of temperature on the energy distribution then consists of the relocation of relatively few electrons from positions just below the Fermi level to the energy region above.

The Fermi-Dirac distribution function fits the properties of metal electrons very well for two reasons: (1) despite the large densities of the electrons in the solid, their Coulomb interaction is small owing to the neutralizing effect of the positive ion lattice; and (2) only a part of the available states is occupied so that electrons are free to assume higher energies under thermal agitation.

Certain additional facts have to be considered to provide the background by which the photoelectric effect in metals and some of its detail features can be explained. First, the electrons—free inside the metal—find themselves bound by a potential energy E_i with respect to vacuum (see insert of Fig. 7.1b). The total energy of an electron is the sum of the potential energy E_i and the kinetic energy E which is distributed according to the Fermi-Dirac law. At $T = 0$, the highest total energy equals $-E_i + E_f$, and it is this amount at least which has to be supplied by a photon to eject an electron. In other words, the threshold energy or work function at absolute zero is given by

$$E_0 = e\varphi = E_i - E_f \tag{7.10}$$

where φ represent the work function in volts as it is also used in the literature. We have therefore the following picture for the wavelength dependence of the photoemissive yield in metals (see Fig. 7.1c):

At long wavelengths, above the threshold $\lambda_c = ch/E_0$, there is no photocurrent at $T = 0$. When the energy is increased to and above the value of the work function $E_0 = h\nu_c$, photocurrent begins to appear, and it will grow as at larger energies a greater population of electrons becomes eligible for ejection. Ultimately, however, a type of volume effect will interfere since shorter wavelengths may penetrate deeper into the metal producing photoelectrons which cannot reach the surface. This behavior, indicated by the solid curve of Fig. 7.1c, applies to absolute zero. At finite temperatures, some electrons reach energies above the Fermi level and thus require less than the absolute threshold energy for their ejection from the metal. In fact, a fraction (which is measurable only at higher temperatures for most metals) reaches the escape energy by thermal agitation alone and is emitted as thermionic current. Thus the photoelectric yield at finite temperatures is typified by a tail (dashed curve in Fig. 7.1c) extending beyond the wavelength of the absolute threshold.

The thermal behavior of photoemission has been accounted for quantitatively by Fowler[4] (see also Hughes and Du Bridge[5]). The starting point of this theory is the Fermi-Dirac distribution function over the various velocity space cells $d\dot{x}\,d\dot{y}\,d\dot{z}$ at $\mathbf{v} = \dot{\mathbf{x}} + \dot{\mathbf{y}} + \dot{\mathbf{z}}$, viz.,

$$dn = 2\,\frac{m^3}{h^3}\,\frac{d\dot{x}\,d\dot{y}\,d\dot{z}}{\exp\left[(\tfrac{1}{2}mv^2 - E_f)/kT\right] + 1} \tag{7.11}$$

It is next assumed that all those electrons can leave a metal in a direction x normal to the surface which have reached a velocity $\dot{x}_0$ large enough so that

$$\frac{m}{2}\,\dot{x}_0{}^2 + h\nu \geqslant E_i \tag{7.12}$$

Integration over all numbers $dn(\dot{x})$ of electrons for which Eq. (7.12) is fulfilled yields a photocurrent I:

$$I = A'T^2\,\frac{h(u)}{(E_i - h\nu)^{\frac{1}{2}}} \tag{7.13}$$

where A' is a constant, $u = h(\nu - \nu_c)/kT$, and

$$h(u) = e^u - \frac{e^{2u}}{2^2} + \frac{e^{3u}}{3^2} - \cdots\,; \qquad u \leqslant 0 \tag{7.14}$$

$$h(u) = \frac{\pi^2}{6} + \frac{u^2}{2} - \left(e^{-u} - \frac{e^{-2u}}{2^2} + \frac{e^{-3u}}{3^2} - \cdots\right); \qquad u \geqslant 0 \tag{7.15}$$

The constant A' may be written as $P \cdot A$, where P represents the combined probabilities of electron excitation, transmission to the surface, etc., and A is the constant contained in Richardson's equation for thermionic current

$$I = AT^2 e^{-eq/kT} \tag{7.16}$$

In fact, Eq. (7.13) goes over into (7.16) for $\nu = 0$ and large T.

Fowler's function was found to be in excellent agreement over a temperature range between 300 and 1000 °K for such metals as silver, gold, and palladium. The temperature dependence of the photoelectric effect for $\nu = \nu_c$ is given simply by T^2 so that the thermal coefficient of photoemission $(\partial I/\partial T)/I$ at the threshold equals $2T/T^2 = 2/T$. This amounts to $0.6\%/°K$ at 300 °K. A somewhat larger effect $(1 - 2\%/°K)$ presumably based on the same mechanism is found for semiconductors such as cesium-antimony (see e.g. Garbuny et al.[6]). The temperature dependence of photoemission has to be taken into consideration, if the work function is derived from yield curves. This can be done either by extrapolating the slope of the photoemission curve (of the type shown in Fig. 7.1c) before it ends in the tail section or, preferably, by a graphical analysis such as based[5] on Eq. (7.13). Table 7.1 lists work functions and

TABLE 7.1

WORK FUNCTIONS AND THRESHOLD WAVELENGTHS OF SOME IMPORTANT METALS[a]
AND THE IONIZATION POTENTIAL OF THEIR FREE ATOMS[b]

Metal	Work functions (V)	Threshold wavelength $\lambda_{\max}$ (μ)	Ionization Potential of Free atoms (V)
Cesium	1.94	0.639	3.87
Rubidium	2.13	0.582	4.159
Potassium	2.25	0.551	4.318
Sodium	2.29	0.541	5.12
Lithium	2.69	0.461	5.363
Barium	2.56	0.484	5.19
Strontium	2.74	0.452	5.667
Calcium	3.20	0.387	6.09
Magnesium	3.67	0.338	7.61
Chromium	4.37	0.284	6.74
Molybdenum	4.20	0.295	7.35
Tungsten	4.54	0.274	8.1
Copper	4.36	0.284	7.68
Silver	4.63	0.268	7.54
Gold	4.80	0.258	9.18
Platinum	5.55	0.223	8.88

[a] From W. Kluge, in Landolt-Börnstein, "Zahlenwerte und Funktionen," Vol. II, Part 6, I. Springer, Berlin, 1959.
[b] From "Handbook of Chemistry and Physics," 44th ed. Chem. Rubber Publ. Co., Cleveland, Ohio, 1962-1963.

ionization potentials for some metals of interest. It is seen that, qualitatively, the work functions behave as the ionization energies with
respect to position of the metals in the periodic table.

7.1.1.2 Quantum Efficiency of Metallic Photoemission

Of further theoretical and even more practical import is the question
concerning the absolute yield of photoemission. In other words, we are
inquiring about the *quantum efficiency*, i.e., the number of ejected
electrons per incident photon at a given frequency. The photoemissive
yield is also measured directly in terms of current per incident radiation
power (amperes/watt) or emitted charge per energy (coulombs/joule) and
by the more subjective unit of current per lumen (the latter being
defined as the flux density produced by a candle at unit distance). These
units are interrelated as follows:

$$1\,\frac{\text{ampere}}{\text{watt}} = 1\,\frac{\text{coulomb}}{\text{joule}} = \frac{1.24\,\mu}{\lambda}\,\frac{\text{electrons}}{\text{quantum}} = 1.61 \times 10^{-3}\,\frac{\text{amperes}}{\text{lumen}_{0.556}} \quad (7.17)$$

where the lumen has been equated to 0.00161 W at 0.556 μ. The determination of the yield thus requires the determination of the incident
radiation power, either by computing the fraction of power received
from a known source such as a black body, or by calibration, e.g., with
a bolometer. This task is difficult only for the ultraviolet wavelength
region where conventional black body cavities radiate little power and
the calibration of bolometers may be unreliable. The quantum efficiency
versus wavelength is shown for several metals in Fig. 7.2, which is the
result of measurements taken by Schulze[7] on photoemissive layers
evaporated onto the inner wall of a glass sphere. Although the yield from
thin films depends on the thickness, which in these experiments may not
have been uniform, certain salient features are apparent which are also
confirmed by other determinations. There is first the dependence of the
wavelength threshold for the various metals on their position in the
periodic system and, correlated with it, also the yield. In the range of
wavelengths shown, the quantum efficiency is generally the smaller, the
higher the work function. The absolute values of the yield are of particular
interest. A comparison with Table 7.1 indicates that for photon energies
well above the absolute threshold the quantum efficiencies of K, Sr, Ca,
and Mg are as low as 10^{-3} and 10^{-4}, and approach 10^{-1} only in the ultraviolet region near 6 eV. The more inert metals of high work function
exhibit still lower yields.

Any theory attempting to account quantitatively for these phenomena
is apt to be quite complex. It must be realized that the approach dis-

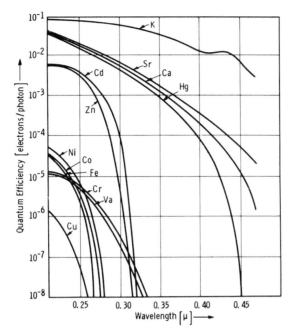

FIG. 7.2. Quantum efficiency of metals versus wavelength of incident light [from R. Schulze, *Z. Physik* **92**, 223 (1934), and W. Kluge, *in* Landolt-Börnstein, "Zahlenwerte und Funktionen," Vol. II, Part 6, I. Springer, Berlin, 1959].

cussed in the preceding section and its result in Eq. (7.13) are valid only near the frequency of the photoelectric threshold and do not shed much light on the absolute yield. For example, the constant A' contains factors of probability which can be determined only, if specific assumptions are made regarding the mechanism of the photoelectric emission process.

However, in setting up a model for the effect in metals, one has to deal with a particular difficulty: a free electron cannot absorb the energy of a photon in the optical range because it thereby would acquire a momentum which the photon cannot supply. This problem exists also for *free-free* transitions in plasmas, and there it finds its answer in the interaction of the electron with transient fields. In other words, momentum can be conserved because more than two particles are involved, to wit, the photon, the electron, and a third center which provides the field. In the simple theory of metals, electrons are assumed to be completely free within the lattice, bound only with respect to vacuum by a potential energy E_i. This condition gives rise to a field at the surface which consequently can supply a site for the generation of photoelectrons.

In the more refined model of metal electrons, a periodic potential is recognized owing to the positive ion lattice. Therefore, photoemission can originate as a *surface effect* as well as a *volume effect* (Tamm and Schubin[8,9]). It is noteworthy that even these basically different classes of phenomena are difficult to distinguish experimentally and still more difficult to assess theoretically. Fan[10] computes the volume effect for metals, especially sodium and potassium, and finds that the order of magnitude of the yield as well as of the wavelength threshold are the same as for the surface effect.

However, certain measurements by Thomas[11] appear to indicate that photoemission in potassium is a volume effect. In this experiment, potassium was evaporated onto a quartz substrate in a high vacuum (10^{-8}–10^{-9} mm Hg). While the metal layer was in the process of building up to thicknesses of several hundred atoms, photoelectric yield measurements were undertaken in response to monochromatic light. Typical curves are shown in Fig. 7.3. Thomas[11] interprets the thickness at which the yield begins to saturate as the *range* of the photoelectrons on the grounds that electrons originating at deeper levels cannot reach the surface. Below the saturation point, the yield is linear with thickness; hence the photocurrent appears to be overwhelmingly due to a volume effect. It is also seen that the saturation thickness increases markedly with wavelengths longer than 0.289 μ, implying decreasing range for higher energies. The interpretation[11,12] of this behavior requires an analysis of the losses which electrons in the pertinent velocity range

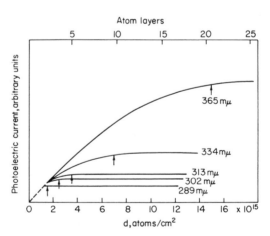

Fig. 7.3. Photoelectric yield versus thickness of potassium layer evaporated on quartz. Thickness at which saturation begins is interpreted as range of electrons as a function of their energy [H. Thomas, *Z. Physik* **147**, 395 (1957)].

suffer as they move through the metal lattice. These interactions may involve the entire lattice by excitation of plasma oscillations [with ν given by Eq. (5.153)] so that internal kinetic energies of the electron near 4 eV are absorbed in a single process; or, at lower velocities, multiple collisions with electrons, impurity centers, phonons, etc., may be dominant.

It is questionable whether these losses are necessarily characterized by an increase of the range for increasing wavelength. In contrast to the alkali metals which show a pronounced *selective* yield maximum at certain wavelengths, the other metals display a photoemission which increases monotonically with frequency. This fact may indicate different range-wavelength relationships for different classes of metals. Suhrmann[13] explains the selective effect in potassium entirely with increasing transmissivity (i.e. decreasing k with increasing ν) and, in fact, offers an alternative explanation for the findings of Thomas, viz. by the change of surface condition and work function for increasing layer thickness. An experimental arrangement designed by the author and used by J. S. Talbot may aid in the clarification of this issue since it substantially separates the effect of work function and of range in photoemission. In this arrangement (see Fig. 7.4), the inner wall of a glass sphere is coated with photoemissive cesium-antimonide ($\lambda_c \approx 0.600\ \mu$) onto which a layer of an alkali metal has been evaporated, e.g. sodium ($\lambda_c \approx 0.541\ \mu$). Thus photoelectrons will be produced almost exclusively in the Cs_3Sb substrate by suitable monochromatic light; they then are partially absorbed by the metal layer, while the rest is collected by the central anode. In the experiment, the metal thickness is depleted at a controlled and measurable rate by immersing into liquid nitrogen a trap attached to the sphere through a well-defined orifice. The photocurrent will then increase exponentially up to a thousand fold its initial value with time (see insert of Fig. 7.4) so that the range of the photoelectrons is given by the corresponding thickness decrease. In this manner, the range of electrons in a cesium layer produced by red light in cesium antimonide was measured as 30 Å (here, too, substantially no electrons are produced in the metal layer).

The various aspects of the surface and volume effects for yield, angle and energy distribution, etc., have been treated in a series of theoretical articles.[10,13–17] To summarize, the photoelectric effect near the threshold is adequately described by the Fowler equation (7.13) with respect to temperature and frequency dependence for most metals (excluding, e.g., the alkalis). Under these conditions, a true surface effect may conceivably be dominant. At higher frequencies, the volume effect will become important and ultimately make the only contribution to the current (whereas in alkali metals this latter situation may exist throughout the

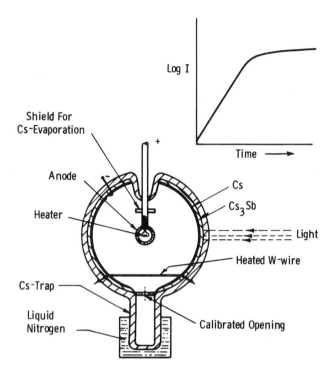

FIG. 7.4. Range determination of photoelectrons from cesium antimonide film in top layer of cesium. After the Cs_3Sb surface has been formed cesium is evaporated onto it. This top layer is in turn depleted by immersing the trap in liquid nitrogen. From the Langmuir current in tungsten wire, produced by ionization of cesium atoms, the rate of thickness decrease is known, so that the range in cesium is obtained from the photocurrent versus time characteristic (see insert).

entire frequency range). The quantum efficiency of the volume effect at a frequency ν for electrons of total energy E_j and momentum vector $\mathbf{k}$ may be written as

$$\eta_{\nu,j,k} = \delta N \cdot B \cdot T \cdot D \tag{7.18}$$

Here, the quantum efficiency is the product of the following four factors:

(1) the number of electrons δN in the energy and momentum range dE_j and $d\mathbf{k}$ ($E_j > E_i - h\nu$);

(2) the transition probability B in this range of energy and momentum;

(3) the probability T of reaching the surface of the metal; and

(4) the probability D of passing through the surface without reflection.

The total quantum efficiency η is obtained by integrating over all allowed energy and momentum values. Equation (7.18) is not limited to metals, but applies also to solids with bound electrons. In these cases, the volume effect is always dominant because momentum can be conserved anywhere within the solid owing to the binding forces.

7.1.1.3 *Energy Distribution of Photoelectrons from Metals*

The statistical distribution of energies with which photoelectrons emerge from a metal represents a problem related to that of the quantum efficiency. The kinetic energy which an electron has left at the surface is given by the excess $h\nu - E_0$ [cf. Eq. (7.1)] minus the effect of collisions in the metal. In the latter, we have also to include the thermal energy kT, although it may not be detectable within the accuracy of such experiments. Thus we should expect that the emerging photoelectrons exhibit kinetic energies between zero and the excess value, the latter increasing linearly with the frequency of the incident light.

The measurement of the energy distribution may be performed in principle with any method of velocity analysis such as that based on deflection. Nevertheless, there exists only one procedure, both simple and reliable, which has been applied almost exclusively for the purpose, viz., the determination by retarding fields (see insert, Fig. 7.5). At a given frequency of light incident on the photocathode, one measures the current resulting from the application of a varying negative potential E which includes the contact potential between the two electrodes and which has, preferably, spherical symmetry. The current-voltage characteristic has typically the shape of the dotted curve in Fig. 7.5. Differentiation of this function yields the value dI/dE, i.e., that portion of photocurrent contained in the potential interval between E and $E + dE$. This represents, of course, the energy distribution which, when normalized, has the profile given by the solid curve in Fig. 7.5 and is then approximately the same for most metals.

7.1.1.4 *The Vectorial Photoelectric Effect*

Before leaving the subject of photoemission in metals, we have to discuss an effect which, although discovered as early as 1894 by Elster and Geitel,[18] has not found a completely satisfactory explanation. In this and other respects, the problem is related to the question of the original sites and energies of the photoelectrons, in particular the question, whether one deals with a surface or a volume effect.

Elster and Geitel found that light obliquely incident on a photosurface

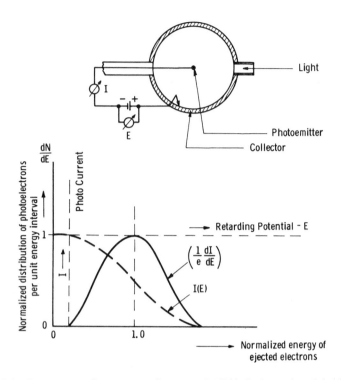

FIG. 7.5.   Photocurrent I versus retarding potential E (dashed curve) and the derivative (solid curve), which yields the energy distribution of photoelectrons from metal surfaces. Insert in upper right corner: arrangement of the retarding potential method for the determination of electron energy distributions.

consisting of a liquid sodium-potassium alloy yielded a much larger photocurrent when the electric vector was contained in the plane of incidence than when it was directed normal to that plane. The ratio of photocurrents for the two polarizations was a function of the angle of incidence and reached a maximum of about 60 for 60°. Obviously, the important magnitude here is the electric vector component normal to the surface. The larger this component, the greater is the photoemissive yield. Indeed, this is exactly what should be expected, at least qualitatively, on the basis of the classical concept that the prospective photoelectrons undergo oscillations under the influence of, and parallel to, the electric field; hence the name *vectorial photoelectric effect*. On the basis of quantum theory, however, the phenomenon is much less plausible in these terms. Furthermore, it is difficult to establish a quantitative agreement of the observed data with this picture.

By its nature, the vector effect requires a smooth, reflecting surface to be observable. This condition can be best achieved with the alkali metals, either in molten or solid form, or evaporated on one of the noble metals. Under such circumstances one also observes a pronounced spectral selectivity, i.e., a maximum of photoemission for certain wavelengths (e.g., at 0.40μ for potassium). The connection between the two types of selectivity may well be tenuous; the alkali metals show a more or less abrupt transparency in the ultraviolet (cf. Section 5.2.4), and thus the decrease of yield at higher energies corresponds to vanishing absorption. Nevertheless, just this fact led Ives and co-workers[19-21] to extensive experimentation with alkali metals on platinum and similar materials and ultimately to an interpretation of the vector effect for these composite cathodes with which the selectivity phenomena can be particularly well observed. The conclusion of Ives is simply that the yield is determined by the intensity of the standing waves resulting by superposition of the incident and reflected beam in the composite metal structure. Indeed, the intensity of light just above the substrate metal for the two directions of polarization agrees quantitatively rather well with the corresponding trends of the photocurrents.[19]

The question whether the yield is determined by the intensity ($\propto E_z^2$) or by the absorption has been answered by experiments for the case of pure molybdenum by Juenker and his co-workers.[22] They observed that the photoelectric emission for light with the electric vector in the plane of incidence follows the computed values of the energy density near the surface at various angles of incidence, but not the power absorbed (see Fig. 7.6a). However, if the electric vector is normal to the plane of incidence, the absorbed power has the same trend as the energy density; hence the normalized yield follows either in the dependence on angle (see Fig. 7.6b). Despite the pronounced dependence of the photoelectric effect in molybdenum on angle of incidence and polarization, no spectral maximum of selectivity was found in this investigation[22].

7.1.2 THE EXTERNAL PHOTOELECTRIC EFFECT IN SEMICONDUCTORS AND INSULATORS

7.1.2.1 *General Relationships*

The existence of a forbidden gap E_g between an almost filled valence band and an almost empty conduction band produces conditions for the photoelectric effect which differ considerably from the behavior of metals. Perhaps the most significant feature of the bound electron structure is the virtual absence of inelastic scattering between electrons for certain

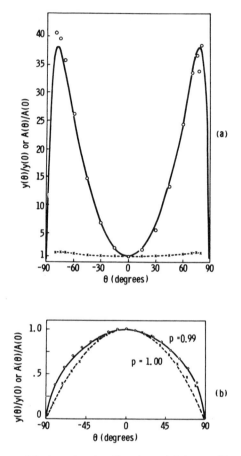

FIG. 7.6.   The vectorial photoelectric effect in molybdenum [D. W. Juenker, J. P. Waldron, and R. J. Jaccodine, *J. Opt. Soc. Am.* **54**, 216 (1964)]. (a) Experimental points represent photocurrent (normalized to perpendicular incidence) as a function of angle, if electric field vector is in plane of incidence. The solid curve is the computed energy density as a function of the angle of incidence. Experimental points establish correlation of yield to energy density rather than to absorbed power (dashed curve). (b) With electric field perpendicular to plane of incidence, experimental points for normalized yield agree with curve which simultaneously represents absorbed power and energy density.

ranges of energy (see Section 7.1.2.2). Thus in contrast to the metallic transport processes, the photoelectron in semiconductors and insulators may have a range of several hundred angstroms, encountering losses only by lattice scattering, i.e., phonon production. Such exchanges involve little energy. Hence, in principle, the photoemissive yield can be high. As we shall see, however, this requires a sufficiently small *electron affinity* E_a , which is the energy difference between the vacuum level and the bottom of the conduction band.

Another important difference lies in frequency response and energy distribution of the photoelectrons as determined by the retarding field method (see insert of Fig. 7.5). These properties, of course, follow from the density of energy states of the electrons within the solids. Figure 7.7 presents a comparison between the conditions in metals, semiconductors, and insulators and the voltage-current characteristics that are to be expected from the model of these three cases. The presentation chosen for Fig. 7.7 is based on the results obtained by Apker and his coworkers,

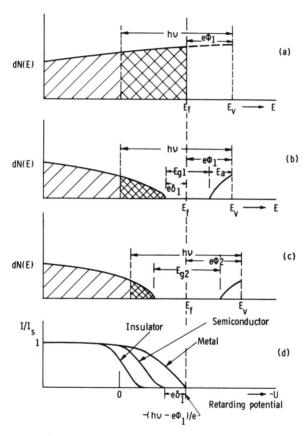

FIG. 7.7. Density of energy states and electron distribution in (a) metals, (b) semiconductors, and (c) insulators, and (d) the respective normalized photoemissive yields at retarding potential. Coincidence of Fermi level and the same *hν* has been assumed for the three cases.

particularly as described in their original paper on photoelectric emission in semiconductors.[23]

It is assumed in Fig. 7.7 that the Fermi levels of the materials coincide and that the work function of the metal and the semiconductor are equal The probability of finding electrons in the conduction band of semi-conductors and insulators because of thermal excitation is negligible for $kT \ll E_g$, and thus photoelectrons from these materials normally origi-nate in the valence band region unless there exist F-centers and other intermediate levels. Whereas (at $T = 0$) the most energetic photoelec-trons in metals come from the Fermi level (cf. Fig. 7.7a and d), in semi-conductors they originate at the top of the valence band (cf. Fig. 7.7b) which lies by an amount $e\delta$ deeper. Thus the retarding potential $-E_t$ at which the onset of photoemission is noticeable differs for the two cases:

$$-E_t = h\nu - e\varphi \qquad \text{(metals)} \qquad (7.19)$$

$$-E_t = h\nu - e(\varphi + \delta) \qquad \text{(semiconductors)} \qquad (7.20)$$

Similarly, the threshold frequency of semiconductors is given by

$$\nu_t = \frac{e}{h}(\varphi + \delta) \qquad (7.21)$$

Since functions for the density of states have different shape for metals and semiconductors, the corresponding voltage-current character-istics vary distinctly in the slope with which the yield rises from zero. However, the saturation point is the same for both cases under the conditions assumed (the electrons in states lying within an energy $h\nu$ below the common vacuum level constitute the maximum that the photo-emitter can supply regardless of the accelerating voltage of the collector, at least within the scope of the phenomena considered here). As in Fig. 7.5, E includes the contact potential with respect to the collecting electrode. The first derivative of the yield curve then is a measure of the energy distribution with which the photoelectrons leave the surface. Aside from their larger gap E_g and the resulting changes for the threshold and saturation points, insulators show the same qualitative behavior as semiconductors (see Fig. 7.7c and d), except for enhanced emission in the presence of F-centers.

7.1.2.2 Phenomena of Semiconductor Emission

The agreement of observed voltage-current characteristics with the theoretically expected profiles (see Fig. 7.7d) is very good within the scope of the assumptions advanced for it. In the work already referred to (Apker[23]), the yield of various metals and semiconductors could be examined by interchanging their position alternatively as photocathodes such that all were observed under substantially the same conditions of vacuum, illumination, and the collecting electrode arrangement. In particular, the behavior of metals and semiconductors could be compared, and where the work functions were approximately equal (as in platinum and tellurium), the pertinent characteristics in Fig. 7.7d could indeed be verified faithfully by the experiment. However, for certain semiconductors, one finds the following behavior:

As the energy $h\nu$ of the incident photon is increased, the energy distribution of the photoelectrons also increases at first, but changes rather abruptly to low values for certain "second" thresholds of frequency. This was first shown by Apker and Taft[24] (see Fig. 7.8) who explained the phenomenon by an inelastic collision process between the photo-

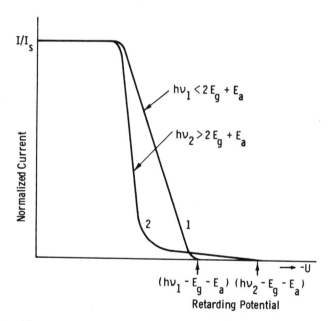

FIG. 7.8. Normalized photoemissive yields of semiconductors in retarding fields for the two conditions that (curve 1) the photon energy is insufficient for pair production across the gap; and (curve 2) large enough for inelastic electron-electron collisions. The second case is characterized by relative scarcity of electrons at higher energies (schematic presentation from measurements of L. Apker and E. A. Taft [*J. Opt. Soc. Am.* **43**, 78 (1953)] on various semiconductors).

electron and an electron in the valence band. More specifically, if the photon energy reaches a value

$$hv = 2E_g + E_a \qquad (7.22)$$

then it is capable of producing a hole-electron pair in addition to the photoelectron.

Because of this possible involvement of two and perhaps more electrons, it is meaningful to subdivide photoemissive semiconductors into two classes[25,26] which in fact represent categories of high and low yield:

(1) $E_g > E_a$ (see Fig. 7.9a). For the condition that the gap is larger than the electron affinity, there exists a relatively large range of excess energy for the photoelectron before it produces a hole-electron pair, viz.

$$E_g + E_a < hv < 2E_g + E_a \qquad (7.23)$$

or

$$0 < hv - (E_g + E_a) < E_g \qquad (7.24)$$

Under these conditions, the main loss mechanism consists of phonon production, amounting in the average to about 0.02 eV per collision so that the range of electrons with velocities of a few electron volts is about 250 Å.[27] If the electron affinity is small enough, the escape probability is high even after pair production. In all, the range of the photoelectrons in this class of semiconductors is of the same order as the penetration depth of the light. Hence, the quantum efficiency may be as

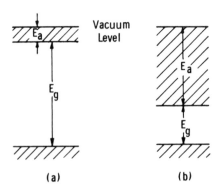

(a)   (b)

Fig. 7.9. Semiconductors with electron affinity E_a (a) smaller (high yield) and (b) larger (low yield) than energy gap E_g .

high as 0.4 electrons per photon, the theoretical maximum after allowance is made for losses due to deflections into the interior. To this class belong photolayers of some alkali-antimonides, e.g., Cs_3Sb and Na_2KSb, or certain I–VIa compounds, e.g., Cs_2Te and Rb_2Te. This is the reason for the practical importance, in particular, of the antimonide photo-surfaces in which high yield is combined with a wavelength threshold location in the visible spectrum. The addition of excess alkali atoms reduces the electron affinity, hence increases the yield as shown in Table 7.2 which is a selection of data presented by Spicer.[26,28]

TABLE 7.2

BAND GAP E_g, ELECTRON AFFINITY E_a, AND PHOTOEMISSIVE YIELD
OF ALKALI-ANTIMONIDE PHOTOCATHODES[a]

Material	E_g (eV)	E_a (eV)	Quantum efficiency (electrons/photons)
Cs_3Sb	1.6	0.45	0.30
Na_2KSb	1.0	1.0	0.30
$[Cs]Na_2KSb$	1.0	0.7	0.40
Na_3Sb	1.1	2.0–2.4	0.02
K_3Sb	1.0	1.1–1.8	0.07
Rb_3Sb	1.0	1.2	0.10

[a] From W. E. Spicer, *J. Appl. Phys.* **31**, 2077 (1960); *Phys. Rev.* **112**, 114 (1958).

(2) $E_g < E_a$ (see Fig. 7.9b). For the condition that the gap is smaller than the electron affinity, inelastic electron-electron scattering with resulting pair production may occur, even if Eq. (7.23) is fulfilled. The explanation for this is that the energy step E_a is not encountered until the electron reaches the surface so that the electron affinity appears as an excess which can be converted within the volume into one or more pair-production energies E_g depending on how much smaller E_g is than E_a. Therefore yield conditions prevail which approach those of the metallic state. The range of the photoelectrons may be 30 Å or less, i.e., by a factor of 10 smaller than the penetration depth of the light. Typical examples of semiconductors in which the electron affinity is much larger than the energy gap are germanium and tellurium. However, some alkali-antimonides also belong to this class with the result that their quantum yields are appreciably reduced from the Class (1) values (see Table 7.2). K_3Sb and Rb_3Sb may be considered as examples for an intermediate state between the two classes.

7.1.2.3 Photoemission from Insulators

Materials with large forbidden gaps are likely to belong to class (1) described in the preceding section. Thus if the problem of the high resistivity is circumvented by evaporating a very thin layer onto a metal substrate, relatively high photoemissive yields may be obtained from such materials as the alkali halides or the alkaline earth oxides.

In addition, there exist certain effects of photoemission enhancement in insulators. These phenomena occur in the presence of intermediate levels, such as F-centers (see Section 4.6.4.5), and secondary processes, such as excitons (see Section 4.6.4.4). The two types of defects may be interdependent. For example, irradiation of potassium iodide with ultraviolet light ($h\nu = 5.66$ eV) produces excitons and—at least partly through them—also F-centers.[29,30] Furthermore, because of their mobility and energy, excitons are extremely efficient in dislodging the electron from its color center position so that it may appear as a photo electron. As a result, radiation in the fundamental absorption band of an originally clear sample may generate a photocurrent which grows to a saturation value during a period of minutes. F-centers, of course, can be produced by other means, such as electron bombardment, and they may be ionized by relatively small energies. Nevertheless, the exciton process is most important because of its relatively high quantum efficiency ($\approx 10^{-2}$–10^{-3} electrons/quantum). With the exciton as intermediary, an electron may in principle be displaced in time and space with respect to the point of incidence of the photon.

7.1.3 Photoconductivity

The variation of electrical resistivity with the intensity of incident light is called photoconductivity, regardless of the sign of such change. The phenomenon is the result of an internal photoelectric effect in which carriers are excited by radiation energy from lower to higher levels. If photoconductivity is to occur, the light quantum must have sufficient energy to cause the transition of a charge carrier from a nonconducting to a conducting state or, more general, from one condition of mobility to another; there must be sufficient mobility in the conducting state; and the number of carriers generated by the radiation must be measurable in comparison to the free carriers normally available in the material. It is for these reasons that photoconductivity, at least in the visible and infrared region, is conspicuous only in a few semiconductors, but not in metals and insulators.

Expressions for the photoconductive response to the incident photon flux are readily derived. If electrons of a density n_e (cm^{-3}) have a drift velocity v_{e_z} under the influence of an impressed field E_z, then $n_e v_{e_z}$ pass through 1 cm^2 of the x-y plane each second. This net motion in the z-direction represents a current density $j_z = n_e \cdot e \cdot v_{e_z} = n_e \cdot e \cdot \mu_e \cdot E_z$, where we have introduced the mobility μ_e (defined as the velocity per unit field). The corresponding relationship is valid also for holes (subscript h) so that we have for the combined conductivity

$$\sigma = \frac{j_z}{E_z} = (n_e \mu_e + n_h \mu_h)e \tag{7.25}$$

Under the effect of incident light, the conductivity can obviously experience a change either by variation in the number of carriers or by a change of their mobility. The latter process, although conceivable, e.g., by transitions of carriers between overlapping bands, has so far played an unimportant role compared with the effects resulting from the photo generation of carriers. The relative variation of conductivity is, therefore,

$$\frac{\Delta\sigma}{\sigma} = \frac{\Delta n_e \mu_e + \Delta n_h \mu_h}{n_e \mu_e + n_h \mu_h} \tag{7.26}$$

When light is incident on the photoconductor, the number of excess carriers produced by the radiation grows at first linearly with time until saturation is reached because of the fact that recombination occurs after a lifetime τ. There results thus a stationary excess which, for instance, in the case of electrons in optically thin samples (uniform absorption rate) amounts to

$$\Delta n_e = G_e \tau_e = \frac{P k \eta_e \tau_e}{h\nu} \tag{7.27}$$

where we have expressed the generation rate G_e in terms of the number of incident photons (given by the ratio of incident power density P to quantum energy $h\nu$), the absorption coefficient k and the quantum efficiency η_e. As we shall show in the following, there are two important alternatives: (1) pair production by photons so that $\Delta n_e = \Delta n_h$; and (2) generation of either electrons or holes from imperfections—in particular, donor or acceptor impurity centers. The relative amount of photoconductivity for the two categories is, according to Eqs. (7.26) and (7.27),

$$\frac{\Delta\sigma}{\sigma} = \frac{P k \eta \tau}{h\nu} \cdot \frac{\mu_e + \mu_h}{n_e \mu_e + n_h \mu_h} \quad \text{(intrinsic)} \tag{7.28}$$

$$\frac{\Delta\sigma}{\sigma} = \frac{P k_i \eta_i \tau_i}{h\nu_i} \cdot \frac{\mu_i}{n_e \mu_e + n_h \mu_h} \quad \begin{array}{l}\text{(extrinsic—electrons} \\ \text{or holes—subscript } i)\end{array} \tag{7.29}$$

In either process, the effect is proportional (in the linear domain) to the number of incident photons, P/hv, the absorption coefficient k, the quantum efficiency of the absorbed photon η and the time constant τ. There are, however, significant differences between the instrinsic and extrinsic type of photoconductivity. As a group, imperfection centers require activation energies which are by an order of magnitude, or more, smaller than the gap energy; hence, they can be sensitive to wavelengths in the medium and far infrared. Intrinsic photoconductors, of course, offer a much larger number of interaction centers than impurity or other defect sites which, of necessity, must be diluted by a factor of 10^4–10^7 in the host substance. This difference makes itself felt to a considerable extent in the quantum efficiency, the absorption coefficient, and therefore the penetration depth. Finally, an important difference exists in the recombination mechanism which, in turn, influences the rise and decay of the response and, in particular, the time constant.

7.1.3.1 Intrinsic Photoconductors

The preceding discussion has brought out that the internal photoelectric effect, operating by means of pair production of a hole and an electron, will require a threshold photon energy which, for a given material, is smaller than that needed to eject external photoelectrons from the valence band, but larger than the activation energy of imperfection photoconductivity. This places the wavelength threshold of most intrinsic photoconductors which are in practical use between 0.5 and 10 μ. Table 7.3 lists the energy and wavelength thresholds, as well as the temperature dependence of the gap, for a number of important photoconductors.

The values given in Table 7.3 are representative of the majority of results cited in the literature; they vary, however, more or less with the structure and the method of preparation. For example, even the first photoconductor known, selenium (discovered by Smith[31] in 1873), has not yielded accurately reproducible results, not the least because of the existence (and co-existence) of several allotropic forms, although systematic purification increases the reproducibility (see Choyke and Patrick[32]). The energy gap varies with temperature and, partly for that reason, also with pressure. The temperature coefficient shown in the last column of Table 7.3 can be positive or negative, but it will be noted that its magnitude is usually in the order of 5×10^{-4} eV/°C (or about $6k$, since according to Table 1.2 the Boltzmann constant k equals 1.38×10^{-16} erg/deg and 1 eV $= 1.60 \times 10^{-12}$ ergs). As might be expected, there exists — at least for some substances — a trend of E_g with the position of

TABLE 7.3

Material	Temp. (°K)	E_g (eV)	λ_m (μ)	dE_g/dT (eV/°C) ($\times$ 10^{-4})
Se	300°	1.8	0.69	−9
Te	300°	0.35	3.5	1.9
Ge	300°	0.81	1.5	−4.5
	77°	0.88	1.4	
Si	290°	1.09	1.1	−4
PbS	295°	0.43	2.9	+4
	90°	0.33	3.8	
	20°	0.30	4.1	
PbSe	295°	0.25	5.0	+4
	90°	0.17	7.1	
	20°	0.15	8.2	
PbTe	295°	0.32	3.4	+4
	90°	0.25	5.1	
	20°	0.21	5.9	
InSb	300°	0.18	6.9	−3.5
	195°	0.19	6.5	
InAs	300°	0.33	3.7	−3.5
InP	300°	1.26	0.98	−4.6
GaSb	300°	0.70	1.8	−4.1
GaAs	300°	1.35	0.92	−4.9
GaP	300°	2.24	0.55	−5.4
Cs_3Sb	300°	1.6		
K_3Sb	300°	1.1		
Cu_2O		2.9		−2
AgI		2.8		

the elements in the periodic table, particularly if the same element is followed through a family of compounds of the same type (see, e.g., the III–V combinations). Such systematic variations, if indeed reliable, facilitate the search for photoconductors of certain desired properties. The most outstanding example of this kind of research is the continuing exploration of semiconductors for their potential as detectors of radiation in the medium and far infrared. Of particular import are the lead salts[33] (PbS, PbTe, PbSe) and the III–V compounds[34] (e.g. InSb). In all, it is seen that the upper wavelength limit accessible to intrinsic photodetectors which are now known is limited to about 7-8 μ, and practical requirements (sensitivity, time constant) reduce the useful limit further to about 6 μ. An exception, however, is the compound mercury telluride (if, in fact, it is intrinsic) with a long-wavelength limit of about 40 μ. HgTe can be mixed with CdTe to form a photoconductor which, depending on composition, has an adjustable threshold response between 0.8 and 40 μ.[35]

In general, intrinsic photoconductors absorb the incident radiation energy (if larger than the gap) within a thickness of 10^2 to 10^4 Å [see, e.g., Eq. (4.108)]. Intrinsic photoconductivity, therefore, is largely a thin film phenomenon, and practical detectors based on this effect are prepared by evaporation, vapor-phase condensation, or other deposition of a thin layer on a suitable substrate. The preparation of intrinsic materials in bulk does not add to the response; in fact, it adds substantially to the noise since competing processes (such as thermal excitation) increase in proportion to the volume. Because of the high concentration of hole-electron pairs which are produced in the thin active layer, recombination is relatively large, the time constant therefore reduced, and the response weakened [cf. Eq. (7.28)]. Thus the photoconductive yield is strongly dependent on the penetration depth and, *cet. par.*, increases with the latter.

The rise and decay of intrinsic photoconductivity with time can be quite complex. In the idealized case that hole-electron pairs are produced solely by the incident radiation, the rate of recombination will be proportional to the square of the carriers present at any instant. Thus the time constant depends on intensity, is different for rise and decay and, at any rate, difficult to define because the growth and decline of the carriers does not follow an exponential law. A simpler situation exists, if only one type of carrier prevails. In that event, the recombination rate is proportional to the number of carriers produced by the light and the time response is in the exponential regime.

7.1.3.2 *Impurity Photoconductors*

Any kind of defect in a solid can give rise to donor or acceptor levels for which the activation energy is smaller than that of the forbidden gap. We shall limit our discussion, however, to extrinsic photoconductors produced by impurity atoms, because only this type of imperfection can be to some extent measured and controlled.

Figure 7.10 shows the energy level diagram for impurity atoms in germanium. The energy values, drawn approximately to scale, are referred to the top of the valence band, or the bottom of the conduction band, whichever is nearer. As to the position of the levels, two distinct groups will be noticed (see also Tables 7.4 and 7.5). The first consists of acceptors (B, Al, Ga, In) and donors (P, As, Sb, Li), all only 0.01 eV above or below the nearest band. The acceptors are elements in column III, the donors, except Li, elements in column V. These impurities have, therefore, either one more (5) or one less (3) electron than required

TABLE 7.4

ENERGY GAPS OF IMPURITIES IN GERMANIUM[a]

Impurity	Donor (D) Acceptor (A)	Eg^b (eV)	λ_{max} (obs.) (μ)
Li	D	−0.0093	
Cu	A	+0.040	29
	A	+0.31	
	D or A	−0.26	
Au	D	+0.053	>15
	A	+0.15	9
	D or A	−0.20	5.5
	A	−0.04	>15
Zn	A	+0.03	>38
		+0.09	
B	A	+0.0104	
In	A	+0.0112	>38
Cd	A	+0.05	
	A	+0.16	
Mn	A	+0.16	3.5
	D or A	+0.37	7.6
Fe	A	+0.34	4.1
	A	−0.27	4.6
Co	A	+0.25	5.5
	D or A	−0.31	4.6
Ni	A	+0.22	5.6
	D or A	−0.30	4.6
Pt	A	+0.04	
	A	+0.25	
	D or A	−0.20	

[a] From R. A. Smith, "Semiconductors." Cambridge Univ. Press, London and New York, 1959; and R. H. Bube, "Photoconductivity of Solids." Wiley, New York, 1960.
[b] The minus sign indicates the level position below the edge of the conduction band; the plus sign, that above the valence band.

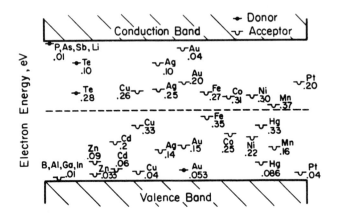

FIG. 7.10. Energy level diagram for impurity atoms in germanium. Energy values indicated are in electron volts and refer to nearest band edge.

by the valency of 4 in the germanium host lattice. The common features of this group are now explained by assuming that the extra electron or hole describe "hydrogenic" orbits[36] around the impurity atom, the difference between this model and the hydrogen atom being the existence of a dielectric constant and the change in effective electron mass. One has for the ionization energies (which represent the theoretical values for the impurity level positions):

$$E_i = \frac{2\pi^2 m^* e^4}{h^2 \epsilon^2} \tag{7.30}$$

where ϵ is dielectric constant of germanium ($\epsilon = 16$), m^* the effective mass of the electron ($m^* \approx 0.25 m_e = 2.3 \times 10^{-28}\,g$), and h and e are Planck's constant and electronic charge, respectively. It thus follows that E_i should be about $1/1000$ of the ionization energy of the hydrogen ground state, viz., in the order of 0.01 eV, independent of the impurity atom itself.

The second group of impurities (Te, Zn, Cd, Cu, Ag, Au, Fe, Co, Ni, Pt) shown in Fig. 7.10 has, in contrast to the first, a great variety of energy states of much larger values which indicate a much tighter binding of the hole or electron to the impurity atom. There exists no quantitative relationship such as Eq. (7.30) for these deep levels. The position of the impurity atoms in the periodic table and the resulting difference in valency from 4 of the host lattice of germanium result in a more complicated multi-level structure.

Gold, e.g., has a single (6s) electron to contribute to holes in the valence band (which can occur at very low temperatures). Thus, it has a donor state of 0.05 eV above the valence band. If the 6s electron, however, is not stripped from the neutral gold impurity, the latter can accept up to 3 additional electrons to accommodate the tetrahedral bonding structure of germanium. There exist therefore 3 acceptor levels of gold, which when filled represent in succession the ionization states Au^-, Au^{2-}, Au^{3-}, respectively. One produces these states successively by the process of "counter-doping," which implies here the admixture of proper amounts of n-type impurities of low electron binding energy (e.g., Sb, As, etc.).

Tellurium with a valency of 6 $(5s^2\,5p^4)$ has two possible donor states. Silver and copper have three acceptor states for the same reason as gold. Similar impurity levels are possible also in silicon (see Table 7.5).

TABLE 7.5
ENERGY GAPS OF IMPURITIES IN SILICON[a]

Impurity	Donor (D) Acceptor (A)	Eg^b	λ_{max} (obs.)
Li	D	−0.033	
Cu	D	+0.24	5.2
	A	+0.49	2.5
Au	A	+0.39	
	D or A	−0.30	
Zn	A	+0.092	4.1
	A	+0.30	4.1

[a] From R. A. Smith, "Semiconductors." Cambridge Univ. Press, London and New York, 1959; and R. H. Bube, "Photoconductivity of Solids." Wiley, New York, 1960.
[b] The minus sign indicates the level position below the edge of the conduction band; the plus sign, that above the valence band.

Thus, there are a great number of energy transitions available which are small enough for photoconductivity in the intermediate and far infrared. However, thermal excitation will compete with radiation at the long wavelengths. A photodetector responding to quanta of 0.01 to 0.3 eV will, of course, register its own thermal excitation as a signal. This means that the detector has to be cooled to temperatures that are the lower, the larger the wavelength limit is which corresponds to the detecting transition. More specifically, photoconductors have to operate at temperatures low enough so that only a small portion of the levels, which are to be left open for the optical transitions, are, in fact, occupied by thermal excitation. In the simplest case, we have for the number of electrons n excited thermally from donor states into the condition band or the corresponding case of holes in the valence band

$$n = N^{\frac{1}{2}} \left(\frac{2\pi m^* kT}{h^2} \right)^{\frac{3}{4}} e^{-eE_g/(2kT)} \tag{7.31}$$

where N is number of impurity atoms. Equation (7.31) is valid for the assumption that $n \ll N$ and that the number of counterdoped impurities is negligible with respect to n. From equations of the type (7.31), cooling requirements can be calculated. Levels in the order of 0.15 eV (e.g., p-type Au-doped germanium) require cooling to liquid nitrogen temperatures or, preferably, $10°$-$20°$K less than that; the second p-level of Zn in Ge at 0.09 eV requires liquid neon or hydrogen; the first cu level at 0.04 eV needs liquid helium; finally, the 0.01 eV levels or those yet lower may require pumping of the liquid helium.

7.1.3.3 Effects related to Photoconductivity

There exist processes which differ from photoconductivity more by the methods of observation than by their operating principle. Some of these are the following:

(1) The photomagnetoelectric effect[37,38] (PME, also PEM-effect) consists of the generation of hole-electron pairs by light in the presence of a magnetic field (see Fig. 7.11). Holes and electrons, given certain initial velocities by the incident photon energy, drift into opposite directions and thereby produce a measurable electric field at right angles to the field (Hall effect). The process is of practical interest in the III–V compounds,[39,40] mainly because of the large diffusion length of electrons in the crystalline samples. Moreover, some III–V compounds have small intrinsic gaps such as InSb (see Table 7.3). For these reasons, as well as the possibility of impurity doping, these materials are of interest as infrared detectors.

(2) Junction effects can have a variety of causes. A photovoltaic effect[41] can be produced by light at the junction of two different materials as the inverse mechanism of recombination radiation (Section 4.6.5.1). The process was, in fact, the first quantum detection effect which was observed and applied, viz., in selenium. Useful detectors can be built on the basis of such mechanisms for response to photons of 1 eV and above.

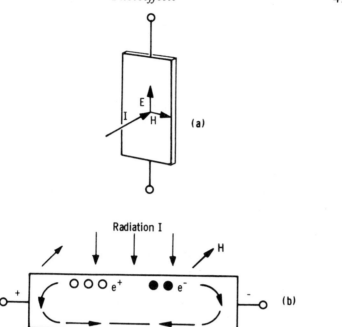

Fig. 7.11. Scheme of the photoelectromagnetic effect. (a) Orientation of propagation direction, magnetic field, and resulting electric field. (b) Cross section of PEM sample, indicating flow of electrons and holes as they drift in the magnetic field.

However, it should be noted that also such low energy gap phenomena as superconductivity have been suggested as possible quantum detection mechanisms.[42]

(3) *Oscillatory Photoconductivity* is a more recently discovered effect which differs from ordinary photoconductivity, not by techniques of observation, but by the involvement of additional basic processes in the semiconductor. The phenomenon consists of the periodic variation of photoconductive response with increasing photon energy at a temperature of a few degrees Kelvin. This was found by Blunt[42a] for a deep acceptor level of Cu in InSb and interpreted as accompanying emission of longitudinal optical phonons in integer numbers at $k = 0$ by Levinstein and his coworkers[42b-c]. This group found the effect also for Au and Ag in InSb as well as for intrinsic transitions in InSb, and in all these cases the spacing between successive minima of response was about 0.024 eV which corresponds to the energy of the eligible optical phonon as determined by other types of measurement. In detail, however, the phenomenon reveals the involvement of yet other processes* as also

---

* These include the periodic foreshortening of the carrier lifetime τ_i [cf. Eq. (7.29)].

shown, e.g., by the oscillatory intrinsic photoconductivity of CdS in the visible region.[42d]

7.2 Fluctuation Processes and Noise

7.2.1 MEAN FLUCTUATIONS OF RANDOM EVENTS

The minimum energy that can be discerned, or the maximum accuracy with which a measurement of energy can be undertaken by a detector, is determined by competing disturbances, called noise. If such perturbations were not always present, arbitrarily small powers could be detected and the precision of measurements would be unlimited. However, basically because matter and energy exist in discrete entities, as atoms, electrons, and quanta, every measurement is a counting process and as such subject to, and limited by, statistical fluctuations. This is the reason for the existence of basic detector noises. It follows that noise theory is, to a large extent, based on statistics. The statistical approach has the advantage of giving diverse fluctuation and noise phenomena a unified theory. There exists, of course, more than one statistic, since the occurrence of events may have restrictions on their independence.

Suppose the number n of certain random (i.e., entirely independent) events is counted, such as the number of radioactive atoms disintegrating within a given time interval, the number of molecules contained in a given volume element at any instant, or the number of births per day among a population of one million. In each case, one will arrive, after a large number of trials, at a mean value $\bar{n}$, and we shall assume that there is no systematic variation of $\bar{n}$, e.g., with time or across a volume. However, even then, in each individual count, a deviation from the mean will be observed, viz.,

$$\varDelta n_1 = n_1 - \bar{n}, \qquad \varDelta n_2 = n_2 - \bar{n}, \qquad \varDelta n_3 = n_3 - \bar{n} \qquad (7.32)$$

These fluctuations $\varDelta n$ will, on the average, have the value 0 (by definition of $\bar{n}$). However, it is possible to make a more meaningful statement regarding the average of the squares of these deviations, the so-called "mean square deviation," viz.,

$$\overline{\varDelta n^2} = \bar{n} \qquad (7.33)$$

Equation (7.33) asserts that the mean square deviation $\overline{\varDelta n^2}$ is equal to the mean of the number n. This relation, which can be derived, e.g., from

Boltzmann statistics, assumes independence of the counted events (Poisson distribution). A consequence of Eq. (7.33) is that while the deviation increases with the square root of the counted number, the relative probable error of the count (i.e., $\Delta n/n \sim 1/n^{\frac{1}{2}}$) will decrease at the same rate. Thus, if a million atoms hit a given surface per second, this number will fluctuate from second to second by the order of 1000 molecules, so that the probable relative error at each individual count is 0.1 %. After 10^4 sec, the rms deviation is 10^5, the relative rms deviation being 10^{-5} or 0.001 %. Thus the accuracy of a measurement or the reciprocal of the detectable size of a quantity will increase with the square root of the observed population, therefore, in general with the square root of the observation time ($\propto 1/\text{bandwidth } \Delta f$).

From the point of view of statistical physics, it is of interest that Eq. (7.33), as well as the mean square deviations belonging to other types of statistics, are equivalent to, and derivable from, the fluctuations of the general thermodynamic quantities. One such basic relationship will be discussed for the energy fluctuations of a subsystem in Section 7.2.6 (cf. Eq. 7.58). From a more practical point of view, it is of interest that the relationships for fluctuations, aside from representing the limits of measurement, also show how information can be derived from noise. Ultimately, this is a consequence of the correlation between the deviation and the mean so that the latter can be derived from the magnitude of noise, if it cannot be determined directly. Typical examples are the determination of carrier lifetimes in solids from their generation-recombination noise behavior[42e-f] and the Hanbury Brown-Twiss experiment (see Section 6.2.5.1).

7.2.2 SHOT NOISE

Equation (7.33) can be at once applied to the derivation of an expression for *shot noise* (other names are "Schottky[43] noise," "shot effect"). Shot noise is associated with the emergence of electrons from an emitter, or their arrival at an electrode, in a manner similar to the fluctuations of the number of raindrops on a roof. When a current i is emitted from a cathode, i/e electrons are generated per second, so that one counts during a time τ

$$\bar{n} = \frac{i\tau}{e} = \frac{i}{2\,\Delta f\,e} \tag{7.34}$$

A very general relationship between τ and Δf is implied here. It results from the fact that the number of elementary measurements processed by an apparatus is equal to the product, $2 \, \Delta f \cdot \tau'$, of the number Δf of periods per unit time counted by the detector, the observation time τ', and a factor of 2 for the number of elementary measurements per period, namely, phase and amplitude.[44–47] Hence, an individual fluctuation Δn is registered during a time $\tau = 1/2 \, \Delta f$. We now apply Eq. (7.34) for the mean square deviation in the number $\bar{n}$ counted in the average during $1/(2 \, \Delta f)$, viz., by combination with Eq. (7.33):

$$\overline{\Delta n^2} = \frac{\overline{\Delta i^2}}{(2e \, \Delta f)^2} = \bar{n} = \frac{i}{2e \, \Delta f} \tag{7.35}$$

Therefore, we obtain for the current fluctuation

$$\Delta i_{\mathrm{rms}} = \sqrt{\overline{\Delta i^2}} = \sqrt{2ie \, \Delta f} \tag{7.36}$$

Equation (7.36) expresses the shot noise current Δi_{rms} and thus the ultimate limit of accuracy with which a current i can be measured at a fixed bandwidth Δf. Again, by setting $\Delta i_{\mathrm{rms}} = i$, we find that $2e \, \Delta f$ is the smallest current detectable with a signal-to-noise ratio of unity.

7.2.3 JOHNSON NOISE

Whereas shot noise results from the statistical fluctuations in the number of electrons leaving or arriving at a surface, Johnson[48] noise is a consequence of the thermal fluctuations of electrons within a resistor. The nature of this perturbation, as well as its derivation, is best argued[49] on the basis of a thermodynamic equilibrium (see Fig. 7.12), although other derivations exist.

Assume that a resistance R is matched to a transmission line which, in turn, is matched to an antenna. This terminology implies that electro-

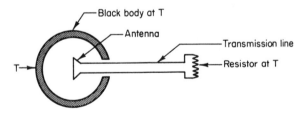

FIG. 7.12.   Equilibrium of power received by microwave antenna and noise generated in impedance under matching conditions with wave guide; for the derivation of Johnson noise.

magnetic modes flowing in either direction through the waveguide are not reflected at the terminals. The antenna is surrounded by a black body at temperature T which is also the temperature for the rest of the apparatus. Now the transmission line will pass a certain frequency interval dv in the microwave region, in which black body radiation makes a contribution which is given simply by the Rayleigh-Jeans formula (see Section 2.3.1)

$$W(v,\, T)\, dv \,=\, 2\pi\, \frac{v^2 kT}{c^2}\, dv \qquad (7.37)$$

The effective area of the antenna is known from antenna theory to be

$$A \,=\, \frac{\lambda^2}{2\pi} \,=\, \frac{c^2}{2\pi v^2} \qquad (7.38)$$

Altogether it follows from Eq. (7.37) and (7.38) that a power

$$P\, dv \,=\, kT\, dv \qquad (7.39)$$

is received by the antenna and, according to the previous assumptions, absorbed without reflection by the resistance R. Thus, to remain at the same temperature T, the resistance has to generate the same power (7.39) within dv. Now this must appear as a fluctuation with a power

$$P_n \,=\, kT\, \Delta f \qquad (7.40)$$

where Δf is the bandwidth over which the noise is measured. Since the resistor R is matched to the transmission line, only one-half of its (open circuit) noise voltage V_n will appear across its terminals. Thus we have, expressing Eq. (7.40) in terms of $\overline{V}_n$ and R,

$$\frac{(\overline{V}_n/2)^2}{R} \,=\, kT\, \Delta f \qquad (7.41)$$

or

$$\overline{V_n^2} \,=\, 4R \cdot kT \cdot \Delta f \qquad (7.42)$$

This is the formula for the mean square voltage of Johnson noise which results from the thermal agitation of carriers in a resistance R.

These noise formulae were derived originally by Nyquist[50] (hence also "Nyquist noise") by the argument, in effect, that there exist $\Delta Z = \Delta v$ modes in a one-dimensional configuration. Each mode carries kT because of thermal electron excitation so that Eq. (7.40) results.

It follows from the arguments presented here that Johnson noise is independent of frequency for those low frequencies for which the Rayleigh-Jeans approximation to Planck's law is valid. A more general expression is given by substituting Eq. (2.38) in place of Eq. (7.37) in Eq. (7.39) so that the Johnson noise voltage at all frequencies is

$$\overline{V_n^2} = 4RkT\,\Delta f\,\frac{hf/kT}{e^{hf/kT} - 1} = 4Rhf\,\frac{\Delta f}{e^{hf/kT} - 1} \tag{7.43}$$

We mention in passing that the balance of power, received and reradiated, is a measure of the temperature of objects scanned by the antenna pattern. This is the basis of a temperature measuring device, viz., the radiometer (Dicke[51]).

7.2.4 Photon Noise

Since radiation is communicated by quanta, there will exist in any measurement of it an uncertainty, or noise, due to the statistical fluctuation in the numbers of quanta arriving at the detector (Einstein[52]). That is to say, one might expect that any measurement of n photons is associated, according to Eq. (7.33), with an uncertainty proportional to the square root of n. However, photons do not obey the classical statistics of Boltzmann which gave rise to the mean square fluctuation law (7.33).

Instead, they follow Bose-Einstein statistics for which the mean square deviation is given by (see Section 7.2.6)

$$\overline{\Delta n^2} = \bar{n} + \frac{\bar{n}^2}{N} = \bar{n}\left(1 + \frac{\bar{n}}{N}\right) \tag{7.44}$$

where $\bar{n}$ is the average number of photons counted in a volume and N is the number of possible energy states or standing waves within an interval $d\nu$ in a volume V [cf. Eq. (2.32)], viz., in vacuo

$$N = g_\nu V = \frac{8\pi\nu^2\,d\nu}{c^3} \cdot V \tag{7.45}$$

The origin of the term $\bar{n}/N$ in Eq. (7.44) is a density bunching effect (Bothe[53]) due to induced emission (Sections 3.1.4 and 3.3.1). Photons emitted in black-body radiation will, of course, undergo positive and negative absorption. The positive effect, which in thermal sources is always larger, does not alter the purely random statistics of the fluctuations. Stimulated emission, however, will trigger a pairing of photons which emerge at the same time and with the same energy and direction.

Now $\bar{n}$, the number of photons contained in volume V is given by Planck's formula for the density of radiation energy [see Eq. (2.37)], viz.,

$$u_\nu \cdot V = \bar{n}h\nu = g_\nu \frac{1}{e^{h\nu/kT} - 1} \cdot h\nu \cdot V = \frac{Nh\nu}{e^{h\nu/kT} - 1} \tag{7.46}$$

where g_ν is given by Eq. (7.45). Introducing these values into Eq. (7.44) one obtains for the mean square fluctuation of the number of photons

$$\overline{\Delta n^2} = \bar{n}\left(1 + \frac{1}{e^{h\nu/kT} - 1}\right) \tag{7.47}$$

or

$$\overline{\Delta n^2} = \bar{n}\left(\frac{e^{h\nu/kT}}{e^{h\nu/kT} - 1}\right) \tag{7.48}$$

The mean deviation of energy is given by multiplying the number $\bar{n}$ of photons by their energy $h\nu$:

$$\overline{\Delta n^2}(h\nu)^2 = \overline{\Delta E_\nu^2} = \overline{E_\nu} \frac{h\nu e^{h\nu/kT}}{e^{h\nu/kT} - 1} \tag{7.49}$$

It is obvious that for high frequencies $(h\nu \gg kT)$ Eq. (7.47) yields the classical mean deviation (7.33). This is the case, approximately, for frequencies higher than those corresponding to about a few microns and $T \sim 10^2$–10^3 °K. On the other hand, for frequencies in the microwave region, where

$$kT \gg h\nu$$

Eq. (7.47) yields

$$\overline{\Delta n^2} = \bar{n}\left(1 + \frac{kT}{h\nu}\right) \approx \bar{n}\frac{kT}{h\nu}$$

The mean square deviation of the radiation energy contained in a volume is then

$$\overline{\Delta E_\nu^2} = E_\nu kT, \qquad kT \gg h\nu \tag{7.50}$$

However, if integrated over the entire radiation spectrum (and if T is not very small), the fluctuation is nearly that computed from classical statistics. One obtains for the power radiated over the whole spectrum from an area A of a black body:

$$\overline{\Delta W^2} = 16\sigma kT^5 A \, \Delta f \tag{7.51}$$

This represents the square of the noise equivalent power at a bandwidth Δf when photon noise is the main limitation (Lewis noise[54]). Measurements of photon noise, particularly with respect to testing the validity of the second term in Eq. (7.47), corroborated the theory (see e.g. Harvit[55]). The validity of Eqs. (7.46)–(7.51) is limited, of course, to black body radiation. The nonthermal photon gas obeys fluctuation laws different to an extent determined by the degeneracy parameter δ (see Section 6.2.5.2).

7.2.5 Temperature Noise

In the phenomenon known as Brownian movement, particles of small mass suspended in a liquid or a gas are seen to undergo random changes of position and velocity. The effect is understood as the result of buffeting by the neighboring molecules which are in irregular thermal motion in the surrounding gaseous or liquid medium. Although the molecules are very much smaller, they share with a particle of mass M an amount of kinetic energy which is for motion in any one direction, according to the equipartition law, equal to

$$\tfrac{1}{2}Mv_x^2 = \tfrac{1}{2}kT \tag{7.52}$$

Basically, this law represents a performance limitation to detectors of very small mass or very small heat capacity, independently of the process by which energy fluctuations are exchanged with the environment. The effect is called temperature[56] (or thermal) noise.

We can compute its size readily by order-of-magnitude considerations. Assume the detector consist of N oscillators each of which have a mean energy kT. We thus define its heat capacity as

$$C = Nk \tag{7.53}$$

We assume further that a certain number of them fluctuate by kT and thereby cause a mean fluctuation $k\,\varDelta T_{\text{rms}}$ of all N oscillators. Thus,

$$(\varDelta NkT)_{\text{rms}} = (Nk\,\varDelta T)_{\text{rms}} \tag{7.54}$$

With the mean square deviation law applied to (7.54)

$$\overline{(\varDelta N)^2} = \bar{N} = \overline{\left(N\frac{\varDelta T}{T}\right)^2} \tag{7.55}$$

we then have, because of Eq. (7.53),

$$\overline{\left(\frac{\varDelta T}{T}\right)^2} = \frac{1}{N} = \frac{k}{C} \tag{7.56}$$

and thus for the mean square of the temperature fluctuation, i.e., the temperature noise

$$\overline{\Delta T^2} = \frac{kT^2}{C} \tag{7.57}$$

7.2.6 THE EINSTEIN-FOWLER EQUATION

So far we have interpreted many noise phenomena in terms of a fluctuation in the number of particles, although some of these processes are more directly explained as fluctuation in energy. The starting point for this is the Einstein[52]–Fowler[57] formula for the mean square deviation $\overline{(\Delta E)^2}$ of the mean energy $\bar{E}$:

$$\overline{(\Delta E)^2} = kT^2 \frac{\partial \bar{E}}{\partial T} \tag{7.58}$$

With

$$\frac{\partial \bar{E}}{\partial T} = C, \qquad C \Delta T = \Delta \bar{E} \tag{7.59}$$

we have at once the formula for the temperature noise as in Eq. (7.57):

$$\overline{(\Delta T)^2} = \frac{kT^2}{C}$$

Another application of the Einstein-Fowler equation is the derivation of photon noise (see Section 7.2.4). We write Planck's equation of the radiation energy contained in a volume V, in terms of number of photons:

$$E_\nu d\nu = \bar{n}h\nu d\nu = \frac{8\pi h\nu^3}{c^3} \frac{V}{e^{h\nu/kT} - 1} d\nu \tag{7.60}$$

Introduction of Eq. (7.60) into Eq. (7.58) yields

$$\overline{(\Delta E^2)} = \overline{\Delta n^2}\, h^2\nu^2 = kT^2 \frac{8\pi h^2\nu^4}{c^3 kT^2} \frac{e^{h\nu/kT}}{(e^{h\nu/kT} - 1)^2} V \tag{7.61}$$

or

$$\overline{\Delta n^2} = \bar{n} \left(\frac{e^{h\nu/kT}}{e^{h\nu/kT} - 1} \right) \tag{7.62}$$

This is identical with the expression (7.48) for photon noise so that with the proper substitutions we come back to Eq. (7.47):

$$\overline{\Delta n^2} = \bar{n} \left(1 + \frac{\bar{n}}{N} \right) \tag{7.63}$$

7.2.7 FLUCTUATIONS DUE TO STRUCTURE

The discussion of noises has dealt so far with fundamental effects. It is characteristic of these basic noises that they can be accurately accounted for in terms of the atomic constants and that they represent in each case the ultimate limitation of performance. By contrast, there are random processes of complex and often not understood origin, and in these cases improved design, processes, and detector material can reduce the disturbance, although usually only by the empirical method of trial and error. Here we are mainly dealing with phenomena based on structural properties of the detector material.

7.2.7.1 Flicker and Current Noises[56]

A certain group of fluctuations is characterized by the fact that they contribute powers to various frequencies, f, which are proportional to $1/f$ or $1/f^2$, or generally $1/f^\alpha$. Such *Flicker effect* occurs for currents across contacts and junctions, although the latter are not necessarily the dominant cause.[58] Flicker noise is, of course, of possible interest to photoemissive detectors, while current noise is an important limitation of photoconductor performance, especially in film deposits. Certain photoconductors of single crystal structure are virtually free of any $1/f$ noise, if proper surface etching and contacting methods are employed.

7.2.7.2 Fixed Pattern Noise

Nonuniformities or graininess of film deposits are important in imaging devices since differences in response across the sensitive two-dimensional detector simulate or mask a signal. Means of fixed pattern cancellation have been pursued with varying success. On the other hand, quality control of the surfaces can, at least in principle, attain any desired degree of perfection.

7.3 Practical Detectors of Optical Radiation

It is possible to classify optical detectors according to a point of view schematically presented in Fig. 7.13. This subdivision is guided by the nature of the spectral range limitation to which the detector is subjected, namely: (a) *The nonselective response*, which is applicable to the entire spectrum (at least, in principle); (b) the *semi-infinite range sensitivity*, for

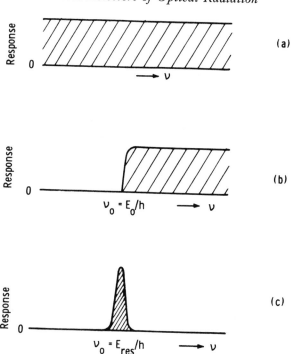

Fig. 7.13. Scheme of detector classification. (a) Response of the nonselective (thermal) detector. (b) The semi-infinite response of the photodetector. (c) Profile of resonance detectivity.

frequencies above a threshold ν_0; and (c) the *resonance response* for a limited region given by the Q-value of the measuring system.

This classification fits quite naturally the principal mechanisms available for the detection of the optical spectrum. Thus class (a) of the nonselective system is represented by the *thermal detectors* which measure the incident power cumulatively, regardless of the size of the quanta contained in the radiation stream. Class (b) encompasses all those sensing effects which require for their activation a threshold energy E_0 of the quantum, i.e., a minimum frequency of the incident light. Finally, class (c) is the domain of the high-Q oscillators which until recently was limited to the reception of radio and microwave frequencies. By convention, one also divides detectors into those operating by (I) *thermal* and (II) *photon* effects. Obviously, our categories merely split (II) into classes (b) and (c). Such classifications are not only differentiations of the underlying operating principles; their chief virtue lies in the indication of the performance that can be expected.

7.3.1 DEFINITION AND EVALUATION OF DETECTOR CHARACTERISTICS

7.3.1.1 *The Role of Performance Figures*

The design of detection systems is based on the performance of the detecting element. One needs, therefore, standards by which the capabilities of the basic components can be measured, not just to compare detectors of one kind, but for the selection of the type of sensing mechanism which best suits the design requirements. One such performance parameter which is useful for the art of measurement in general is the noise equivalent power (NEP). This is a figure of merit which still depends on certain operating conditions such as the receiving area of the detector, bandwidth, nature of the spectrum, etc., which have to be specified along with the NEP. The manner in which the NEP depends on these variables follows functional relationships which is the same for many (but not all) types of detectors. Definitions and standards have therefore been introduced which refer specifically to conditions under which some of these variables are set equal to unity. Clearly defined, such figures of merit describe the performance of a detector for at least one operating point, although not necessarily for the entire operating region.

7.3.1.2 *Responsivity*

A term which has no reference to noise limitations, *responsivity* is defined as the detector output per unit power of incident radiation. Thus, it is measured as microvolts/microwatt, as milliamperes/watt, or as the quantum efficiency, viz., number of transition events per incident quantum. The significance of this performance figure is the information it yields regarding the amount of amplification or magnification needed.

7.3.1.3 *Noise Equivalent Power (NEP)*

The smallest power detectable with a signal-to-noise ratio of unity is called the *noise equivalent power*. For any detector, the NEP will depend on all the characteristics of the incident radiation, and thus tight specifications are needed. As an example, most infrared detectors are evaluated with a black body at $500°K$. The NEP is, furthermore, often dependent on chopping frequency and it is proportional to the square root of the bandwidth (at least for the case of a Poisson distribution). Conventionally, one refers the NEP to unit bandwidth and cites the frequency at which the detector has been tested, usually either 90 or 900 cps. Symbolically, this is written as NEP (500, 900, 1), denoting the black-body temperature, the chopping frequency, and the noise bandwidth, respectively.

7.3.1.4 *Detectivity*

As a figure of merit, the noise equivalent power suffers from the drawback that large values correspond to low performance. For this reason, Jones[56] introduced the term *detectivity* which is defined as the reciprocal of the NEP. The detectivity is therefore an inherent characteristic of the detector:

$$D = 1/NEP \qquad (7.64)$$

7.3.1.5 *The D* Detectivity*

Since the NEP is often proportional to the square root of the area of the detector, Jones introduced the symbol D^* for the detectivity referred to unit area:

$$D^* = \frac{\sqrt{A}}{\text{NEP}} \quad [\text{cm } W^{-1} \text{ sec}^{-\frac{1}{2}}] \qquad (7.65)$$

7.3.1.6 *The Measurement of Performance Figures*

The figures of merit introduced so far can all be determined provided a known amount of radiation power can be applied to the detector. The most reliable source is the black body. The fraction P_i of power received by the detector is determined by geometrical considerations (see Chapter 2 which also gives the radiation formulae). Assuming the validity of the square law relationship between NEP and bandwidth Δf, one determines the noise equivalent power from the signal-to-noise ratio S/N by

$$\text{NEP} = \frac{P_i}{(S/N) \times (\Delta f)^{\frac{1}{2}}} \qquad (7.66)$$

Signal and noise are separately evaluated by the arrangement shown schematically in Fig. 7.14. The chopper will intercept the radiation from the opening of the black body and generate a fixed frequency. The shape of the chopper teeth may introduce harmonics which should be considered when studying the frequency behavior. The noise bandwidth is determined by the wave analyzer. Both signal and noise are observed for a certain amount of time and introduced into Eq. (7.66).

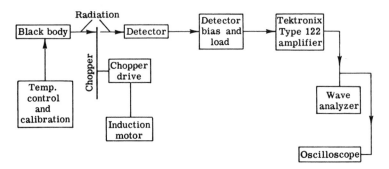

Fig. 7.14.   Block diagram of detector test arrangement.

7.3.1.7 *The Measurement of Time Constants*

The apparatus shown in Fig. 7.14 is, in principle, capable of determining time constants of rise and decay by varying the chopping frequency and evaluating the frequency response. The latter is given by

$$S \approx \frac{1}{(1 + \omega^2\tau^2)^{\frac{1}{2}}} \tag{7.67}$$

from which the time constant τ can be derived. The drawback of this method is the limitation to relatively long time constants, since the measurement of times less than 1–10 μsec requires excessive chopper speeds. The methods of scanning the image of a slit over a detector by means of a rotating mirror improves on this limit. Figure 7.15 shows a method in which the beam from the source is repeatedly reflected between a set of stationary mirrors and the faces of a multisided rotating mirror (Garbuny *et al.*[59]). Time constants in the order of 10^{-8} sec can be determined with this latter method.

7.3.2 Thermal Detectors

Until relatively recent times, the methods available for the detection and measurement of infrared radiation in contrast to the rest of spectrum were solely based on thermal conversion processes. In all of these "thermal detectors," the incident energy is absorbed by a black layer on the detector, converted into heat and evaluated by the resulting temperature increase. The mechanism of absorption and the principle of reducing the thermal mass to a minimum is common to all thermal detectors;

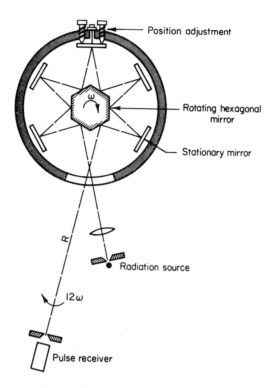

FIG. 7.15. Generator of very fast light pulses [M. Garbuny, T. P. Vogl, and J. R. Hansen, *Rev. Sci. Instr.* **28**, 826 (1957)].

they differ, however, in the means by which the temperature is measured. We shall cite in the following three major groups of thermometric mechanisms, although there exist variations for each of these.

7.3.2.1 *The Bolometer*

The most important quantity in bolometers is the thermal coefficient, C, of the resistivity ρ:

$$C = \frac{1}{\rho}\frac{d\rho}{dT} \tag{7.68}$$

Radiation incident on a bolometer element then will produce a temperature change ΔT with a resulting change of resistance R:

$$\Delta R = RC\,\Delta T \tag{7.69}$$

This resistance change is measured in a bridge arrangement as a change in potential which is suitably amplified (see, e.g., Smith et al.[59a], p. 91). For metals,

$$\rho = \rho_0(1 + \alpha t) \qquad (7.70)$$

where ρ_0 is the resistivity at $t = 0\,°C$ and t is in $°C$. In this equation, ρ is linear with t over a wide temperature range ($\alpha = 0.0030$ for Pt and Ag, 0.00344 for Al, 0.006 for Ni and Fe) so that C in Eq. (7.68) is inversely proportional to temperature (in $°K$) over certain temperature ranges. The temperature behavior of intrinsic semiconductors is largely determined by the thermal excitation of carriers across a forbidden gap. Conductivity σ and resistivity ρ in semiconductors are given by equations of the form

$$\sigma \sim e^{-eE_g/kT}, \qquad \rho \sim e^{eE_g/kT} \qquad (7.71)$$

Here, the effect of temperature on mobility and factors containing T or T^2 in front of the exponential are ignored as unimportant in comparison to the exponential. The C-value of semiconductors, then, assumes an inverse square law relationship with respect to temperature:

$$C \approx \frac{eE_g}{k} \cdot \frac{1}{T^2} \qquad (7.72)$$

The C-values of semiconductors are in the order of 10 to 100 times larger than those of bulk metals. In this respect, bolometers made of a combination of transition-metal oxides have proven very effective ("thermistors").

Although the C-values for metals were seen to be only in the order of $0.3–0.6\%/°C$, evaporated metal films may have considerably larger thermal coefficients of resistivity. This is true especially in the transition region of very small thickness (see Fig. 7.16).

The phenomena of superconductivity have been utilized for bolometric purposes in the laboratory by Andrews and his co-workers.[60] The superconducting "transition region," in which the resistivity of superconducting materials decreases from finite values to zero, lies for most material at a few degrees Kelvin and has a width of $\sim10^{-2}\,°K$, or less, depending on the purity of the metal. It is obvious that the C-value can assume very large values in superconductivity. Furthermore, the transition region of most metals is low enough for the specific heat to obey the T^3-law so that very high speeds of thermal response can be expected from superconducting bolometers. Nevertheless, certain difficulties have so far denied the practical use of this method. First,

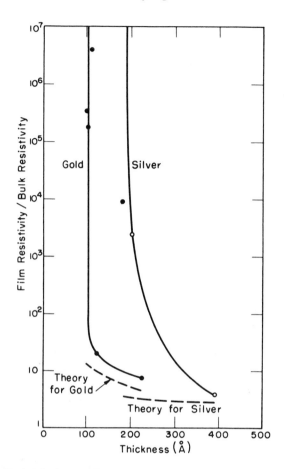

FIG. 7.16. Resistivity increase in metal films. Values normalized to bulk turn abruptly toward infinity at a thickness for which electron micrographs show the onset of island structure (R. C. Ohlman, private communication).

there is the need for liquid helium, which must be pumped for transition temperatures below 4.2 °K. Even the highest transition regions, such as that for niobium nitride, require, at least, liquid hydrogen for operation at 14–18 °K.* Another difficulty attending practical use of super-conducting bolometers is the instability which is associated with the extremely narrow transition region of most materials.

---

* The materials with highest transition temperatures at zero magnetic field, known at present, are V_3Si (17°K) and Nb_3Sn (18°K).

7.3.2.2 Thermocouple Detectors

Another important thermometric process is the Seebeck effect (1822). In this phenomenon, the junction between two different metals, A and B, . will develop a potential U with respect to the return junction B-A, if it is held at a different temperature. One has, therefore, a relation of the form

$$U = \pi_{AB}(T_2 - T_1) \tag{7.72}$$

The detailed theory of the thermocouple detector indicates the desirability of high electrical, but low thermal, conductivity of the junction materials so as to reduce ohmic losses as well as those by heat conduction to the cold junction. In metals, the ratio of thermal conductivity, K, and electrical conductivity, σ, is, of course, constant according to the Wiedemann-Franz law:

$$\frac{K}{\sigma T} = L \tag{7.73}$$

Thus, it appears that the Wiedemann-Franz law restricts the possibilities of manipulating thermocouple efficiencies.

For most metal combinations, the Seebeck constant π is in the order of 10–$100 \; \mu V/°C$. Semiconductors may exceed this range, but the value L in Eq. (7.73) may then be unfavorably large. In practice, thermocouples employing junctions between bismuth and antimony and bismuth and tellurium have been used successfully for bolometers.

7.3.2.3 The Pneumatic Detector (Golay Cell)

Golay and his co-workers[61,62] applied the principle of the gas thermometer to the detection of radiation, particularly in the infrared, although his device, known as the *Golay cell*, has been applied successfully also from the ultraviolet to the millimeter wave region. The basic element of the detector is a small volume of gas bounded by a radiation absorbing film and a flexible mirror (see Fig. 7.17). When radiation is incident on the black film, heat is communicated to the gas which then expands and thereby bends the silvered membrane. The latter is part of an optical system by which the image of a grid is superposed on the grid itself so that slight changes in the degree of register have a large effect on the monitoring light measured by a photocell.

For spectrometer applications, the Golay cell is very useful because of its sensitivity, high for a thermal device (NEP $< 10^{-10}$ W $\times$ sec$^{1/2}$). The ultimate limit of accuracy is, in fact, given by temperature noise.

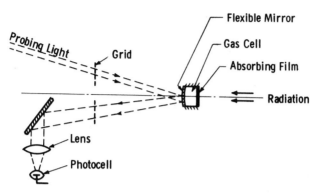

FIG. 7.17.   Scheme of the Golay cell.

However, its operation, combining optical and mechanical features, is sensitive to vibrations (microphonics) and somewhat slow (10^{-3}–10^{-1} sec).

Film Blacks. The thermal detectors described so far have in common the need for the conversion of incident radiation power into a temperature signal. The highest values for the absorption of light in the visible and infrared are achieved with films which operate, in a sense, as black bodies in their microstructure. Such blackening has been used for a long time with lamp soot (and other finely distributed forms of carbon particles) which absorbs infrared up to about 10 μ. Powdered glass is an effective infrared absorber for long wavelengths. These layers are deposited on substrates either directly or by means of a binding vehicle such as lacquer. A rather large thermal mass is, however, an inherent property of such deposits. To avoid this difficulty, similar, but more efficient structures were suggested first in 1930[63,64] in the so-called film "blacks."

Such layers are formed by evaporating metals at pressures in the order of 1 mm Hg of nitrogen or hydrogen. While evaporation of metals in high vacua leads to highly reflecting deposits, metal atoms surrounded by a buffer gas at 1 mm have a mean free path of only about 10^{-2} mm and as a result condense while diffusing through the evaporation space. They then settle on exposed surfaces as feathery, often velvet, black deposits. Under the electron microscope these layers appear as crystallites in the order of 10^{-6} cm diameter arranged in chain-patterns, perhaps because of electrostatic forces. A great number of metals and semiconductors are suitable for this process. Gold black is the most efficient among the diffusion evaporates.[65] The ratio of absorption achieved per unit thermal mass (heat capacity per unit area) is orders of magnitude larger in gold black than in other "black" deposits.

7.3.3 PHOTODETECTORS

The practical detectors belonging to Class b, responding selectively to frequencies above a characteristic threshold, all operate by means of one or the other quantized mechanism discussed in Section 7.1. As a group, *photodetectors* have the advantage of higher sensitivity and shorter response time over the thermal detectors, but they either do not operate at all at the longer wavelengths or they exact there the penalty of elaborate cooling means.

Thus photoemission requires activation energies above 1 eV (wavelengths below 1.2 μ); in fact, only one type of compound, viz., cesium-silver oxide, can accomodate this region of the near infrared. On the positive side, however, is the enormous versatility of the photoelectric effect, its speed of response—so high that it defies measurement at least in its normal process—and especially its quantum efficiency. The latter, for reasons outlined in Section 7.1, is in the visible region highest for certain semiconducting compounds. A long sustained effort of industrial laboratories has been expended on the development of photocells which approach as closely as possible the theoretical limit of yield, and this goal has been virtually achieved with the mixed alkali antimonides. At best, however, photoemission is a phenomenon of small currents; to avoid Johnson noise under such conditions, the photoelectric effect is usually combined with secondary emission in the *photomultiplier* principle by which the number of electrons originally ejected by light is augmented in 6 to 14 stages by *dynodes*. It can then be shown that shot noise will be approached as the irreducible lower limit.

The role that is played in the visible and ultraviolet region by photoemissive cells, is filled in the infrared by photoconductors. As pointed out before, their quantum yield may be unity and their speed of response is always higher than that of the thermal detectors, typically 10^{-6}–10^{-3} sec for intrinsic materials, 10^{-6}–10^{-9} sec for impurity photoconductors.

7.3.3.1 *Conditions of Operation*

An important method of operating a detector particularly in the infrared is by intercepting (chopping) the incident radiation at a given frequency and amplifying the signal with a suitable band pass amplifier. This reduces false signal contributions due to dc drift, radiation background, etc., considerably, namely to that Fourier component which these noises have in the admitted frequency interval. The signal developed by a photoconductor (as well as by a bolometer, photocell, etc.) is measured as the voltage drop across a load resistor. Specifically, if a battery of voltage V_0 is connected in series with the photoconductor R and a load resistor R_L, the voltage drop V_L equals

$$V_L = V_0 \frac{R_L}{R + R_L} \qquad (7.74)$$

so that a variation dR yields a signal

$$dV_L = (-)V_0 \frac{R_L}{(R + R_L)^2} dR = (-)V_0 \frac{RR_L}{(R + R_L)^2} \frac{dR}{R} \qquad (7.75)$$

It was seen in Section 7.1 that photoconductivity is based on changes Δn in the number n of carriers, due to incident radiation, with resulting changes in the conductivity

$$\sigma = ne\mu \qquad (7.76)$$

where e and μ are the carrier charge and mobility, respectively. Furthermore, σ is related to R through the dimensions of the sensitive element l and A:

$$R = \frac{l}{A\sigma} = \frac{l}{ne\mu A} \qquad (7.77)$$

The signal voltage is then, after inserting Eqs. (7.77) and (7.75),

$$\Delta V = V_0 \frac{RR_L}{(R + R_L)^2} \frac{\Delta n}{n} \qquad (7.78)$$

If G is the generation rate of carriers (which is, of course, proportional to the radiation intensity), and τ their lifetime with respect to recombination, we have for the number n' of extra carriers the differential equation

$$\frac{dn'}{dt} = G - \frac{n'}{\tau} \qquad (7.79)$$

the integral of which is

$$n' \equiv \Delta n = G\tau(1 - e^{-t/\tau}) \qquad (7.80)$$

Hence, the equilibrium number of radiation generated carriers is

$$\Delta n = G\tau \qquad (7.81)$$

This equilibrium condition is reached only if the radiation flux remains constant during a time large compared with the lifetime τ. For $t \ll \tau$, $\Delta n = Gt$, as seen from Eq. (7.80). In all, the response to a sinusoidally varying radiation intensity is given (writing V_S for ΔV) by

$$V_S = V_0 \frac{RR_L}{(R + R_L)^2} \frac{G_{rms}\tau}{n} \frac{1}{(1 + \omega^2\tau^2)^{1/2}} \qquad (7.82)$$

7.3.3.2 *Noise Characteristics*

A discussion of the fundamental noise sources has been given in Section 7.2. Which type of noise will be limiting, depends not only on the detector, but for a given detector also on the operating conditions. Bolometers are limited, in general, by Johnson noise, but at low frequencies $1/f$ noise can be the main perturbation. Lead sulfide, in its spectral region the most sensitive detector, is limited usually by $1/f$ noise; at higher frequencies or lower operating temperatures, its limitation is reported to be the fundamental photon noise. Whenever photon energies are large compared to kT of the detector, the basic limitation will be that of rms fluctuations in the number of signal carriers produced by the radiation. Examples are the infrared photoconductors with impurity gaps of less than 0.1 eV which are operated at liquid helium temperatures. These detectors are "background limited," i.e., their basic noise level is determined by the photon fluctuations of the "warm" background. Thus the noise output depends on the unshielded angle of view. Other examples are the photoelectric detectors in the visible and ultraviolet. It is usually possible to compute noise power and voltage for a given detector by means of the noise formulae given in Section 7.2 within a factor of 2 or 3.

In doped germanium crystals, for example, the dominant noise at intermediate temperatures (liquid nitrogen) is a variation of shot noise, called generation-recombination noise (see e.g. Van Vliet[66]). Its origin lies in the mean square deviation of the number of carriers which hover between generation and annihilation and are counted during the observation time $t_0 \cong 1/f$. Using the symbols of Eq. (7.82), the rms noise voltage due to this cause is given by

$$(\overline{V_n^2})^{1/2} = 2V_0 \frac{RR_L}{(R + R_L)^2} \frac{\tau^{1/2} \Delta f^{1/2}}{n^{1/2}} \frac{1}{(1 + \omega^2\tau^2)^{1/2}} \tag{7.83}$$

where n is the mean of the total number of carriers in the crystal. In case of gold-doped germanium, the theoretical values (7.83) agree well with observation. Typical NEP values averaged over the radiation profile of a black body at 500 °K are 5×10^{-11} W-sec$^{1/2}$ for Au-doped germanium, (p-type) while the equivalent value for thermal detectors is 10^{-8}–10^{-9} W-sec$^{1/2}$. Equation (7.83) can be used as the basis for determining τ from the noise (cf. Section 7.2.1).

Figure 7.18 compares the typical spectral response of infrared detectors for the range up to 15 μ. The sensitivity is expressed in terms of the detectivity D^* which has been introduced in Section 7.3.1.5. It is seen that the photoconductors near their spectral peak perform better than

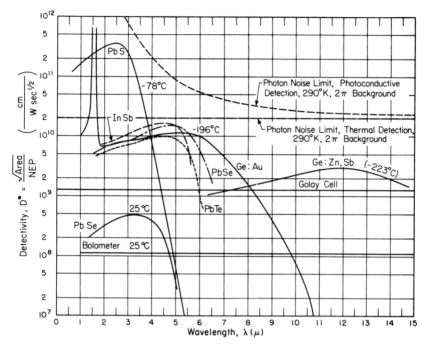

FIG. 7.18. Comparison of detector performance.

even the best thermal detectors. On the other hand, photoconductors are limited to spectral regions of a few octaves. The long wavelength limit is determined, of course, by the energy gap involved in each case. There is also a gradual decline of detectivity for short wavelengths, and this for several reasons. First, there are fewer quanta per unit energy as we progress to lower values on the wavelength scale. Also, as the energy of incident photons exceeds the threshold of pair production, the depth of the absorption region shrinks, and as the interaction space is reduced in thickness, the recombination rate increases with unfavorable results for the detectivity. This behavior is not noticeable in impurity photoconductivity which is characterized by absorption coefficients typically of 1 cm^{-1} so that, in fact, crystals of millimeter or centimeter length have to be grown to provide a sufficient number of active centers. All imperfection photoconductors have, of course, also an intrinsic region which is only slightly affected by the presence of the defects. Here, the transition to the intrinsic domain is marked by the sharp increase of detectivity as shown, for example, in Fig. 7.18 by the peak of the Ge:Au curve. Background noise limits the detectivity as indicated by the dashed curve for 2π angle of view into an environment at room temperature. This curve will be shifted toward the upper right corner of the figure, if the detector is surrounded by a shield cooled to low temperatures. The performance data presented in Fig. 7.18 can, and often have been, exceeded by selected samples so that the limit of photon noise has been approached asymptotically.

7.3.3.3 Time Constants

The simplest condition for the rise and decay of carriers exists when the number of electrons or holes produced by radiation is negligibly small in comparison with the number of existing recombination centers. It is for this case that Eq. (7.79) is valid with its solutions (a) for rise of photocurrent:

$$n_r' = G\tau(1 - e^{-t/\tau}) \qquad (7.84)$$

(b) for decay of photocurrent:

$$n_d' = G\tau e^{-t/\tau} \qquad (7.85)$$

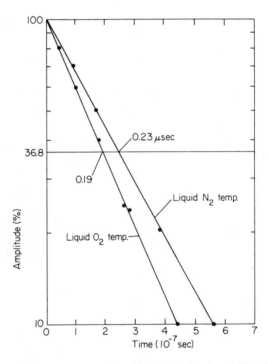

Fig. 7.19. Decay of photoconductivity in Au-doped Ge with time.

The time constant τ is simply given by

$$\tau = \frac{1}{n_c \cdot \sigma \cdot v} \qquad (7.86)$$

where n_c is the number of recombination centers per cubic centimeter, v the carrier velocity (cm sec^{-1}) and σ the cross section (cm^2) of the recombination process.

Figure 7.19 shows typical photocurrent decay curves for gold-doped (p-type) germanium[67] at two different temperatures. It is seen that the decay is exponential, as given by (7.85); that values in the order of 10^{-7} sec are measured; and that the time constant is shorter for higher operating temperatures.

There also exist, in photoconductivity, conditions in which the number of radiation-produced carriers is not much smaller than the number of recombination centers. In such cases, the response of current to time will be more complicated in that rise and decay do not follow symmetric functions, and two different time constants have to be defined.

7.3.4 RESONANCE DETECTORS

Although radiation in the radio and microwave range is almost exclusively detected by tuned oscillators, resonance receivers in the optical spectrum have made their appearance only in rather recent times. There lies, in fact, little advantage in the sharply tuned, high-Q detection of thermal sources: the signal received is proportional to Q^{-1}, the noise only to $Q^{-1/2}$ [see e.g. Eq. (7.36)], hence the signal-to-noise ratio of continua detection is inversely proportional to $Q^{1/2}$. If, however, the source is nearly monochromatic, then it is important to match the emitted spectrum by the response of the receiver, and the signal-to-noise ratio will indeed increase in proportion to $Q^{1/2}$.

Nevertheless, there are occasionally reasons to use resonance detectors even for thermal radiation—if only for the inherent interest which such detection processes possess or, as described in the next paragraph, because the wavelength region requires it.

7.3.4.1 *The Radiometer*

Since thermal radiation extends into the microwave region, a microwave antenna will communicate to its transmission line an amount of noise characteristic of the thermal radiation it intercepts.[51] Therefore, the noise communicated by the antenna to the transmission line may be used as a measure for the temperature of the object or objects in the lobe field of the antenna pattern. This, in short, represents the operating principle of the radiometer. The radiation power available to the antenna within the frequency interval $d\nu$ is given by

$$P_\nu = W_\nu(\nu, T) \cdot A \tag{7.87}$$

where the power radiated W_ν (ν, T) is given by the Rayleigh-Jeans radiation law (see Section 2.3.1). It has been shown in connection with the derivation of the Johnson noise formula in Section 7.2.3 that the power received by the system is

$$P_\nu \, d\nu = kT \, d\nu \tag{7.88}$$

Thus, the antenna noise power equals the Johnson noise, viz., it is equal to kT per unit frequency.

A typical radiometer is shown schematically in Fig. 7.20. The main components are an antenna, a mixer with local oscillator, a wide band amplifier, and a narrow bandwidth amplifier following the second detector. The minimum detectable temperature follows from considerations of mean deviations, as discussed in Section 7.2:

$$\Delta T = \frac{TF}{\sqrt{\Delta\nu \cdot \tau}} \tag{7.89}$$

where TF is the product of antenna temperature and noise figure. The product of observation time τ and bandwidth $\Delta\nu$ represents the number of pulses counted during one observation. Equation (7.89) states that the probable relative error $\Delta T/T$ is, except for the noise factor F, equal to reciprocal square root of the number of counts (cf. Section 7.2.1). To gain an idea of the capabilities of the radiometer, assume $T = 300\,°\mathrm{K}$, a noise factor $F = 10$, a bandwidth of 10^7 cps

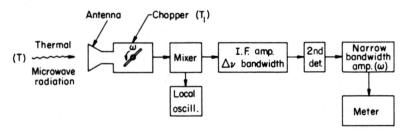

Fig. 7.20.   Scheme of the radiometer. [R. H. Dicke, *Rev. Sci. Inst.* **17**, 268 (1946)].

and an observation time of 1 msec (corresponding to a scanning rate of 1000 cps). It follows from Eq. (7.89) that under these conditions a minimum temperature difference of 30 °K can be detected. Actually Eq. (7.89) presents a simplification which ignores additional noise sources.

However, despite its relatively low sensitivity, the radiometer is a practical device rather than of mere academic interest. Microwaves, particularly those of wavelength larger than one centimeter, penetrate the atmosphere even in the presence of rain and clouds, with relatively little disturbance. Hence, all-weather operation of thermal detection is possible with the radiometer.

7.3.4.2 *Cascade Transition Detectors*

A principally different method of detecting radiation has the following general features:

Assume a system capable of going through various energy states E_0 to E_4 in succession. One of these transitions shall require a rather sharply defined quantum $h\nu_0$, while the rest of the energy requirements may be supplied by auxiliary means such as pumping light. Finally, one of the transitions serves to give an observable effect. This cascade sequence of transitions can then take place only if an optical quantum $h\nu_0$ has been received. It is necessary that the temperature is sufficiently low and the incident radiation sufficiently monochromatic, so that altogether no disturbing influences can compete with the signal produced by the incident radiation. One such scheme has been proposed by Bloembergen.[68] The scheme postulates the existence of narrow energy levels, such as presented by the ions of rare earths embedded as impurities in host lattices. The operation of this proposed so-called *quantum counter* rests on the conditions that a transition $E_1 \to E_2$ occurs if, and only if, a quantum energy $h\nu_0 = E_2 - E_1$ is incident. Only under this conditions will an auxiliary light source (optical pump) of high intensity provide a further transition $E_2 \to E_3$. This process is then made observable by recombination $E_3 \to E_1$ or $E_3 \to E_4 \to E_1$.

Obviously, the transition $E_3 \to E_1$ might be of a different kind, e.g., corresponding to a photoconductive or photoemissive excitation and be made observable as such. The main advantage of cascade schemes is so-called *quantum-mechanical amplification*, in that the absorption of an infrared quantum triggers a quantum transition of much larger energy. This then may be easier to observe than the direct process of low energy transition. Nevertheless, the relative fluctuations in the number of such excitations due to GR noise or photon fluctuations, hence the signal-to-noise ratio at this level, are not improved by such means. Moreover, the limitation to narrow spectral regions (high Q operation) and complexity make these schemes inferior to existing means of thermal source detection, unless special requirements exist.

7.3.4.3 Photoelectric Mixing

In Section 6.2.5.3 a process was described in which light waves differing in frequency by $\Delta\nu$ would produce a photoemissive current modulated at $\Delta\nu$.[69-70] This fact permits the application of a technique known in radio engineering as *heterodyning*. The application of the scheme to the optical region would provide for a local laser beam which is mixed in the photoemitter with the incident radiation. The differential frequencies can then be analyzed by means of conventional microwave or radio circuitry.

7.3.4.4 Cyclotron Resonance Detector

The principle of transitions between Landau levels has been discussed already for the case of emission in Section 6.3.2.4. The inverse process of absorption has made possible a unique detector of infrared radiation in which electrons or holes are held in quantized cyclotron orbits by a magnetic field. Incident photon energy of

$$h\nu = \frac{eH}{2\pi m^*} h \tag{7.90}$$

corresponds (in first order) to the level spacing of the various possible orbit energies and is thus able to raise carriers to positions of changed mobility.

The detector is tunable by a variation of the magnetic field H, and the operating range can be further controlled by a choice of materials with suitable effective carrier mass m^*. For example, germanium ($m^* = 0.135m_e$) requires $H = 1.46 \times 10^5$ G, and indium antimonide ($m^* = 0.0157m_e$) $H = 1.7 \times 10^4$ G to achieve a photoconductive response resonant at 100 μ.

Although processes of this kind had been discussed and proposed before,[71,72] the first tunable cyclotron resonance detector was successfully demonstrated by Brown and his co-workers.[73] Indium antimonide was used at liquid helium temperatures in combination with a superconducting magnet. The detector had a D^* of about 5×10^{10} [cm W^{-1} sec$^{-1/2}$], a time constant of less than 10^{-6} sec, and was tunable in a range between

60 and 300 μ. Since very high fields appear feasible with superconducting magnets, a yet wider spectral region seems to be accessible to this nearly ideal resonance detector. Thus with the development, in addition, of photoconductors operating with hydrogenic impurities in the millimeter region,[74] the entire optical spectrum has been opened up to practical processes of detection.

REFERENCES

1. A. EINSTEIN, *Ann. Physik* [4] **17**, 132 (1905); [4] **20**, 199 (1906).
2. H. HERTZ, *Ann. Physik* [3] **31**, 983 (1887).
3. See e. g. C. KITTEL, "Introduction To Solid State Physics," 2nd ed., p. 256. Wiley, New York, 1956.
4. R. H. Fowler, *Phys. Rev.* **38**, 45 (1931).
5. A. L. HUGHES and L. A. DU BRIDGE, "Photoelectric Phenomena." McGraw-Hill, New York, 1932.
6. M. GARBUNY, T. P. VOGL, and J. R. HANSEN, *J. Opt. Soc. Am.* **51**, 261 (1961).
7. R. SCHULZE, *Z. Physik* **92**, 223 (1934).
8. I. TAMM and S. SCHUBIN, *Z. Physik* **68**, 97 (1931).
9. I. TAMM, *Phys. Rev.* **39**, 170 (1932).
10. H. Y. FAN, *Phys. Rev.* **68**, 43 (1945).
11. H. THOMAS, *Z. Physik* **147**, 395 (1957).
12. H. MAYER and H. THOMAS, *Z. Physik* **147**, 419 (1957).
13. R. SUHRMANN *in* "Der Lichtelektrische Effekt und seine Anwendungen" (H. Simon and R. Suhrmann, eds), pp. 85-92. Springer, Berlin, 1958.
14. R. E. MAKINSON, *Phys. Rev.* **75**, 1908 (1949).
15. M. J. BUCKINGHAM, *Phys. Rev.* **80**, 704 (1950).
16. H. B. HUNTINGTON, *Phys. Rev.* **89**, 357 (1953).
17. H. PUFF, *Physica Status Solidi* **1**, 636 and 704 (1961).
18. J. ELSTER and H. GEITEL, *Ann. Physik* [3] **52**, 433 (1894); [3] **55**, 684 (1895); [3] **61**, 445 (1897).
19. H. E. IVES, *Phys. Rev.* **38**, 1209 (1931).
20. H. E. IVES and H. B. BRIGGS, *Phys. Rev.* **38**, 1477 (1931); **40**, 802 (1932).
21. H. E. IVES and T. C. FRY, *J. Opt. Soc. Am.* **23**, 73 (1933).
22. D. W. JUENKER, J. P. WALDRON, and R. J. JACCODINE, *J. Opt. Soc. Am.* **54**, 216 (1964).
23. L. APKER, E. TAFT, and J. DICKEY, *Phys. Rev.* **74**, 1462 (1948).
24. L. APKER and E. TAFT, *J. Opt. Soc. Am.* **43**, 78 (1953).
25. E. A. TAFT, H. R. PHILIPP, and L. APKER, *Phys. Rev.* **110**, 876 (1958).
26. W. E. SPICER, *J. Appl. Phys.* **31**, 2077 (1960).
27. J. A. BURTON, *Phys. Rev.* **72**, 531 (1947).
28. W. E. SPICER, *Phys. Rev.* **112**, 114 (1958).
29. L. APKER and E. TAFT, *Phys. Rev.* **79**, 964 (1950).
30. L. APKER and E. TAFT, *Phys. Rev.* **81**, 698 (1951).
31. W. SMITH, *Nature* **7**, 303 (1873).
32. W. J. CHOYKE and L. PATRICK, *Phys. Rev.* **108**, 25 (1957).

33. See, e.g., R. A. SMITH, *Advan. Phys.* (*Phil. Mag. Suppl.*) **2**, 321 (1953).
34. H. WELKER and H. WEISS, *Solid State Phys.* **3**, 1 (1956).
35. W. D. LAWSON, S. NIELSEN, E. H. PUTLEY, and A. S. YOUNG, *J. Phys. Chem. Solids* **9**, 325 (1959).
36. W. SHOCKLEY, "Electrons and Holes in Semiconductors." Van Nostrand, Princeton, New Jersey, 1950.
37. I. K. KIKOIN and M. M. NOSKOV, *Physik. Z. Sowjetunion* **5**, 586 (1934).
38. T. S. MOSS, *Proc. Phys. Soc.* (*London*) **B66**, 993 (1953).
39. S. W. KURNICK and R. N. ZITTER, *J. Appl. Phys.* **27**, 278 (1956).
40. C. HILSUM, D. J. OLIVER, and G. RICKAYZEN, *J. Electron.* **1**, 134 (1955).
41. J. J. LOFERSKI and P. RAPPAPORT, *in* "Methods of Experimental Physics" (L. Marton, ed.), Vol. 6, Part B, p. 365. Academic Press, New York, 1959.
42. E. BURSTEIN, D. N. LANGENBERG, and B. N. TAYLOR, *Phys. Rev. Letters* **6**, 92 (1961).
42a. R. F. BLUNT, *Bull. Am. Phys. Soc.* [2] **3**, 115 (1958).
42b. W. ENGELER, H. LEVINSTEIN, and C. STANNARD, JR., *Phys. Rev. Letters* **7**, 62 (1961).
42c. H. J. STOCKER, C. R. STANNARD, JR., H. KAPLAN, and H. LEVINSTEIN, *Phys. Rev. Letters* **12**, 163 (1964).
42d. Y. S. PARK and D. W. LANGER, *Phys. Rev. Letters* **13**, 392 (1964).
42e. A. VAN DER ZIEL, *J. Appl. Phys.* **24**, 1063 (1953).
42f. W. BEYEN, P. BRATT, H. DAVIS, L. JOHNSON, H. LEVINSTEIN, and A. MACRAE, *J. Opt. Soc. Am.* **49**, 686 (1959).
43. W. SCHOTTKY, *Ann. Physik* [4] **57**, 541 (1918).
44. G. GRAU, *Z. Angew. Physik* **17**, 21 (1964).
45. N. WIENER, *Acta. Math.* (*Stockholm*) **55**, 117 (1930).
46. A. KHINTCHINE, *Math. Ann.* **109**, 604 (1934).
47. D. K. C. MACDONALD, *Phil. Mag.* [7] **40**, 561 (1949).
48. J. B. JOHNSON, *Phys. Rev.* **32**, 97 (1928).
49. R. H. DICKE, R. BERINGER, R. L. KYHL, and A. B. VANE, *Phys. Rev.* **70**, 340 (1946).
50. H. NYQUIST, *Phys. Rev.* **32**, 110 (1928).
51. R. H. DICKE, *Rev. Sci. Instr.* **17**, 268 (1946).
52. A. EINSTEIN, *Physik. Z.* **10**, 185 (1910).
53. W. BOTHE, *Z. Physik* **20**, 145 (1923).
54. W. B. LEWIS, *Proc. Phys. Soc.* (*London*) **59**, 34 (1947).
55. M. HARVIT, *Phys. Rev.* **120**, 1551 (1960).
56. R. C. JONES, *Advan. Electron.* **5**, 68 (1954); *Advan. Electron. Electron. Phys.* **11**, 87 (1959).
57. R. H. FOWLER, "Statistical Mechanics" 2nd ed. Macmillan, New York, 1936.
58. J. B. JOHNSON, *Phys. Rev.* **26**, 71 (1925).
59. M. GARBUNY, T. P. VOGL, and J. R. HANSEN, *Rev. Sci. Instr.* **28**, 826 (1957).
59a. R. A. SMITH, F. E. JONES, and R. P. CHASMAR, "The Detection and Measurement of Infrared Radiation." Oxford Univ. Press (Clarendon), London and New York, 1957.
60. D. H. ANDREWS, R. M. MILTON, and W. DESORBO, *J. Opt. Soc. Am.* **36**, 518 (1946).
61. H. A. ZAHL and M. J. E. GOLAY, *Rev. Sci. Instr.* **17**, 511 (1946).
62. M. J. E. GOLAY, *Rev. Sci. Instr.* **18**, 347 and 357 (1947).
63. A. H. PFUND, *Rev. Sci. Instr.* **1**, 397 (1930).
64. H. C. BURGER and P. H. Van CITTERT, *Z. Physik* **66**, 210 (1930).
65. L. HARRIS, R. T. MCGINNIES, and B. M. SIEGEL, *J. Opt. Soc. Am.* **38**, 582 (1948).
66. K. M. Van VLIET, *Proc. IRE* **46**, 1004 (1958).
67. T. P. VOGL, J. R. HANSEN, and M. GARBUNY, *J. Opt. Soc. Am.* **51**, 70 (1961).
68. N. BLOEMBERGEN, *Phys. Rev. Letters* **2**, 84 (1959).

69. A. T. FORRESTER, R. A. GUDMUNDSEN, and P. O. JOHNSON, *Phys. Rev.* **99**, 1691 (1955).
70. A. T. FORRESTER, *J. Opt. Soc. Am.* **51**, 253 (1961).
71. H. J. ZEIGER, C. J. RAUCH, and M. E. BEHRNDT, *J. Phys. Chem. Solids* **8**, 496 (1959).
72. D. W. GOODWIN and P. H. HOWARD, *J. Appl. Phys.* **32**, 2056 (1961).
73. M. A. C. S. BROWN, M. F. KIMMETT, and V. ROBERTS, *Proc. IRIS* **8**, 117 (1963), unclassified.
74. E. H. PUTLEY, *Proc. IEEE* **51**, 1412 (1963).

AUTHOR INDEX

Numbers in parentheses are reference numbers and indicate that an author's work is referred to although his name is not cited in the text. Numbers in italic show the page on which the complete reference is listed.

445

SUBJECT INDEX

452

Q

Q- switching, 362
Q-value, 13, 15
Q'-value, 120
Quadrupole moment, 70, 95, 132
  transitions, 183ff
Quantization
  of energy levels, 21
  of hydrogen, 149
  of oscillator model, 42
Quantum, 21
  concepts in Bohr atom, 80
  counter, 439
  efficiency
    of metallic photoemission, 390ff
    of various photoemitters, 391, 403
  electronics, 325ff
  numbers, 150ff
  statistics, 46, 205, 319, 384
Quantum mechanics
  energy states, 85
  statistics of photon correlations, 319
  transition probabilities, 91ff
  wave equations, 84
Quartz
  absorption spectrum, 279, 280
  frequency doubling, 373 ff
  optical indices, 289
Quenching of luminescence, 384

R

R-band, 223
Racah notation, 180, 345
Radial probability function, 150
Radiance, 29
  of plasmas, 56
Radiant flux, 30
  density, 29
Radiation
  and structure, 3, 6, 8ff
  classical pattern, 71ff
  equilibrium, 34ff, 89
  field of classical oscillator, 69ff
  gamma—, 3
  laws of thermodynamic origin, 33ff
  power, 29ff, 37
    of classical oscillator, 72
    total from oscillating charge, 74

pressure, 37
quantized—, 21
Radiationless transitions, 210, 213, 230
Radiator configurations, 360
Radiometer, 437
Raman effect, 199
  stimulated, 378
Random directions of spontaneously emitted
    photons, 91
Range of photoelectrons, 392ff
  in semiconductors, 398
Rare earths, 176, 210
  as laser materials, 340ff, 371
  energy levels, 216ff
  line widths, 136
Rate equation, 332
Rayleigh-Jeans law, 42, 45
Recoil shift, 125
Recoilless photon exchange, 127
Recombination in solids, 228
Reduced mass, 82
  of rotator, 191
Reflection
  at boundary, 254ff
  at normal incidence, 255
  at oblique incidence, 258
  coefficient, 256
  complete, 272
  total, 259
Reflectivity, 252, 256, 282
  comparison, theory and experiment, 264
  definition, 282
  in reststrahlen region, 285ff
  of superconductors, 294ff
  spectra and structure, 286ff
Refraction, 257ff
Refractive index, 259
Relaxation
  oscillations, 362
  time of metal electrons, 292
Resolving power
  of Fabry-Pérot, 173
  of grating, 172
  of prisms, 172
Resonance
  detectors, 437ff
  energy transfer, 183, 345
  imprisonment, 104ff, 110
    in gas lasers, 347, 350
  line, 99

profile
of absorption, 78
of maser oscillation, 333
Spectroscopic terms, levels, and states, 113
Spectrum
electromagnetic-1ff
infrared—, 2
optical, 5ff
ultraviolet, 2
visible, 2
Spikes in laser operation
by relaxation, 363, 365ff
giant—, 363
Spin
and magnetic moment, 159
fine structure, 155ff
in helium, 181
—orbit interaction with field, 162
orientation, 157
Spontaneous emission, 89, 91
transition probability, 89ff
Standing waves
in cube, 18
in electron orbits, 83
in photoemitter, 397
number of, 19
of vibrating string, 17
Star diameter determinations, 311, 319
Stark broadening, 130
States, 113
density, 16
Stationary solution of oscillator equation, 270
Statistical theory of pressure broadening, 132ff
Statistical weight, 55, 144
Statistics of quanta, 40ff
Stefan-Boltzmann law, 36ff
Stern-Gerlach experiment, 160
Stimulated emission
classical, 77
interpretation, 90
produced in Raman scattering, 378
quantized process, 89
Structure and radiation, 8ff
Submillimeter region
detectors, 437, 440, 441
transparent materials, 279, 283ff
Sum rule
Burger-Dorgelo, for intensities, 114

f-values, 98ff, 115, 119
in dispersion, 272, 273
Sun radiation
dark edge effect, 30, 65
spectra, 65, 66
Superconductors, 293ff, 428
Surface effect of photoemission, 392
Systems for lasers, 359

T

Tellurium photoconductivity, 407
as impurity, 411
Temperature
determination from populations, 123
determination from line intensities, 101, 138
determination from line shape, 123
limits to practical radiators, 50
negative, 326
noise, 420
of arcs, 54
of flames and explosions, 53, 54
Tensors for polarization, 374
Term scheme, 113ff, 154ff
designations, 154
for LS coupling, 158
for molecular spectra, 188
for multiple electrons, 175
of atom in plasma, 57
Thallium-bromide-iodide transmittance (see KRS — 5)
Thermal detectors, 426
Thermal effect
on energy gap, 406, 407
on photoemission, 387ff
on resistivity, 428
Thermal equilibrium and thermal radiation, 54
Thermal radiation, 28ff
coherence properties, 314
from plasmas, 53ff
from solids, 51ff
Planck curves, 44
Thermalizing of electrons 346
Thermionic current, 388
Thermistor, 428
Thermocouple detectors, 430
Thermodynamics
for the derivation of Kirchhoff's law, 34, 35